CONCEP'
MATHEMATICAL
MODELING

WALTER J. MEYER

Professor of Mathematics
Adelphi University

DOVER PUBLICATIONS, INC.
Mineola, New York

To my children
Alex and Fred

Bibliographical Note

This Dover edition, first published in 2004, is an unabridged republication of
the work originally published by McGraw-Hill, Inc., New York, 1984.

Library of Congress Cataloging-in-Publication Data

Meyer, Walter J. (Walter Joseph), 1943-
    Concepts of mathematical modeling / Walter J. Meyer.
        p. cm.
    Originally published: New York : McGraw-Hill, 1984.
    Includes bibliographical references and index.
    ISBN-13: 978-0-486-43515-2 (pbk.)
    ISBN-10: 0-486-43515-6 (pbk.)
    1. Mathematical models. I. Title.

QA401.M515 2004
511'.8—dc22

                                                            2004047763

Manufactured in the United States by LSC Communications
43515607   2022
www.doverpublications.com

CONTENTS

PREFACE

Thou hast ordered all things
in measure and number and weight.
Ch. 11, verse 20, Book of Wisdom

This book consists of a number of more or less independent sections designed to illustrate the most important principles of the mathematical modeling process. It is intended for an introductory undergraduate-level course. A student with a good knowledge of calculus and a little probability and matrix theory would find all of it accessible. If one wishes to avoid the sections involving probability, there is still a semester's worth of material here. The sections are organized according to a definite point of view, but the text can also be treated as a sampler with success (especially the sections in the first three chapters).

The basic premise of this book is that all mathematics students need an understanding of mathematical modeling, even those who won't become practitioners of the craft. My aim is to show students that mathematical modeling is interesting and useful and to illustrate, without undue technicality or specialization, some of the main themes and methods of the subject. A course based upon this book would be a good springboard to advanced studies in aspects of modeling which are touched upon lightly here: optimization, computer simulation, hands-on experience with substantial projects, statistics, etc.

The first time I taught mathematical modeling, my notes could have been entitled "Models I Happen to Like." It didn't seem sensible to publish anything so purely personal, and so I began thinking harder about how to teach mathematical modeling. One approach is to organize by mathematics: first we study differential equations models in various fields, then we study probability models in various fields, etc. Another approach is to organize by subject matter: first we study population and various mathematical models for it, then we study political science and various mathematical approaches to it, etc. Excellent books of these types have been written; so I have chosen a third approach for this book: to display aspects of the *process* of modeling. There is a lot to this process besides knowing mathematics and the subject matter of the model. There are strategies, attitudes toward data, choices to be made about mathematical tools— in short, a lot of "know-how." Put another way, this book concerns the question: "What does 'modeling' mean in the phrase 'mathematical modeling'?"

By covering sections out of order, an instructor can easily replace my organization of the material with a more mathematical organization. For example, a substantial part of a course could be devoted to applications of difference and differential equations by starting with the long Section 2 in Chapter 5 and adding to it Sections 3 and 4 of Chapter 1 and Section 5 of Chapter 3. Likewise, a course segment on probabilistic models could use the long Section 3 of Chapter 5, together with Section 6 of Chapter 1, Section 2 of Chapter 2, and Section 1 of Chapter 5. One could assemble a minisegment on applications of geometry from Section 1 of Chapter 2 and Sections 6 and 7 of Chapter 3. A unit on optimization could be based on Chapter 4 and Section 7 of Chapter 1.

For a subject matter organization, one could devise a long course segment dealing with population.

As an aid to instructor and student, each section is preceded by an abstract and statement of prerequisites for that section. This should help the instructor make use of the considerable flexibility of the book.

Answers or hints are provided for selected exercises (those marked by dots in the margin).

There is no shortage of good mathematical models to use in writing a textbook. The challenge seems to be in choosing a good set of constraints to guide the selection process. Some of the constraints (besides personal taste) which I set for myself are:

1. The level should be fairly elementary.
2. The sections should be largely independent.
3. There should be variety in subject matter (physical, biological, social sciences, and operations research), mathematical tools, and historical eras from which the applications are drawn.
4. Certain important subjects, namely difference equations and optimization, which often do not find a natural home in the curriculum for mathematical science majors, should appear.
5. I have tried to emphasize "classic" models and to exclude models of narrow or passing interest, no matter how novel they may be. Hopefully, the models treated here will merit inclusion in the curriculum for many years.

I owe special thanks to the following people who reviewed this book when it was in manuscript form: Martin Braun, Queens College; James P. Jarvis, Clemson University; William F. Lucas, Cornell University; Joseph Malkevitch, York College (CUNY); Rochelle Meyer, Nassau Community College; Herman F. Senter, Clemson University; Donald R. Sherbert, University of Illinois at Urbana-Champaign; Robert L. Wilson, Jr., Washington and Lee University; and Matthew Witten, Illinois Institute of Technology. Others who have influenced or assisted me include: James Frauenthal, Bell Labs; Noreen Goldman, Princeton University; T. N. E. Greville, University of Wisconsin; Frank Hoppensteadt, University of Utah; Nathan Keyfitz, Harvard University; Dick Montgomery, Southern Oregon State University; Fred Pohle, Adelphi University; Bill Quirin, Adelphi University;

and H. A. Sulz, State University of New York at Buffalo (SUNY). Joseph Mal-kevitch deserves special mention: his friendship, advice, and assistance over many years have shaped my thinking and left many marks throughout this book. The entire COMAP organization deserves credit for encouraging mathematical modeling and helping to create a need for books like this. Special thanks are also due to my patient family and my excellent typists, Virginia Armbruster, Marie Glass, Kathleen Haese, and Lenore Nemirow.

Walter J. Meyer

CHAPTER ONE

THE SCOPE OF MATHEMATICAL MODELING

1 MODELS, MATHEMATICAL AND OTHERWISE

Abstract This section introduces the concepts of mathematical and non-mathematical models in a nontechnical way. Our examples show that mathematical models are often, but not always, better; that nonmathematical models may evolve into mathematical ones; and that experimental work may be needed to provide data for mathematical models.

Prerequisites None.

No human investigation can claim to be scientific if it doesn't pass the test of mathematical proof.

Leonardo Da Vinci

Mathematical modeling is an attempt to describe some part of the real world in mathematical terms. It is an endeavor as old as antiquity but as modern as tomorrow's newspaper. It has led to some good mathematical models and some bad ones, which are best forgotten. Sometimes mathematical models have been welcomed with great enthusiasm—even when their value was uncertain or negligible; other times good mathematical models have been greeted with indifference, hostility, or ridicule. Mathematical models have been built in the physical, biological, and social sciences. The building blocks for these models have been taken from calculus, algebra, geometry, and nearly every other field within mathematics.

In short, mathematical modeling is a rich and diverse activity with many interesting aspects. The aim of this book is to display by examples some of the many facets of mathematical modeling.

1

But before we plunge into this, it seems only fair to say something about models of a nonmathematical nature. In ordinary language the word "model" has many meanings. What we will mean by it is this.

Definition

A model is an object or concept that is used to represent something else. It is reality scaled down and converted to a form we can comprehend.

For example, a model airplane, made of wood, plastic, and glue, is a model of a real airplane. Another example is the idea that, in politics, public opinion is like a pendulum because it changes periodically from left- to right-wing ideas then back again in a way which reminds us of a pendulum swinging back and forth. In our terminology we would say that a pendulum is a model for public opinion.

A model aiplane and pendulum are physical objects; so they are not mathematical models.

Definition

A mathematical model is a model whose parts are mathematical concepts, such as constants, variables, functions, equations, inequalities, etc.

Example 1 that follows illustrates the differences between mathematical and nonmathematical models. In this example the mathematical model is, in many ways, superior to its nonmathematical counterpart. The other examples in this section also illustrate the great value of mathematical models. But we shall see that nonmathematical models have value as well. Among other things, they often stimulate the development of mathematical models.

Example 1 Aircraft Flight

To find out how an aircraft will behave in flight, we could make a physical model of the aircraft and test it under various weather conditions. There are a great many things one might want to know: Is the plane stable in the air? How fast can it go? How steeply can it climb? Etc. To focus our discussion, let's consider the question of how great the lift force on the plane is when it takes off.

The lift force is the force pushing up on the wings. This force is largely what determines how steeply the plane can climb. If we did experiments with a physical model, we could find out almost anything we wanted to know about it. For example, we could discover that the lift force was dependent on how fast the plane was moving. By flying the plane at different speeds, we could make a table of values relating lift force to velocity and a graph of this table of values that might look like Figure 1.

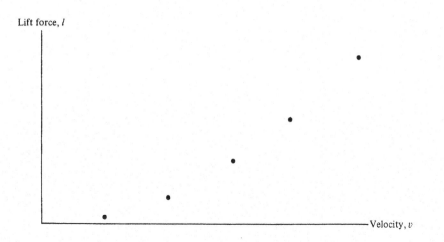

Figure 1 Points are plotted from a table of values obtained from wind-tunnel experiments. The different points represent trials at different speeds.

But there is an entirely different approach to this problem, one based on a mathematical model. This mathematical model consists of a single equation which relates the lift force to other factors. It is

$$l = C_l \frac{\rho}{2} s v^2 \tag{1}$$

where l = lift force

C_l = a certain numerical value called the lift coefficient whose exact value depends on the shape of the plane

ρ = density of the air

v = velocity of the plane

s = total surface area of the tops of the wings

We can estimate s from the blueprints of the plane we propose to build. ρ is a measurement we can make in the atmosphere. (It may differ a little from one airport to another.) C_l is a number which differs from plane to plane and is a little hard to estimate for a plane that has not yet been built and tested. But there are methods that yield reasonable estimates; so let's assume C_l is known. Then the product $C_l(\rho/2)s$ in Equation (1) becomes a known constant. If we call this constant a, then Equation (1) becomes an equation linking only two variables, l and v:

$$l = a v^2 \tag{2}$$

Using this equation, we can generate the graph shown in Figure 2 with a moment's worth of calculation and plotting.

Which approach is better, experiments on the physical model or predictions from the mathematical model? Building physical models is time consuming: it might take days to make a good model plane. It's also expensive. In both these

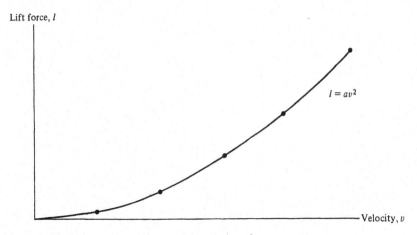

Figure 2 The curve is plotted from the formula $l = av^2$.

respects the mathematical model is superior. But it has another advantage. It tells us things that our physical experiments don't.

For example, suppose we wanted to increase the lift force on the plane. Figure 1, which is based on our experiments, shows that lift increases as velocity increases. A larger velocity will probably do the trick; so we might try outfitting the plane with a more powerful engine. But Figure 1 doesn't tell us exactly what velocity we need to achieve a given lift force. By using Equation (2), we can find exactly what value of v we need to achieve a given value of l. This is merely a matter of solving for v and substituting the value of l desired.

If we go back to Equation (1), we also discover there is another means of accomplishing our goal, namely, to increase s by making the wings bigger. To discover this from our physical experiments would be impossible without building a series of additional physical models with various wing sizes. Thus our mathematical model is cheaper, faster, and more versatile than experimentation on a physical model of a plane.

Of course there is much more to aircraft performance than lift force. There is air drag, stability in flight, and a host of other factors to be considered. In principle, each of these factors *could* be modeled mathematically in an equation. The whole collection of these equations would be a mathematical model for aircraft performance. Such giant models have been built and are being continually improved. Although they are still far from perfect, they are slowly replacing physical models and wind-tunnel experiments in the aircraft industry. In 1976 the American Society of Mechanical Engineers' winter meeting featured a symposium on the question of whether computer-implemented mathematical models of aircraft performance might soon make wind-tunnel experiments obsolete.

One thing one must keep in mind about mathematical models is that they don't arise out of thin air. They need to be discovered first, and this takes time and ingenuity. The present-day availability of good mathematical models for aircraft design is a far cry from the situation that prevailed in the early history of aviation. For example, in 1879 the Aeronautical Society of Great Britain could report:

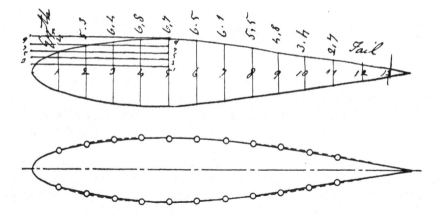

Figure 3 *Above*: Sir George Cayley's sketch of the cross section of a trout. [*From "Aeronautical and Miscellaneous Note-Book (ca. 1790–1826) of Sir George Cayley," Cambridge University Press, 1933.*] *Below*: A comparison of Cayley's trout section with modern low-drag airfoil sections. Circles indicate trout; ——— N.A.C.A. 63A016; ------- LB N-0016. (*Reprinted from T. von Kármán: "Aerodynamics: Selected Topics in the Light of their Historical Development." Copyright 1954 by Cornell University. Used by permission of the publisher, Cornell University Press.*)

"Mathematics up to the present date have been quite useless to us in flying." A cynic might reply that nothing else was much use either. The first sustained flight by the Wright brothers didn't take place until years later in 1903.

Developing a mathematical model requires not only time and ingenuity, but data as well. In the case of aeronautics, a good many of the data required are obtained from wind-tunnel experiments. Thus physical experiments and non-mathematical models play a role in giving birth to mathematical models. Our emphasis on the value of mathematical models does not mean that nonmathematical models have only slight value.

In the case of aircraft flight, there is a striking example of a nonmathematical model that turned out to be beautifully simple and fairly effective. In the early nineteenth century, long before anyone had built a working airplane, Sir George Cayley was concerned about finding a good shape for the cross section of a wing (Figure 3) that would minimize air drag. Lacking a good mathematical model and being unable to carry out the necessary experiments, Cayley hit upon the interesting notion of making the wing cross section in the shape of a trout. His reasoning was that the water drag on the trout swimming through water was analogous to the air drag on a wing traveling through air. A further assumption of Cayley's nonmathematical model was that Mother Nature, through the mechanisms of evolution (the struggle for existence and survival of the fittest) would have already hit upon a good design.

Example 2 Chemistry

In 1786 the philosopher Immanuel Kant asserted that chemistry could never become a science. It would be hard to find a more spectacularly wrong prediction in all of the history of human thought. But we should be kind to Kant because in his time chemistry truly seemed like a random collection of recipes and rules of thumb.

One-hundred years later chemistry was well on its way to becoming a science. Specifically, in 1887 the chemist Clemens Winkler provided dramatic evidence of the value of Dmitri Mendeleev's Periodic Table of the Elements (see Section 5 of Chapter 2). This is an important milestone because it showed that there was order and logic behind the jumble of facts which chemists had collected up until that time.

Yet another hundred years or so brings us to the modern era, in which parts of chemistry are well on their way to becoming a mathematical science in the sense that mathematical models are playing a vital role in the process of chemical discovery. A case in point is the determination of the exact spatial relationships of the atoms making up a molecule by William Nunn Lipscomb, who won the Nobel prize in chemistry in 1976. Lipscomb won his Nobel prize in part for discovering new types of molecules in the borane family. Before many of these molecules were actually found in nature, Lipscomb had predicted their existence with the aid of higher mathematics (and a children's construction toy called *D-stix*).

Lipscomb's own words give testimony to the growing role of mathematics in chemistry:

> To give you an idea of how radically chemistry has changed over the years, I can say that I spend practically no time in a laboratory. At the Gibbs Laboratory we do no chemical analysis and practically no synthesis. But by contrast I probably use more time on Harvard's computers and other computers available to us than any individual—huge chunks of computer time.
>
> It's a far cry from my own antecedents. I became interested in chemistry when my mother gave me an A. C. Gilbert chemistry set. But today the real research is in ideas—even intuition— expressed usually, in the special languages with which mathematicians and computers communicate.

Example 3 Paranoia

One of the most influential models of the twentieth century is also one of the least mathematical: Sigmund Freud's model of the personality as divided into three warring factions—the id, the ego, and the superego. His model contains more than just this three-way division; it also contains a lot of assumptions about how and why these parts of the personality compete for control. None of it is mathematical, and, partly for this reason, it is vague and hard to test but easy to argue about. Nevertheless, even Freud's greatest critics agree that he captured important germs of truth in his model.

Will these germs of truth ever be transformed into a good mathematical model? At the present time there is no satisfactory mathematical model of an entire normal human personality. Some would say that there never can be. After all, would it ever be possible to measure the strength of a feeling such as anxiety? It is worth noting that Freud himself thought of anxiety in quantitative terms.

Although he had no instrument or procedure to measure it, he described it as something which could, in principle, be precisely measured.

Even though there is not yet a satisfactory mathematical model of a whole normal human personality, there is a good model of a small part of a sick human personality. This is a mathematical model, called *PARRY*, devised by K. M. Colby to imitate the conversation of a paranoid personality. The model is in the form of a computer program, which can hold a conversation with an ordinary human being (in written form via computer terminal) about its psychological difficulties. The entire conversation is supposed to resemble a session between a paranoid human being and a therapist.

The model has the capability of "understanding" English well enough that it can devise appropriate responses to most remarks the therapist might make. These responses are not random however. The model has built into it a large repertory of beliefs and feelings, which revolve around betting on horse races, dishonest bookmakers, the Mafia, and other related topics. The model has the capability of generating a very large number of different conversations in response to the various things the human conversational partner might say. The conversations which the model can engage in may be thought of as the fruits of the model (rather like weather reports are fruits of a single mathematical model of the atmosphere).

The model itself is the computer program, i.e., the calculations and instructions that determine how PARRY will respond to a given remark of the therapist. The response at any point depends on two factors: the last few exchanges in the conversation and the emotional state the model has generated for itself at that point. This emotional state is portrayed in the model by numbers which keep track of specific emotions, such as fear, anger, and mistrust. These numbers are constantly updated according to mathematical formulas which take into account the remarks each side makes in the conversation.

How realistic is PARRY? Experiments indicate that it is realistic enough to fool professional therapists. A group of psychiatrists was asked to interview a group of paranoid "patients" via a teletype machine and without personal contact. Although none of them knew it, only some of them were dealing with real patients; the others were conversing with PARRY. None of the psychiatrists who conversed with PARRY suspected that they were not dealing with a human being. In another stage of the experiment, typewritten transcripts of the conversations were sent to other psychiatrists, who were then asked to decide which conversations involved PARRY and which involved human paranoids. These psychiatrists showed no ability to discriminate between PARRY and human paranoids. The number of successful identifications was about what would be expected through chance guessing.

Although PARRY is very successful at what it does, it must be kept in mind that what it does is very limited. It can't say anything intelligent about baseball, mathematics, or other subjects outside the repertory it is programmed for. If you mention the Easter Bunny, it will probably respond that it's all a racket controlled by the Mafia. If you ask it to do a math problem, it might simply refuse. If you tell it a joke, it won't laugh; nor can it tell you one back. It is also very bad at making logical connections. The reason it fools the psychiatrists is that psychotherapy sessions are narrow in exactly the way PARRY is narrow. Psychotherapy sessions do not normally include jokes, tests on arithmetic or logic, or conversation about irrelevant subjects such as the Easter Bunny. In addition, therapists who believe they are talking to paranoids are prepared for the conversations to be slightly weird. Consequently, if the computer program performs poorly now and then, the poor performance can easily be mistaken for mental illness.

Perhaps you are wondering what conversation with PARRY is like. Here is an example. The uppercase letters show the responses of PARRY, and the lowercase letters show the human being's part of the conversation.

Why are you in the hospital?
I SHOULDN'T BE HERE.
Who brought you here?
THE POLICE.
What trouble did you have with the police?
COPS DON'T DO THEIR JOB.
What do they do instead?
COPS ARREST THE WRONG PEOPLE.
They arrest the right ones sometimes.
THE LAW SHOULD GET THOSE ITALIAN CROOKS.
Are you Italian?
MY BACKGROUND IS BRITISH, BUT I WAS BORN IN THIS COUNTRY.
Do you know any Italian crooks?
I TRY TO AVOID THE UNDERWORLD.
Are you afraid of the underworld?
THE MAFIA ARE THE WORST GANGSTERS.
Are you connected with the Mafia?
YOU KNOW, THEY KNOW ME.
How do they know you?
THEY SPY ON ME.
Why do they?
THEY WANT TO GET FALSE EVIDENCE TO FRAME ME.
What do they have against you?
THEY ARE OUT TO GET ME.
What for?
I HAVE INCRIMINATING EVIDENCE AGAINST THE MAFIA.
Are you sure of all this?
YOU DON'T BELIEVE ME?
I am doing my best.
(SILENCE ON THE "PATIENT'S" PART)
(Silence on the psychiatrist's part)
CAN'T YOU THINK OF A DECENT QUESTION TO ASK?

You seem hostile to me.
SOME OF YOUR REMARKS ANNOY ME.
I am sorry.
FORGET THE EXCUSES, YOU GUYS ARE ALL ALIKE.

Example 4 Shortest Paths

Suppose you have a network of roads and intersections, as shown in Figure 4. The numbers on the line segments represent the lengths, in miles, of the road segments. This picture is an example of a weighted graph. This weighted graph may be thought of as a mathematical model of the real network of roads. We might use this model to find the shortest path from one intersection to another.

The weighted graph is not the only way to model the road network mathematically. Indeed, for some purposes it may not be a very good model. For example, if we wanted to feed this information to a computer, we couldn't just show the computer the weighted graph because most computers can't see. They have nothing analogous to the human eye which they could use for visual processing. What we could do instead is to describe the road network with a distance matrix.

Definition

The distance matrix of a road network is a square matrix with as many rows and columns as intersections. If we denote the ith-row, jth-column entry by a_{ij}, then the following rule shows how a_{ij} is determined:

$$a_{ij} = \begin{cases} \text{the length of the segment joining intersections } i \text{ and } j \\ \infty \text{ if there is no segment joining intersections } i \text{ and } j \\ 0 \text{ if } i = j \end{cases}$$

Finally, we could build a nonmathematical model of the road network out of buttons and string: let the buttons represent the intersections and the strings represent the road segments. Furthermore, cut each string to a length proportional to the length of the road segment it represents. (For example, we might let each mile be represented by 1 inch of string.)

A common mathematical problem is to find the shortest path from one intersection to another in a road network. Which of our three models is best for working this out? In the example of Figure 4 trial and error on the weighted

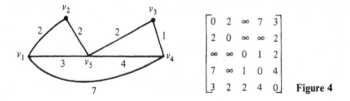

Figure 4

graph seems easy enough. But suppose we had a network with 10, 20, or 50 inter-sections? For graphs of this size the eye becomes confused and trial and error is not efficient. It is possible to get the answer from our distance-matrix model by doing some numerical manipulations on the matrix (relying on some theorems in matrix algebra which we won't mention). But the string-and-button model also gives a quick, easy to understand, and foolproof solution. Grasp the "start" and "end" buttons in opposite hands and pull in opposite directions until some set of string segments tightens up into a straight line leading from the start button to the end button. Thus, each of the three models can be useful in its own way.

Example 5

Figure 5 shows a schematic diagram of the human eye, which is meant to illustrate how the eye muscles move the eyeball around in its socket. Two of the muscles are shown as shaded areas. Within these shaded areas are springs and dashpots (shock absorbers). Naturally the eye muscles don't really have springs or shock absorbers inside them. These mechanisms are used to represent, or to model, the

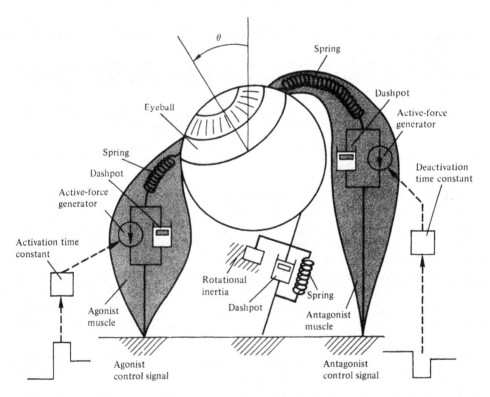

Figure 5 Schematic diagram of the human eye. (*From " The Trajectories of Saccadic Eye Movement" by A. Bahill and L. Stark. Copyright © 1979 by Scientific American, Inc. All rights reserved.*)

muscle. Presumably the authors did this because the average person is familiar with springs and shock absorbers and how they work, and this familiarity will help people understand how the eye muscles work. The spring and shock-absorber arrangement can be thought of as a simple model for the muscle. But it's not a mathematical model because springs and shock absorbers are not mathematical concepts.

This model was devised for purely illustrative purposes. For scientific purposes there is a mathematical model available which is more useful. One part of this model is an equation (the Fenn-Hill-Katz equation) which describes exactly how the "damping force" (the amount of shock absorption) depends on the speed with which the eyeball is moving. Thus an equation takes the place of a physical mechanism. Another equation describes how the spring works. For more details, read "The Trajectories of Saccadic Eye Movement" by Bahill and Stark in the January 1979 *Scientific American*.

Example 6 Electricity

The history of scientific theories of electricity is an interesting illustration of the transition from nonmathematical to mathematical models. One of the earliest ideas about electricity was that it was generated by the action of two fluids. These corresponded to what we nowadays call + and − *charge*. This two-fluid model made it easy to accept the fact that electricity could flow from one point to another since everyone was familiar with the fact that fluids could flow.

Benjamin Franklin made a significant modification by suggesting that only one fluid was involved. This theory eventually became the generally accepted one, but only after years of argument.

These ideas about fluids can be said to be nonmathematical models of electricity. By modern standards they seem primitive, but in their time they were useful because they did explain some of the facts and suggest important experiments to perform.

While the arguments raged about how many fluids there were and their nature, some men (notably Henry Cavendish, Charles de Coulomb, and George Ohm) were doing experiments and devising instruments that measured electricity precisely and provided the beginnings of a mathematical theory. However, this did not occur until the end of the eighteenth century, and it wasn't till many years later that the nonmathematical fluid model became entirely obsolete among researchers. Those of us who deal with electricity only on a practical level have never quite given it up. We speak of current as "juice" and think of voltage as analogous to pressure and amperage as analogous to the volume of fluid flowing.

There are some spectacular examples of men who made great practical advances in electricity without understanding it mathematically. One is the physicist Michael Faraday, to whom we owe inventions such as the dynamo, a mechanism presently used to generate most of the world's electricity. Faraday thought of all forms of energy radiations (like heat or radio signals) as flowing through narrow curved tubes. He probably didn't really think it worked that way but just found it a good model. Instead of contemplating the equations governing the radiations, he thought about the sizes and shapes of these imaginary tubes. And it worked! As the physicist Hermann von Helmholtz says of Faraday,

> It is in the highest degree astonishing to see what a large number of general theorems, the methodical deduction of which requires the highest powers of mathematical analysis, he found by a kind of intuition, with the security of instinct, without the help of a single mathematical formula.

Another striking example of the power of nonmathematical thought is the career of Thomas Edison. Edison was basically a genius tinkerer and knew hardly any mathematics. Happily, this did not prevent him from inventing a prodigious number of electrical devices. Nevertheless, toward the end of his career the state of the art had passed him by. The story is told that an assistant approached him for technical advice and was told to consult a member of the staff who had a better command of mathematics. Edison said, "He knows far more about [electricity] than I do. In fact, I've come to the conclusion that I never did know anything about it."

One of the particular reasons for Edison's difficulties was the increasing use of alternating current, which brings with it engineering problems requiring more mathematics for their solution. At about this time Edison sold his business interests in the electric power industry to the General Electric Company. The technical side of General Electric was under the guidance of a German immigrant with a solid grasp of mathematics, Charles Proteus Steinmetz. Steinmetz makes an apt comparison to Edison. Edison won fame for a device, the electric light bulb, which is basically simple and required mainly lots of persistence for its invention. Steinmetz became famous (among scientists) for a mathematical formula for the energy dissipated in iron as its magnetic field changes owing to an alternating current. The man in the street has seen many of Edison's light bulbs but has probably never seen Steinmetz's formula, and, partly for this reason, Edison is the more famous of the two. But Edison's inventions could not be so widespread and effectively used today without Steinmetz's mathematical work, based on mathematical models of electricity.

BIBLIOGRAPHY

Boden, Margaret: "Artificial Intelligence and Natural Man," Basic Books, New York, 1977. A nontechnical survey with a description of PARRY and other exciting developments in computer science.

Kármán, Theodore, von: "Aerodynamics," Cornell University Press, Ithaca, N.Y., 1954. An excellent exposition designed for the nontechnical reader, equipped only with high school algebra.

Kershner, William: "Student Pilot's Flight Manual," Iowa State University Press, Ames, Iowa, 1968.

Kuhn, Thomas S.: "The Structure of Scientific Revolutions," 2d ed., University of Chicago Press, Chicago, 1970. A thin but important book on how science develops. The history of the theory of electricity is one example discussed.

2 STEPS IN BUILDING A MATHEMATICAL MODEL

Abstract Three steps in mathematical modeling are discussed: formulation, mathematical manipulation, and evaluation. Two examples are used: Galileo's work on gravity and modern research on industrial productivity.

Prerequisites Elementary differential equations, logarithms, and exponential functions.

Modeling cannot be done mechanically. Nevertheless, there are some guidelines for how to go about it. We can divide the modeling process into three main steps: formulation, mathematical manipulation, and evaluation. Formulation can, in turn, be divided into three smaller steps.

Formulation

1. *Stating the question.* The question we start with is often too vague or too "big." If it's vague, make it precise. If it is too big, subdivide it into manageable parts.
2. *Identifying relevant factors.* Decide which quantities and relationships are important for your question and which can be neglected.
3. *Mathematical description.* Each important quantity should be represented by a suitable mathematical entity, e.g., a variable, a function, a geometric figure, etc. Each relationship should be represented by an equation, inequality, or other suitable mathematical assumption.

Mathematical Manipulation

The mathematical formulation rarely gives us answers directly. We usually have to do some mathematics. This may involve a calculation, solving an equation, proving a theorem, etc.

Evaluation

In deciding whether our model is a good one, there are many things we could take into account. Chapter 3 deals with many of these criteria in detail. Obviously, the most important question concerns whether or not the model gives correct answers. If the answers are not accurate enough or if the model has other short-comings, then we should try to identify the sources of the shortcomings. It is possible that mistakes have been made in the mathematical manipulation. But in many cases what is needed is a new formulation. For example, it may be that some quantity or relationship which we neglected was more important than we thought. After a new formulation, we will need to do new mathematical manipulations and a new evaluation. Thus, mathematical modeling can be a repeated cycle of the three modeling steps, as shown in the flowchart of Figure 1.

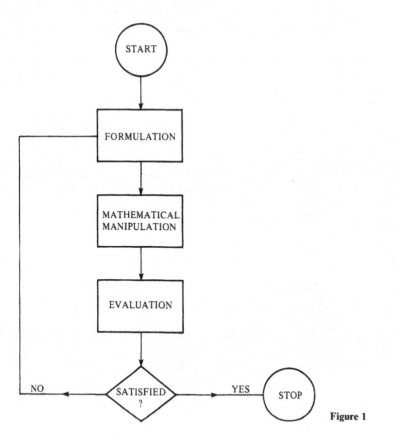

Figure 1

Illustration 1 Galileo's Gravitation Models

One of the oldest scientific investigations was the attempt to understand gravity. It gives a nice illustration of the steps in modeling.

Formulation

1. *Stating the question.* "Understanding gravity" is too vague and ambitious a goal. One of the first more specific questions asked about gravity was: *Why* do objects fall to the earth? Aristotle's answer was that objects fall to the earth because that is their natural place and every object seeks its natural place. Unfortunately, the question of why objects fall, along with Aristotle's answer, never led directly to any useful science or mathematics.

   Around the time of Galileo (1564–1642) people began asking *how* gravity worked instead of *why*. For example, Galileo wanted to describe the way objects gain velocity as they fall. (Think of how a roller-coaster car goes slowly when it starts down, but continually picks up speed.) This gain in velocity is called *acceleration*. Two particular questions Galileo asked were:

a. What formula describes how a body gains velocity as it falls?

b. What formula describes how far a body falls in a given amount of time?

2. *Identifying relevant factors.* Galileo decided to take into account only distance, time, and velocity. There are other factors Galileo might have considered relevant. He might have considered the weight, shape, and density of the object. Does a bowling ball fall faster than a football? He might have considered air conditions. Does it matter if the wind is blowing or what the barometric pressure is? The first assumption Galileo made was:

Assumption 1 If a body falls from rest, then its velocity at any point is proportional to the distance already fallen.

As we shall see and Galileo eventually discovered, this is wrong. But before discarding it, we will complete the modeling process based on this incorrect assumption.

3. *Mathematical description.* We will set up a distance scale to measure an object's fall and use the variable x to measure distances along this scale. At a certain instant in time, say $t = 0$, we begin observing the object's fall. For convenience, let $x = 0$ at this initial point of observation. Let $x(t)$ denote the distance the object has fallen after t seconds. See Figure 2.

Between the time instants t_0 and $t_0 + h$ the distance traveled is $x(t_0 + h) - x(t_0)$. Since the time elapsed is h seconds, the average velocity is

$$\frac{x(t_0 + h) - x(t_0)}{h}$$

(The word "average" is used because there are many examples of motion—the roller coaster for example—for which the velocity is changing from instant to instant.) If we let $h \to 0$, this quotient approaches dx/dt evaluated at t_0, a quantity we call the *instantaneous velocity* at t_0. We sometimes denote this $v(t_0)$ instead of $(dx/dt)|_{t_0}$.

The same ideas apply to acceleration. Between time instants t_0 and $t_0 + h$ the change in velocity is $v(t_0 + h) - v(t_0)$. Since the time elapsed is h seconds,

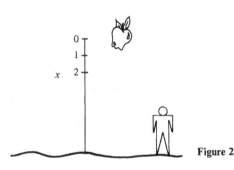

Figure 2

the average change in velocity, i.e., the average acceleration, is

$$\frac{v(t_0 + h) - v(t_0)}{h}$$

As we let $h \to 0$, we obtain $(dv/dt)|_{t_0}$, which is the same as $d(dx/dt)/dt = d^2x/dt^2$ evaluated at t_0. This is called the *instantaneous acceleration* at time t_0.

The mathematical description of Assumption 1 is

$$\frac{dx}{dt} = ax \tag{1}$$

where a is a constant yet to be determined.

It should be noted that we are now departing from historical accuracy because calculus and differential equations hadn't been invented in Galileo's time. From now on we are telling the story as if a modern mathematical modeler were carrying out the investigation.

Mathematical manipulation We solve Equation (1) as follows:

$$\frac{dx}{dt} = ax$$

$$\frac{dx}{x} = a\, dt$$

$$\int \frac{dx}{x} = \int a\, dt$$

$$\ln(x) = at + c$$

$$e^{\ln(x)} = e^{at+c}$$

$$x = e^c e^{at}$$

$$x = ke^{at} \tag{2}$$

We can evaluate k like this: at $t = 0$ the object is at rest; so $x = 0$ when $t = 0$. Making these substitutions in Equation (2) gives

$$0 = ke^0 = k$$

Consequently Equation (2) becomes

$$x = 0 \qquad \text{for all } t \tag{3}$$

This says that the object will never move, no matter how long we wait.

Evaluation Since the conclusion is completely absurd and there are no mistakes in the mathematical manipulation, we need a reformulation. Galileo eventually came to this conclusion also.

Formulation (again) Galileo's new formulation is the same as before except that Assumption 1 is replaced by

Assumption 1* If a body falls from rest, then its velocity at any point is proportional to the time it has been falling. In particular, for each second of fall, the object gains an extra 32 feet/second in velocity.

The mathematical description of this assumption is

$$\frac{dx}{dt} = 32t \tag{4}$$

Mathematical manipulation From Equation (4) we obtain

$$dx = 32t \, dt$$

$$\int dx = \int 32t \, dt$$

$$x = 16t^2 + c$$

When $t = 0$, we have $x = 0$. Substituting these values gives $c = 0$; so

$$x = 16t^2 \tag{5}$$

Evaluation This law of falling bodies agrees well with observations in many circumstances. Section 3 deals with cases for which this model needs further refinement and which involve more cycles through the three steps of model building.

Equations (4) and (5) can be used to solve problems like the following.

Example 1 (a) An object falls, starting at rest, for 2 seconds. How far does it fall and what is its velocity after 2 seconds?

(b) How long does it take an object to fall 144 feet?

Solutions (a) From Equation (4) we find that after 2 seconds

$$\frac{dx}{dt} = 32 \cdot 2 = 64 \text{ feet/second}$$

From Equation (5) we find

$$x = 16 \cdot 2^2 = 64 \text{ feet}$$

(b) From Equation (5) we obtain

$$144 = 16t^2$$

$$t^2 = 9 \text{ seconds}$$

$$t = 3 \text{ seconds}$$

A slight modification of this model deals with the situation where the body does not start at rest, that is, where it has an initial velocity v_0. Galileo assumed that the gains in velocity due to gravity are simply added on to this initial velocity. In place of Equation (4) we have

$$\frac{dx}{dt} = 32t + v_0 \tag{6}$$

When we carry out the integration of Equation (6), we get

$$x = 16t^2 + v_0 t \tag{7}$$

We now turn to an entirely different illustration of the three-step modeling process.

Illustration 2 The Manufacturing Progress Curve

If you have a complicated job to do and you have to do it many times, you'll probably get better at it. Partly this is a matter of learning: practice does make perfect. But ingenuity will also play a role: you will invent shortcuts, devise new tools to assist you, etc. The same is true for a team of workers and managers assembling complex products, like airplanes or automobiles. T. P. Wright studied this in aircraft assembly plants in 1936 and proposed the following model. Since then, it has been used in many branches of manufacturing.

Formulation

1. *Stating the question.* How does the time required to produce an airplane of a given type depend on the number of planes of that type already produced?
2. *Describing relevant factors.* We consider only the time for assembly and the number already assembled. All other factors are ignored. For example, we don't consider whether workers are given cash incentives for efficient work, even though this may be relevant. Wright's data seemed to support the following.

Assumption When the number of planes is doubled, the time for production decreases to about 80 percent of its former value.

3. *Mathematical description.* Let $T(x)$ be the time required for the xth plane. Wright's assumption means that, if the first plane took 100,000 worker-hours, then the second would take 80,000. The fourth would need 64,000. In general, if the time for the first plane is T_1, then the following table describes the situation:

Plane no. x	1	2	4	8	$\cdots$
Hours T	T_1	$(0.8)T_1$	$(0.8)^2 T_1$	$(0.8)^3 T_1$	$\cdots$

The formula we are looking for appears to be

$$T = (0.8)^n T_1 \tag{8}$$

where n is the number of doublings, starting with the number 1, required to obtain x. The formula doesn't involve x directly. In addition it doesn't allow us to compute T for 3 planes, 5 planes, or any number of planes not a power of 2. Some mathematical manipulations are needed to remedy these shortcomings.

Mathematical manipulation The number of doublings to reach x, starting with 1 is, by definition, $\log_2 x$. Thus Equation (8) can be rewritten

$$T(x) = 0.8^{\log_2 x} T_1$$

Taking base-2 logarithms, we get

$$
\begin{aligned}
\log_2 T(x) &= (\log_2 x)(\log_2 0.8) + \log_2 T_1 \\
&= \log_2 x^{\log_2 0.8} + \log_2 T_1 \\
&= \log_2 T_1 x^\alpha
\end{aligned}
$$

where $\alpha = \log_2 0.8 = (\log_{10} 0.8)/(\log_{10} 2) = -0.322$. We can remove the logs from the last equation to obtain

$$
\begin{aligned}
T(x) &= T_1 x^\alpha \\
&= T_1 x^{-0.322}
\end{aligned} \tag{9}
$$

This is called the *manufacturing progress curve* for airplanes, and 80 percent is called the *progress rate*. Figure 3 shows the general shape of such a curve.

Example 2 A new assembly process for airplanes, which uses more automated machine assembly and less human labor, is introduced. Consequently the opportunities for learning decline and the progress rate shifts from 80 to 90 percent. Assume the first plane made with the new process took 100,000 hours.
(a) What is the equation for $T(x)$?
(b) How long does it take to produce the hundredth plane?

Solution (a) From Equation (9) we obtain

$$T(x) = 100{,}000 x^\alpha$$

where $\alpha = \log_2 0.9 = (\log_{10} 0.9)/(\log_{10} 2) = -0.152$.

(b) $$
\begin{aligned}
T(100) &= 100{,}000(100)^{-0.152} \\
&= 49{,}659 \text{ hours}
\end{aligned}
$$

Evaluation Equation (9) gives a pretty good fit to the data available for airplanes. For products other than airplanes, the same basic model seems to apply, but the improvement rate is not necessarily 80 percent. Consequently, the exponent in Equation (9) will not necessarily be -0.322 for other products.

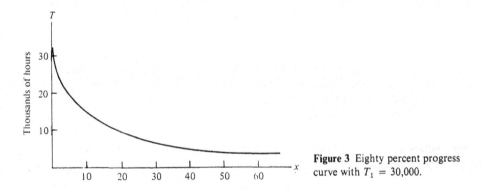

Figure 3 Eighty percent progress curve with $T_1 = 30,000$.

However Equation (9) has one theoretical consequence which may not be entirely realistic. Because

$$\lim_{x \to \infty} x^{-0.322} = 0$$

we must conclude from Equation (9) that the time to produce an airplane will get as close to 0 as we like if we produce enough of them. Here is a reformulation of the model designed to overcome this objection.

Formulation (again) Everything is the same as in our first formulation, but now we assume that there is a certain minimum number of worker-hours needed to assemble a plane under ideal circumstances (when workers and managers have learned all that can be learned about how to do the work, when the best tools and procedures have been devised, etc.). In any given assembly, the time used is this minimum time T_m plus "excess time" T_e. T_m is always the same, but T_e will decrease to zero as x increases:

$$T(x) = T_m + T_e(x)$$

Finding a reasonable formula for $T_e(x)$ requires a close look at the data. See, for example, R. V. Conway and A. Schultz, The Manufacturing Progress Function, *The Journal of Industrial Engineering*, February, 1959. A reasonable first try might be to let $T_e(x)$ have the form earlier obtained for $T(x)$ so that

$$T(x) = T_m + T_0 x^\alpha \tag{10}$$

where T_0, T_m, and α are constants that depend on the product involved.

EXERCISES

• **1** A golf ball is dropped from a resting position, and it falls for 3 seconds.
    (a) How far has it fallen?
    (b) What is its velocity after 3 seconds?

2 Suppose a golf ball is thrown downward (instead of being dropped from rest) with an initial velocity of 8 feet/second.

(a) How far has it fallen after 3 seconds?

(b) What is its velocity after 3 seconds?

● **3** An object falls x feet, starting from rest.

(a) Find a formula for t, the time required, in terms of x.

(b) How long does it take an object to fall 64 feet if it starts from rest?

(c) How fast is the object going after it has fallen 64 feet?

4 Two balls are dropped from rest, the first from a height of 16 feet and the second from 32 feet.

(a) What is the ratio of times required for them to hit the ground?

(b) What is the answer if the heights are h and $2h$?

● **5** Show that, if x is the distance a body dropped from rest falls, then

$$\frac{dx}{dt} = 8\sqrt{x}$$

6 We showed that Galileo's first model was wrong for a body falling from rest. If the initial velocity is $v_0 \neq 0$, what equation would we obtain from Galileo's first model to express x in terms of t, a, and v_0?

7 Which equation for gravity is easier to test experimentally: $v = 32t$ or $x = 16t^2$? Explain what measuring devices would be used.

● **8** Suppose a ball is thrown *upward* with an initial upward velocity of 32 feet/second. Set up a suitable equation linking v and t and then obtain one linking x and t. (Think about whether you want to keep positive values on the distance axis pointing down, as we have done so far. If you wish, you can reverse the distance axis.)

9 A ball starts from rest and rolls down an inclined plane which makes an angle of θ with the ground. Build a reasonable mathematical model to describe how the velocity and distance traveled vary with time.

10 You are going to test the equation $x = 16t^2$. You start your stopwatch and at the same time shout to your friend standing on the roof of a building 32 feet above you. Upon hearing you, your friend drops a ball. When you hear it hit the ground, you stop your stopwatch and record the time t'. This is not exactly the time required for the body to fall because it also includes delays for human reaction time (suppose reaction time is $\frac{1}{4}$ second) and the time for the sound of the shout to travel.

(a) How would you correct t' to give the true time of descent?

(b) Can you redesign the experiment to make these factors irrelevant?

● **11** The first Boeing 707 required about 150,000 worker-hours of labor. If the progress rate is 80 percent, find the number of worker-hours needed for the thirty-second plane. Use Equation (9).

12 Suppose the tenth refrigerator produced took half as long as the first. What is the progress rate? Use Equation (9) with the appropriate α.

13 Suppose $T = 10,000$ hours and the progress rate is 70 percent. Plot a graph of $T(x)$ for x in the range from 1 to about 30.

● **14** What manufacturing progress function $T(x) = T_1 x^\alpha$ describes the table of values below?

Unit number	1	2	4	8
Hours of labor	32,000	25,600	20,480	16,384

15 If T_1 is the time required for the first airplane and Equation (10) is the manufacturing progress curve, express T_1 in terms of T_0 and T_m.

16 Can you find a manufacturing progress function of the form $T(x) = T_1 x^\alpha + T_m$ that gives a good fit to the data in Exercise 14?

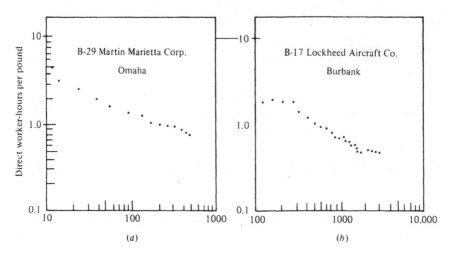

Figure 4 Log-log plots of worker-hours (vertical) versus quantity. (*Redrawn after Miguel Reguero,* " *An Economic Study of the Military Airframe Industry*," *Wright-Patterson Air Force Base, Ohio, Dept. Air Force, pp. 231–235, October, 1957.*)

17 How do you think Equation (9) would change if an aircraft manufacturer offered financial incentives to workers and managers for more efficient work and for making useful suggestions about new procedures?

18 Explain how it can be that Equation (9) gives a good fit to actual data even though it predicts that the time for assembly approaches 0 in the limit.

19 When data are collected and plotted for manufacturing progress curves, T and x are often replaced by their logs, $y = \log T$ and $z = \log x$. Figure 4a shows such a plot, in which the points are nearly on a straight line, say $y = mz + b$. What equation involving T and x could be deduced from this straight-line relation?

20 Suppose the plot of y versus z (see Exercise 19) shows a leveling off, as in Figure 4b. Is this consistent with an equation of the form $T(x) = T_1 x^\alpha + T_m$, $T_m \neq 0$?

Computer exercise

21 Write a program to compute the *total* time required to build any given number of planes, based upon an equation of the form $T(x) = T_1 x^\alpha$. Inputs to the program should be: the time for the first plane, the rate r (from which α is computed), and the desired number x. Output should be the total time.

BIBLIOGRAPHY

For information about Galileo's second model, based on Assumption 1*, consult any comprehensive calculus text or basic physics text.

Andress, F.: The Learning Curve as a Production Tool, *Harvard Bus. Rev.*, p. 87, January, 1954. Contains case studies of the various uses of the manufacturing progress curve.

Koyre, Alexandre: "Galileo Studies," Humanities Press, Atlantic Highlands, N.J., 1978. Chapter 1 of Part II has a discussion of Galileo's first model based on (the mistaken) Assumption 1.

Sherbert, Donald R.: "The Manufacturing Progress Curve," UMAP Module 511, available from COMAP, 271 Lincoln St., Lexington, MA 02173. A very readable and slightly more extensive account than the one given here.

Wright, T. P.: Factors Affecting the Cost of Airplanes, *J. Aeronaut. Sci.*, vol. 3, p. 122, 1936. This is the first published work on the manufacturing progress curve.

3 APPROXIMATE AND LIMITED MODELS— GRAVITY, ROCKETS, AND RAINDROPS

Abstract Mathematical models usually describe only very limited aspects of the world around us and, even then, only give approximately correct answers. Often the limitations and approximations are acceptable for our purposes. If they are not, we can sometimes build a new and better model by adding certain refinements. In this section, we illustrate these points by examining the limitations of Galileo's model of gravity. We study a number of replacements. Each of these alternatives is useful in some circumstances but inappropriate in others.

Prerequisites Formulating differential equations and understanding their meaning. (Solving differential equations is not required, except for certain exercises and one simple occurrence in the text.) Scientific notation. A hand calculator will probably be helpful.

In this section we study falling bodies. The notation is the same as in Section 2. We use t to measure time, in seconds, and x to measure distance, in feet. The distance axis is vertical, and its positive axis points toward the earth; i.e., as a body falls, x increases. (It would be possible to have the axis point up instead, and we shall do this later when discussing rockets.) x is a function of t and will often be written $x(t)$. Likewise, the velocity of a body v varies with time and we may denote it $v(t)$.

Model 1 (Galileo's Model)

Historians of science often assert that the first truly modern scientific mind belonged to a brilliant (but allegedly cantankerous) Italian professor of mathematics named Galileo. He devised the first good mathematical model of gravity, based on the following assumption:

An object, falling to the earth from a *moderate height* and *subjected only to the force of gravity*, gains an extra 32 feet/second in velocity for each second in which it falls. In other words, its acceleration is a constant value of 32 feet/ second. This is true no matter what the weight of the object is.

The italicized phrases are escape clauses, and we shall soon see their significance.

We can express Galileo's assumption with any one of the following equivalent equations:

$$v(t) = v(0) + 32t \tag{1}$$

$$\frac{dx}{dt} = v(0) + 32t \tag{2}$$

$$\frac{d^2x}{dt^2} = 32 \tag{3}$$

where $v(0)$ denotes the initial velocity.

Example 1

A golf ball is at rest at $t = 0$ and dropped to the earth at that instant. The experimenter times its fall and discovers that it takes 1.5 seconds to reach the earth. How fast is it going when it lands?

In this problem $v(0) = 0$. Therefore, Equation (1) implies

$$v(1.5) = 0 + (32)(1.5) = 48 \text{ feet/second}$$

Can we check the prediction of this example experimentally? It's possible but not easy because measuring the velocity of a golf ball when it hits the earth is difficult. Fortunately, we can derive some consequences of our model which are easier to test. Integrating both sides of Equation (2) gives

$$x(t) = v(0)t + 16t^2 + c \tag{4}$$

To evaluate c, substitute $t = 0$ and obtain $c = x(0)$, i.e., the distance traveled after 0 seconds. This is 0. So $c = 0$, and Equation (4) becomes

$$x(t) = v(0)t + 16t^2 \tag{4'}$$

Example 2

How far has the golf ball of Example 1 traveled after 1.5 seconds?

Using Equation (4') and recalling that $v(0) = 0$, we immediately obtain $x(1.5) = 16(1.5)^2 = 36$ feet. This conclusion is testable with relatively simple equipment—a tape measure and stopwatch. A fancier way to do this experiment involves flash photography. Beginning when the ball is dropped, the camera shutter clicks at equal time intervals, but always on the same frame of the film. The various positions of the ball are frozen on the same photograph. Figure 1 shows a drawing of what such a photo might look like. Measurements on the photo can be converted to actual distances by multiplying by a suitable scaling factor. Table 1 shows data collected and scaled up in this way from a photograph

Table 1

Ball position	Distance fallen, centimeters
1	7.70
2	16.45
3	26.25
4	37.10
5	49.09
6	62.18
7	76.36
8	91.58
9	107.89
10	125.34

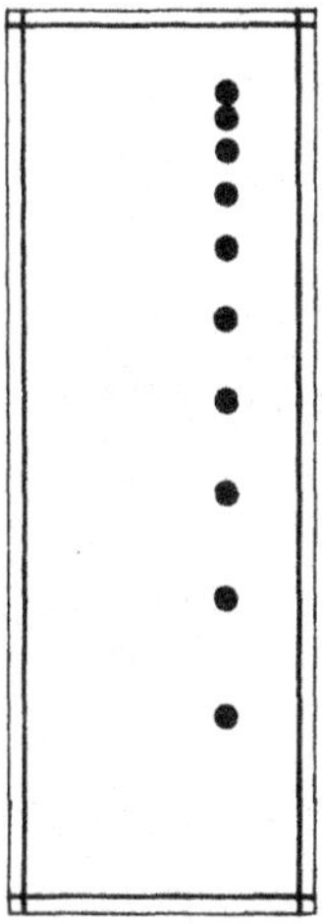

Figure 1

where the camera clicked every $\frac{1}{30}$ second. If you convert centimeters to feet (2.54 centimeters = 1 inch), you will see that these data agree pretty well with model 1.

The next example, however, shows an experiment we could do that would contradict model 1.

Example 3

A raindrop, beginning at rest, falls from a cloud 1024 feet above the ground. How long does it take to reach the ground?

If we set $x(t) = 1024$ in Equation (4′) and solve for t, we obtain

$$1024 = 16t^2$$

$$t = 8 \text{ seconds}$$

However, if we actually performed the experiment, we would discover two things that contradict model 1: first, that the weight of the raindrop makes an important difference in the time it takes to fall and, second, that the fastest time (for the largest raindrop) is about 40 seconds—5 times as long as predicted by model 1 in the above calculation.

What's gone wrong? The problem with applying model 1 is that it is only valid if the object is subjected only to the force of gravity. (Recall that this was one of our escape clauses.) In the case of our raindrops, the force of gravity is opposed by a significant amount of air drag—a lucky thing for us or we might be killed by falling raindrops. Air drag is also present when we drop a golf ball, but it is smaller in relation to the force of gravity because of the greater density of the golf ball and the shorter distance of fall.

The previous example shows clearly that we need a new model to cope with raindrops. Actually we will develop two in detail: one for very small droplets and another for large ones.

Model 2 (Stokes' Law)

For spherical droplets falling in motionless air and having a diameter $D \leq 0.00025$ feet, the acceleration due to gravity [the 32 in Equation (3)] is opposed by an amount proportional to the velocity of the raindrop, specifically by an amount equal to $(0.329 \times 10^{-5}/D^2)dx/dt$. While we are at it, we will also replace the gravitational acceleration of 32 with a more accurate 32.2. Thus,

$$\frac{d^2x}{dt^2} = 32.2 - \frac{0.329 \times 10^{-5}}{D^2}\frac{dx}{dt} \tag{5}$$

Instead of trying to solve this differential equation (see Exercise 8 for hints on solving it), we shall show that it predicts something drastically different from the predictions of model 1: the existence of a *terminal velocity*, that is, a velocity that is an upper bound to how fast the body can go at any time during its fall.

Theorem 1

Model 2 implies a terminal velocity v_{term}.

Proof By setting the right side of Equation (5) equal to zero, we discover a value for dx/dt at which $d^2x/dt^2 = 0$, namely,

$$\frac{dx}{dt} = \frac{32.2 \times 10^5}{0.329}D^2$$

If the droplet ever achieves this velocity, then $d^2x/dt^2 = 0$, which means the rate of change of the velocity is zero. This means the body continues at this velocity. Therefore $v_{term} = (32.2 \times 10^5/0.329)D^2$.

Actually, although we shall not prove it, a droplet falling according to Equation (5) never quite reaches its terminal velocity but gets closer and closer to it. Unless its fall is interrupted by hitting the ground, the velocity eventually becomes so close to v_{term} that, for practical purposes, we consider it equal to v_{term}.

Furthermore, clouds are sufficiently high and a water droplet gets close to its terminal velocity quickly enough that it is not a bad assumption to suppose that the droplet travels at its terminal velocity for its whole trip.

Example 4

Find the terminal velocity of a drizzle drop with diameter $D = 0.00025$ feet. Compare it to the terminal velocity of a fog droplet with one-tenth of that diameter ($D = 0.000025$).

We substitute the value of D in Equation (5), set $d^2x/dt^2 = 0$, and solve for dx/dt. The value which results is v_{term}. For the drizzle drop,

$$v_{term} = \frac{32.2(0.00025)^2}{0.329} \times 10^5$$

$$= 0.612 \text{ feet/second}$$

a bit more than 7 inches/second. For the fog droplet,

$$v_{term} = \frac{32.2(0.000025)^2}{0.329} \times 10^5 = 0.00612 \text{ feet/second}$$

This is one one-hundredth the velocity of the drizzle drop. At this slow rate, droplets need about 165 seconds to fall 1 foot. This, of course, corresponds exactly to our experience of fog: it hardly seems to be falling at all; mostly it just appears to hang around. Indeed we often notice fog lifting. This is because its rate of fall is so slow that the merest updraft will overcome the rate of fall.

In order to study the effects of gravity on larger raindrops, we now introduce yet another model.

Model 3 (Velocity-Squared Model)

For spherical raindrops falling in still air and having diameter $D \geq 0.004$ feet, the acceleration due to gravity is opposed by an amount proportional to the square of its velocity, specifically an amount equal to $(0.000460/D)(dx/dt)^2$; Thus Equation (3) is replaced by

$$\frac{d^2x}{dt^2} = 32.2 - \frac{0.000460}{D}\left(\frac{dx}{dt}\right)^2 \tag{6}$$

This model, like model 2, predicts a terminal velocity which we can obtain by setting $d^2x/dt^2 = 0$ and solving for dx/dt. The result is

$$v_{term} = \sqrt{\frac{32.2D}{0.000460}} = 264\sqrt{D} \tag{7}$$

Example 5

For a raindrop of diameter $D = 0.004$ feet, find the terminal velocity. Also, find how long it takes to reach the ground if it starts its descent in a cloud 3000 feet high.

From formula (7), $v_{term} = 264\sqrt{0.004} = 16.7$ feet/second. Observe how much faster this is than the drizzle drop and fog droplet discussed in Example 4.

To find how long it takes to fall 3000 feet, we make the assumption that it reaches its terminal velocity nearly instantaneously. So we can suppose it goes at

16.7 feet/second for the whole distance. Then the time is $3000/16.7 = 180$ seconds, which is 3 minutes.

The alert reader may have noticed that, although we have given models for small ($D \leq 0.00025$) and big ($D \geq 0.004$) drops, we have given no model for medium-sized ($0.00025 \leq D \leq 0.004$) drops. There is a more general model, which applies to drops of any diameter and incorporates both models 1 and 2. However, as we shall see, the general model is inconvenient to use. Here is a brief description of it.

Model 4 (General Air-Drag Model)

For a spherical raindrop of diameter D the differential equation describing the fall function $x(t)$ is

$$\frac{d^2x}{dt^2} = 32.2 - \frac{0.000920C(D\,dx/dt)(dx/dt)^2}{D} \tag{8}$$

C is a function of the product $D\,dx/dt$, which we will call the *drag-coefficient function*. A glance at Figure 2 shows that the dependence is not a simple one, which could be expressed by a convenient formula. Note that the figure actually shows the graph of log C plotted against log ($D\,dx/dt$).

In studying the general air-drag model, it is convenient to distinguish four separate ranges of values for $D\,dx/dt$.

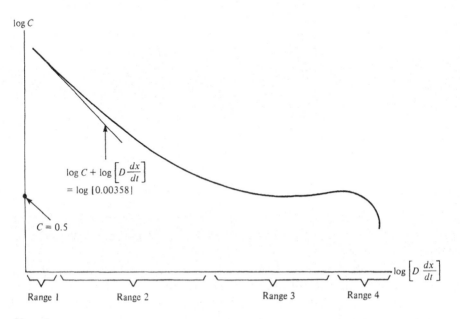

Figure 2

Range 1 (small values of $D \, dx/dt$). Here the plot of $\log C$ versus $\log (D \, dx/dt)$ is linear, with approximate equation

$$\log C + \log \left(D \frac{dx}{dt} \right) = \log 0.00358$$

From the laws of logarithms, we obtain the equivalent equation

$$C = \frac{0.00358}{D \, dx/dt}$$

Substituting this into Equation (8) gives Equation (5).

Theory and experiment both show that, when $D \leq 0.00025$ feet, the raindrop spends all its time in range 1, so that Equation (5) applies to its entire fall. See Exercise 4.

Range 2. In this range there is no convenient equation relating C and $D \, dx/dt$. For this reason, we cannot simplify Equation (8) to get a convenient differential equation. We could find an inconvenient one, but for our present purposes there would be little learned in doing this; so we shall not take the trouble.

Range 3. Here $\log C$ is approximately constant, which means C is approximately constant, with a value about 0.5. Substituting this in Equation (8) gives Equation (6).

Theory and experiment show that, when $D \geq 0.004$ feet, the raindrop spends nearly all its time in range 3, so that Equation (6) can be assumed to apply to its entire fall.

Range 4. Here again we can't obtain a convenient formula for C as a function of $D \, dx/dt$. However, it can be shown that this range never applies to raindrops. The reason is that a raindrop would have to be very large to fall in range 4 and raindrops never get that large because they split when their diameters reach about 0.02 feet.

Before leaving the general air-drag model, it should be noted that it is not as general as it could be. Equation (8) applies only to a raindrop falling through air in the lower atmosphere. It would not apply if the falling object were denser than a raindrop, for example, a speck of dust or a small pebble. Equation (8) may also not apply if the medium in which the object falls were different. For example, water offers more resistance to a pebble falling through it than air does. Pancake batter would offer even more. The physical characteristics of the medium that determine the resistance it offers are its density ρ_m and its viscosity (a measure of its stickiness) μ_m. Here is a generalization of Equation (8) that takes into account the density of the falling object ρ_o, the density of the medium ρ_m, and the viscosity of the medium μ_m:

$$\frac{d^2x}{dt^2} = 32.2 - \frac{3\rho_m C(D \, dx/dt)}{4\rho_o D} \left(\frac{dx}{dt} \right)^2 \qquad (9)$$

Table 2 Experimentally determined terminal velocities for rain-drops of various sizes

Drop diameter (feet)	Terminal velocity (feet/second)	Drop diameter (feet)	Terminal velocity (feet/second)
0.00033	0.89	0.00852	24.82
0.00066	2.36	0.00918	25.64
0.00098	3.84	0.00984	26.43
0.00131	5.31	0.01049	27.08
0.00164	6.75	0.01115	27.67
0.00196	8.10	0.01180	28.20
0.00230	9.41	0.01246	28.59
0.00262	10.72	0.01311	28.95
0.00295	12.03	0.01377	29.25
0.00328	13.21	0.01443	29.44
0.00393	15.21	0.01508	29.61
0.00459	16.95	0.01574	29.73
0.00525	18.52	0.01639	29.80
0.00590	19.96	0.01705	29.90
0.00656	21.28	0.01770	29.96
0.00721	22.62	0.01836	30.03
0.00787	23.84	0.01902	30.06

As before, C is a function whose graph can be described as in Figure 2. Its exact shape will depend on the values of ρ_m and μ_m.

Model 5 (Inverse-Square Law)

Our discussion of models 2, 3, and 4 is meant to show how important air drag can be and how far wrong one can go if one uses model 1, which assumes air drag is not present, in cases where air drag is substantial. However, it would be wrong to think that air drag is all that prevents Equation (3) of model 1 from being exactly correct. Even for objects falling in a vacuum, where clearly there is no drag of any sort, Equation (3) is false. The reason is that the acceleration which the earth's gravitation causes in a raindrop or other object is not a constant, but varies with the distance between the object and the earth. We shall describe this in model 5.

Model 5 will differ from the other models in yet another way: it asserts that gravitational attraction is something not only the earth possesses. Instead, every object in the universe exerts a gravitational attraction on every other object. For small objects, like raindrops, we do not notice its power to attract other objects. Basically, the reason is that a raindrop is very small and the gravitational attraction an object can exert is proportional to its mass. Since the earth has a very large mass, its gravitation has effects we notice, namely objects falling to the earth.

The particular unit of mass we will use is called the *slug*. One slug is an amount of mass that would weigh 32.2 pounds on earth. This definition does not mean

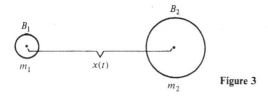

Figure 3

that mass and weight are more or less the same. If an object weighs 32.2 pounds on a certain scale on earth and we take that object and the same scale into outer space, the object will now be weightless, even though it still has 1 slug of mass. If we take it to the moon instead, it will have weight but less than on earth. (Do you remember how easily astronauts were able to jump around on the moon?)

Let B_1 and B_2 be bodies with masses m_1 and m_2 respectively and let $x(t)$ denote the distance between their centers of mass at time t (Figure 3). Then each body attracts the other and, if no other forces are acting, their motions are described as follows:

1. The instantaneous acceleration of B_1 is in a straight line toward B_2 and, at a time t, has magnitude $Gm_2/x(t)^2$, where G is a constant called the *gravitational constant*.
2. The instantaneous acceleration of B_2 is in a straight line toward B_1 and, at time t, has magnitude $Gm_1/x(t)^2$, where G is the gravitational constant.
3. The value of the gravitational constant depends on the units of measurement chosen. If we measure mass in slugs, distance in feet, and time in seconds, then $G = 34.3 \times 10^{-9}$ (approximately).

There are two important differences between this model and model 1:

1. Each object moves toward the other.
2. The accelerations are not constant, but vary with the distance between the bodies and get larger as the bodies get closer.

These differences from model 1 may be negligible or considerable, depending on the masses of the bodies and their distance apart. To illustrate this, we shall consider three examples of objects being attracted to the earth: a raindrop, a rocket ship, and a comet. In these examples we pay no attention to air drag. This is done for simplicity, not realism.

Example 6

Raindrops are too small to move the earth appreciably, as we now show. Let m_e be the mass of the earth and m_r the mass of a typical raindrop. The exact values of m_e and m_r won't concern us, but we will need their ratio m_e/m_r, which is about 6×10^{31}, an enormous number. Let a_e denote the acceleration of the earth due to the attraction of the raindrop. We will show that this is negligible in comparison to a_r, the acceleration of the raindrop produced by the attraction of the earth. Specifically, we will show that the ratio a_e/a_r is negligible.

Model 5 implies

$$a_e = G \frac{m_r}{x(t)^2} \quad \text{and} \quad a_r = G \frac{m_e}{x(t)^2}$$

Therefore

$$\frac{a_e}{a_r} = \frac{m_r}{m_e} = \frac{1}{m_e/m_r}$$

$$= \frac{1}{6 \times 10^{31}} = 1.7 \times 10^{-32}$$

This is indeed a negligibly small number. This value implies $a_e = a_r(1.7 \times 10^{-32}) = 32.2(1.7 \times 10^{-32}) = 5.47 \times 10^{-31}$ feet/second^2, an acceleration so small it could not be perceived or measured.

Example 7

We will now show that raindrops are too close to the earth to change their acceleration much as they fall.

Let R denote the radius of the earth, approximately 2.09×10^7 feet. Let h be a variable which denotes the height of the raindrop above the surface of the earth. We will compare the raindrop's acceleration at the start of the drop's fall, at $h = 5000$ feet, say, with its acceleration when it hits the ground, that is, when $h = 0$. If a_h denotes the acceleration at h feet, so that, for example, a_0 denotes the acceleration at the instant the raindrop hits the ground, model 5 yields

$$a_h = G \frac{m_e}{(R + h)^2} \quad \text{and} \quad a_0 = G \frac{m_e}{R^2}$$

The fractional increase is

$$\frac{a_0 - a_h}{a_h} = \frac{a_0}{a_h} - 1$$

$$= \left\{ \frac{G(m_e/R^2)}{G[m_e/(R + h)^2]} \right\} - 1$$

$$= \left(\frac{R + h}{R} \right)^2 - 1$$

$$= 2\frac{h}{R} + \frac{h^2}{R^2}$$

If the initial height is 5000 feet, h/R is $(5 \times 10^3)/(2.09 \times 10^7) = 2.39 \times 10^{-4}$. Thus

$$\frac{a_0 - a_{5000}}{a_{5000}} = 2(2.39) \times 10^{-4} + (2.39)^2 \times 10^{-8}$$

$$= 4.78 \times 10^{-4} + 5.71 \times 10^{-8}$$

This is a very small fractional increase. When converted to a percentage, it is less than one-twentieth of 1 percent. For most research involving raindrops this small percentage increase in acceleration is safely ignored.

If we wish to compute the value of the raindrop's nearly constant acceleration, we need to know that $m_e = 4.10 \times 10^{23}$ slugs (approximately). Then

$$a_0 = \frac{34.3 \times 10^{-9} \times 4.10 \times 10^{23}}{(2.09)^2 \times 10^{14}}$$

$$= 32.2 \text{ feet/second}^2$$

which agrees with the number we have been using.

The moral of the last two examples is that the complications of model 5 are irrelevant for the study of raindrops. In the next example we show that one of these complications is important in rocketry.

Example 8

An old saying asserts that "what goes up must come down." But man has propelled rocket ships into space which have never returned to earth, a clear contradiction of the proverb. Actually, the proverb would be exactly and universally true if model 1 were exactly and universally true. The reason the proverb fails and a rocket can "escape" from the gravitational attraction of the earth is that the truth about gravity is better captured by model 5 than model 1.

Suppose a rocket leaves the earth with an initial velocity v_0 and $x(t)$ is its distance from the center of the earth at t seconds after blast-off. In our simplified example, after the initial thrust from the firing of its engines, there are no further forces acting on the rocket beside the earth's gravitational attraction. (In reality there will be other forces: air resistance; the gravitational attraction of the sun, moon, and other heavenly bodies; and, finally, the rocket possibly having a second or third stage to provide additional acceleration midway in its flight.)

If the rocket returns to earth, it must change its direction (from up to down) at some time, say t'. At that instant it is neither going up nor down; so its velocity dx/dt is 0.

Model 1 implies

$$\frac{dx}{dt} = v_0 - 32t$$

[This equation looks different from Equation (2). The reason for the minus sign on the 32t term is that the positive direction of the x axis is now pointing up instead of down.] Clearly, if $t = v_0/32$, we will have $dx/dt = 0$. Further, when $t > v_0/32$, we will have $dx/dt < 0$, which means the rocket is now coming down. Thus, model 1 implies the following more precise version of the proverb: "What goes up starts to come down after $v_0/32$ seconds."

Model 5 implies

$$\frac{d^2x}{dt^2} = -G\frac{m_e}{x^2}$$

Multiplying by dx/dt gives

$$\frac{d^2x}{dt^2}\frac{dx}{dt} = -Gm_e x^{-2}\frac{dx}{dt}$$

Integrating both sides with respect to t produces

$$\frac{1}{2}\left(\frac{dx}{dt}\right)^2 = Gm_e x^{-1} + c \tag{10}$$

At $t = 0$, $dx/dt = v_0$ and $x = R$. Making these substitutions gives $c = v_0^2/2 - Gm_e/R$. Thus Equation (10) becomes

$$\left(\frac{dx}{dt}\right)^2 = v_0^2 - \frac{2Gm_e}{R} + \frac{2Gm_e}{x(t)} \tag{11}$$

If $v_0 \geq \sqrt{2Gm_e/R}$, the right side of Equation (11) is positive; so dx/dt will never be 0. The critical value $\sqrt{2Gm_e/R}$ is called the *escape velocity*, denoted v_{esc}. Its numerical value is

$$v_{esc} = \sqrt{\frac{2Gm_e}{R}}$$

$$= \sqrt{\frac{2 \times 34.3 \times 4.10 \times 10^{-9} \times 10^{23}}{2.09 \times 10^7}}$$

$$= 36{,}684 \text{ feet/second}$$

which is about 7 miles/second. Therefore, the proverb we can deduce from model 5 is: "What goes up must come down, unless it starts going up faster than about 7 miles/second."

The conclusion of the last example is that at least one of the differences between models 5 and 1 is highly important for rocket ships. What about the other difference, the fact that not only does the earth attract the rocket, but the rocket also attracts the earth? It is easy to calculate, following the same procedure used for the raindrop, that the rocket's gravitational pull on the earth is negligible.

Is anything large enough to exert a measurable pull on the earth? Objects in our everyday experience on earth are far too small. The sun is large enough and close enough to the earth to exert a measurable force of attraction. This force is what keeps the earth in orbit around the sun instead of moving off in a straight-line path. The moon and the planets provide only minor perturbations of this basic orbit. The only other heavenly bodies worth considering are comets.

Could the earth be wrenched out of its orbit by a close encounter with a comet? To answer this question, we need to be a little more precise about what we mean. What does it mean to be "wrenched out of its orbit"? Any passing comet (or body of any size) will jiggle the earth's orbit slightly. Mostly these changes are so small as to be unmeasurable. In fact, there does not appear to be any case in recorded history where a comet has caused a change in the earth's orbit which was large enough to be detected.

In 1770 Lexell's comet passed within about 8×10^9 feet of the earth. This is about one sixty-second of the distance from the earth to the sun. During its approach, astronomers were unable to determine its mass and were in some suspense about it because they knew that, if the mass had been as large as 10^{17} slugs, it would have produced a measurable change in the earth's orbit. However, no measurable change occurred. Scientists concluded that Lexell's comet must have been smaller than 10^{17} slugs. But it is interesting to note that there are comets of about this size, although most comets range in size between 10^{11} and 10^{17} slugs. So, in a sense, Lexell's comet was a lucky close encounter. Perhaps someday there will be a close encounter with one of the larger comets which will not turn out so lucky.

Let us now summarize the various models we have discussed and put them in perspective. A good place to start is Galileo's model (model 1), which is mathematically the simplest and easiest to work with. Unfortunately, we pay the following prices for this simplicity.

1. It only applies to objects falling to the earth.
2. It only applies to objects falling from moderate heights because it ignores the inverse-square relationship of gravitation to distance (height).
3. It only applies to fairly dense objects because it ignores air drag.

We have improved Galileo's model in two directions (see Figure 4): in one direction, we have added the effects of air drag; in the other, we have incorporated the inverse variation of gravitational attraction with the square of the distance. Table 3 shows how these improved models compare when evaluated against the shortcomings of Galileo's model and in terms of how easy or hard it is to solve the differential equation resulting from the model.

As Table 3 and Figure 4 make clear, we have not discussed any models that take into account both air drag *and* the inverse-square law. Why not build a completely general model, which would take these and all other factors into account? Then, instead of choosing our model to fit the particular circumstances, we could use our general model for any situation. Such models have been devised, but they are unwieldy because they require a great deal of calculation. Whenever possible, scientists prefer something simple. (There is a joke about this: a great

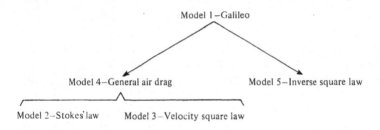

Figure 4

scientist, after a lifetime of studying the complexities of the natural world, remarks about the Creation, "If the good Lord had consulted me, I would have recommended something simpler.")

Unfortunately, not all scientific problems are amenable to simplification. A case in point is the calculation of the trajectories of artillery shells and rockets. Here one often needs to use both the inverse-square law and calculation of air drag. Furthermore, models 2, 3, and 4 for air drag are not adequate since they assume a spherical shell traveling through air which has uniform density and no wind currents. We need new and more complex equations that take into account the shape of the shell, wind patterns, density of the air at various altitudes, and so on. The situation became so complicated that, during and after World War I, a large number of American mathematicians worked on ballistics for the U.S. Army and produced books of firing tables for every combination of atmospheric factors and types of guns and shells. Often a single trajectory would involve about 12 hours of calculation if done by a human being. The automation of this process became the immediate motivation for, and the initial task of, the world's first electronic digital computer.

Table 3

	Convenience of differential equation	Incorporates resistance of medium (air drag)	Handles attracting bodies other than earth	Inverse square
Model 1: Galileo	Good	No	No	No
Model 2: Stokes' law air drag	Moderate	Yes	No	No
Model 3: Velocity-squared air drag	Moderate	Yes	No	No
Model 4: General air drag	Poor	Yes	No	No
Model 5: Inverse square	Moderate	No	Yes	Yes

EXERCISES

● **1** (a) How far will a golf ball, starting from rest, fall in 5 seconds?

(b) How many seconds will it take a golf ball to fall 64 feet if it starts from rest?

(c) How many seconds will it take a golf ball, starting from rest, to attain a velocity of 160 feet/second.

2 (a) Suppose an object, beginning with a velocity of 0 at $t = 0$, falls in accordance with model 1. How fast is it going after falling 16 feet? 100 feet? [*Hint*: First find the time elapsed. Use Equations (4') and (1).]

(b) Find a formula expressing $v(t)$ as a function of $x(t)$ for objects obeying model 1.

● **3** (a) Calculate terminal velocities of raindrops with the following diameters: $D = 0.00005, 0.00010, 0.00015$, and 0.00020. (Think carefully about which model to use.) Use these calculations to make a graph of v_{term} plotted against diameter. What kind of curve do these points lie on?

(b) Calculate terminal velocities of raindrops with the following diameters: $D = 0.005, 0.006, 0.007, 0.008, 0.009$, and 0.010. [Should you use the same model as in part (a)?] Use these calculations to make a graph of v_{term} plotted against diameter. Use the same set of axes as in part (a) to make comparison convenient.

4 Range 1 in Figure 2 extends from $D\,dx/dt = 0$ to about 6×10^{-8}. Find the largest value of D for which v_{term} is small enough to make Dv_{term} be in range 1. (It follows then that, all the while such a raindrop is accelerating up to this velocity, it is in range 1. Furthermore, any smaller raindrop will fall in this range—the Stokes' law range—for its entire descent.)

5 For D values less than or equal to the value found in Exercise 4, make a graph of v_{term} as a function of D. On this same graph, plot the experimentally determined values from Table 2.

● **6** Suppose a certain-size raindrop spends its entire fall in range 1 of Figure 2 and reaches its terminal velocity so soon after starting its fall that we are willing to assume that it spends its entire fall at this velocity. Under these assumptions, what is the formula for $x(t)$, the distance covered after t seconds of fall? (Your formula can involve D as well as t.)

7 Do Exercise 6 under the assumptions the drop spends practically its entire fall in range 3 and at its terminal velocity.

8 In model 2 (Stokes' law), if we set $v = dx/dt$, the differential equation becomes $dv/dt = 32.2 - av$, where $a = 0.329 \times 10^{-5} \times D^{-2}$. This leads to the integration problem

$$\int \frac{dv}{32.2 - av} = \int dt$$

(a) Perform the integrations and show that $v = 32.2a^{-1}(1 - e^{-at})$. In evaluating the constant of integration, assume $v = 0$ when $t = 0$.

(b) An approximation sometimes used for e^x, when x is small, is $e^x = 1 + x$. Substitute this in the formula found in part (a) and show that model 1 results.

(c) Replace v by dx/dt in the formula of part (a) and then solve for x as a function of t.

9 (a) Let $t \to \infty$ in the formula of part (a) of Exercise 8 to find a formula for v_{term}. How does it compare with the one in the next? Show that we may write the formula for v as $v/v_{term} = 1 - e^{-at}$.

(b) Use the formula for v/v_{term} to calculate the t value at which a raindrop with $D = 0.00025$ feet reaches 99 percent of its terminal velocity.

(c) Repeat part (b) for $D = 0.000025$.

10 Set $v = dx/dt$ in the differential equation for model 3 and obtain an integration problem involving the variable v. Show by integration that

$$\frac{dx}{dt} = v = \sqrt{\frac{32.2}{r} \frac{e^{bt} - 1}{e^{bt} + 1}}$$

where $r = 0.00046/D$ and $b = 2\sqrt{32.2r}$. Use the initial condition $v = 0$ when $t = 0$ to evaluate the constant of integration.

11 Solve the differential equation found in Exercise 10; i.e., find x as a function of t. Use the initial condition $x = 0$ when $t = 0$. (See the next exercise for the answer.)

12 Verify that the differential equation for model 3 has as a solution

$$x = \frac{1}{2r} [2 \log(1 + e^{bt}) - bt - 2 \log 2]$$

where $r = 0.00046/D$ and $b = 2\sqrt{32.2r}$.

● **13** (a) Let $t \to \infty$ in the formula for v found in Exercise 10 and thereby find a formula for v_{term}. Use this to find a formula for v/v_{term}.

   (b) Use the formula for v/v_{term} to find the value of t at which v is 99 percent of v_{term}.

14 In Example 6, find how small x [also denoted $x(t)$] would have to be in order for the raindrop, in the absence of other forces, to cause the earth to be accelerated by 32 feet/second^2. Explain why such a value of x is not physically possible.

15 Calculate the exact value of a_{5000} in Example 7.

● **16** Suppose a spacecraft has a mass of 7000 slugs. What acceleration would the craft cause on the earth, in the absence of other forces, if it were 10^7 feet above the earth's surface? What would the acceleration be if the craft were 1 foot above the earth's surface?

17 For a rocket with initial velocity $v_0 < \sqrt{2Gm_e/R}$, show that there will be a maximum distance from the earth, i.e., the function $x(t)$ has a maximum. Find the value of this maximum.

● **18** The weight of a body on earth is defined to be its mass times the acceleration that the earth would produce on the body if it were dropped in a vacuum, isolated from other forces, right at the earth's surface. The weight of a body on the moon is the mass times the acceleration when dropped at the moon's surface. If the earth's radius and density are R_e and ρ_e while the moon's radius and density are R_m and ρ_m, express the ratio of a body's weight on the moon to its weight on earth. (*Hint*: Use the inverse-square law to express the accelerations in terms of m_e and m_m, the masses of the earth and moon respectively, and m, the mass of the body, which remains the same on earth as it is on the moon.)

Computer exercises

19 Table 1 contains observed values. Write a computer program to compute the theoretical distances according to model 1. The method is shown in Example 2. Use the program to compute distances at multiples of $\frac{1}{30}$ second and compare them to the actual values in the table. (Be sure to convert your theoretical values to centimeters for comparison with the table.)

20 Write a computer program to compute the terminal velocity of a raindrop when the diameter is given as input. Restrict the inputs to ranges 1 and 3 and have the program use the appropriate model. Compare the results to Table 2. Experiment to see if any of the models used by your program give decent results for range 2.

BIBLIOGRAPHY

Binder, R. C.: "Fluid Mechanics," Prentice-Hall, Englewood Cliffs, N.J., 1973. Technicalities about drag in air and other fluids.

Blanchard, D. C.: "From Raindrops to Volcanoes," Anchor Books (Doubleday), New York, 1967. This delightful elementary book contains no mathematics, but may inspire some mathematical thoughts. Our exercises on raindrops were motivated by material in this book.

Gunn, Ross, and Gilbert D. Kinzer: The Terminal Velocity of Fall for Water Droplets in Stagnant Air, *J. Meteor.*, vol. 6, p. 243, 1949.

Shapiro, Ascher H.: "Shape and Flow," Anchor Books (Doubleday), Garden City, N.Y., 1961. The story behind air drag is here in a form requiring little mathematics, but some physics, for complete comprehension.

4 MACRO AND MICRO POPULATION MODELS I—EXPONENTIAL GROWTH

Abstract for Sections 4, 5, and 6 Models can be built at different levels of detail. A "macro model" is like a low-power telescope: it takes in a big picture, but not much fine detail is visible. A "micro model" is like a high-power telescope: you see the fine structure, but the field of view is narrow. We illustrate this in the next three sections by presenting three models for population projection, each occupying a different position on the macro-micro spectrum. The first can be thought of as an answer to the question of how many people there will be in the future. The second, the Leslie-matrix model, can be regarded as an answer to the question of how many people of various ages there will be in the future. The last, the sex and family-planning model, can be considered an answer to the question of how many people there will be in your family in the future.

Prerequisites The first of the three sections requires only high school algebra. The second requires understanding matrix multiplication. Some exercises use a bit more linear algebra. The third section requires some probability theory: the independence rule and expected values.

How many people will there be in the population of a certain country in 10 years? How many births? In this section we build the simplest possible model for answering these questions.

We will use the variable t to measure time in years, with $t = 0$ denoting the present. Let $P(t)$ denote the size of the population at time t. Let $B(t)$ be the number of births in the 1-year interval between time t and $t + 1$ and let $D(t)$ denote the number of deaths between t and $t + 1$. The main assumption of our model is that certain rates stay the same. The rates we have in mind are defined as follows:

Definition

1. $B(t)/P(t)$ is called the *birth rate* for the time interval t to $t + 1$.
2. $D(t)/P(t)$ is called the *death rate* for the time interval t to $t + 1$.

Assumptions

1. The birth rate is the same for all intervals. Likewise, the death rate is the same for all intervals. This means that there is a constant b, called the birth rate, and a constant d, called the death rate so that, for all $t \geq 0$,

$$b = \frac{B(t)}{P(t)} \quad \text{and} \quad d = \frac{D(t)}{P(t)} \qquad (1)$$

2. There is no migration into or out of the population; i.e., the only source of population change is birth and death.

As a result of Assumptions 1 and 2 we deduce that, for $t \geq 0$,

$$
\begin{aligned}
P(t + 1) &= P(t) + B(t) - D(t) \\
&= P(t) + bP(t) - dP(t) \\
&= (1 + b - d)P(t)
\end{aligned}
\tag{2}
$$

Setting $t = 0$ in (2) gives

$$
P(1) = (1 + b - d)P(0)
\tag{3}
$$

Setting $t = 1$ in Equation (2) and substituting Equation (3) gives

$$
\begin{aligned}
P(2) &= (1 + b - d)P(1) \\
&= (1 + b - d)(1 + b - d)P(0) \\
&= (1 + b - d)^2 P(0)
\end{aligned}
$$

Continuing this way yields

$$
P(t) = (1 + b - d)^t P(0)
\tag{4}
$$

for $t = 0, 1, 2, \ldots$. The constant $1 + b - d$ is often abbreviated r and called the *growth rate* or, in more high-flown language, the *Malthusian parameter*, in honor of Robert Malthus who first brought this model to popular attention. In terms of r, Equation (4) becomes

$$
P(t) = P(0)r^t \qquad t = 0, 1, 2, \ldots
\tag{5}
$$

$P(t)$ is an example of an *exponential function*. Any function of the form cr^t, where c and r are constants, is an exponential function.

Example 1

Suppose the current population is 250,000,000 and the rates are $b = 0.02$ and $d = 0.01$. What will the population be in 10 years? Formula (4) immediately gives

$$
\begin{aligned}
P(10) &= (1.01)^{10}(250,000,000) \\
&= (1.104622125)(250,000,000) \\
&= 276,155,531.25
\end{aligned}
$$

Naturally, this result is absurd, since one can't have 0.25 of a person. This is a good illustration that the fundamental assumptions of the model are not exactly true, but only approximately. Thus the predictions of the model are only approximations.

Example 2

How many years will it take for the population of Example 1 to double its initial size?

We seek a value of t for which $P(t)/P(0) = 2$. This requires that

$$\frac{(1.01)^t P(0)}{P(0)} = 2$$

Therefore t must satisfy $(1.01)^t = 2$. $t = \log 2/\log 1.01 \approx 69.66$ years.

Let us go further and ask how long the next doubling will take. We want to find k so that $P(69.66 + k)/P(69.66) = 2$. This yields

$$\frac{(1.01)^{69.66+k} P(0)}{(1.01)^{69.66} P(0)} = 2$$

Canceling terms brings us to the same equation we had before: $(1.01)^k = 2$. We conclude that each doubling requires the same period of time.

This is contrary to our common experience of how things increase. For example when you were 2 years old, it took 2 years for you to double your age. But then you were 4, and the next doubling took longer, namely, 4 years. Or suppose you have saved \$100 toward a car and add to your savings by \$50 per month. At first your savings will double in 2 months. But the next doubling takes longer, 4 months. Most of our experience with growth involves examples like this, where the doubling time is not constant. Experts on population have often claimed that population growth problems are especially difficult for the man in the street to understand because population does not grow in ways we are familiar with.

Finally we turn to the question of how many births there will be in the 1-year period from t to $t + 1$. Since $B(t) = bP(t)$, Equation (4) implies

$$B(t) = b(1 + b - d)^t P(0) \tag{6}$$

Example 3

Using the data of Example 1, how many births will occur between $t = 10$ and $t = 11$? Formula (6) gives the answer as

$$B(10) = (0.02)(1.01)^{10}(250,000,000)$$
$$= 5,523,110.6$$

How good is our model? Figure 1 shows the U.S. population between 1900 and 1970 and the best fit by an exponential function. It's not a bad fit, but it certainly is far from exact. Actually, it appears that the population could best be fit by two exponentials, one covering the first three points and the other covering the last four. The transition between the two occurs in the depression decade from 1930 to 1940, when economic circumstances altered birth rates. This example illustrates the great flaw of our model: the assumption of constant rates doesn't

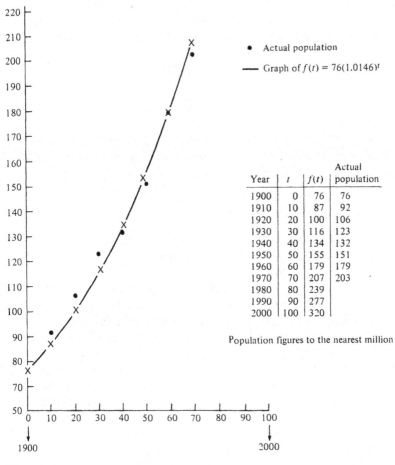

Year	t	$f(t)$	Actual population
1900	0	76	76
1910	10	87	92
1920	20	100	106
1930	30	116	123
1940	40	134	132
1950	50	155	151
1960	60	179	179
1970	70	207	203
1980	80	239	
1990	90	277	
2000	100	320	

Population figures to the nearest million

• Actual population

— Graph of $f(t) = 76(1.0146)^t$

Figure 1

hold exactly. The virtue of the model is that it is simple and its assumption of constant rates is sometimes close enough to the truth for some practical purposes.

EXERCISES

● 1 Suppose the current population is 1,500,000, $b = 0.03$, and $d = 0.01$. What is the population in 5 years according to Equation (5)?

2 Suppose the current population is 1,000,000, $b = 0.025$, and $d = 0.01$. What was the population last year according to the Malthusian model?

● 3 What can you conclude about $P(t)$ if the birth rate equals the death rate?

4 Is it possible, in the Malthusian model, for the population to decline? Explain how.

● 5 Show that, for any fixed number of years, say k, the percent by which the population increases in k years is a function of k, b, and d alone—it does not depend on the population size at the start of the k-year period.

6 (a) Let t^* be the doubling time, that is the period of years required for the population to double its size. Show that $t^* = \log 2/\log (1 + b - d)$.

(b) Find the constant and linear terms of the Taylor series of the function $\log (1 + x)$ evaluated around $x = 0$. Use this approximation to replace the denominator of the formula of part (a). Then show that (approximately) $t^* = 0.7/(b - d)$.

● **7** The crude birth rate b^* differs from our birth rate b in that births are computed per 1000 people in the population at the middle of the year in which those births occurred. Thus $b^* = 1000B(t)/P(t + \frac{1}{2})$. Find the formula relating b^* to b and d.

8 Find a formula for $D(t)$ in terms of $P(0)$, b, and d.

9 (a) Equation (4) is claimed to be valid only for $t = 0, 1, 2, \ldots$. What, if anything, could you say about $P(1\frac{1}{2})$? What is the relation between $P(1\frac{1}{2})$ and $P(2\frac{1}{2})$?

(b) Can you find a formula for $P(t + \varepsilon)$ in terms of t and $P(\varepsilon)$?

(c) Is there anything in our assumptions that would enable us to find a relation between $P(\frac{1}{2})$ and $P(0)$?

10 Suppose a single bacterium splits in two after 1 hour ($t = 1$). After another hour ($t = 2$) each of these splits in two, and the process continues like this. What are b and d, and what is the formula for $P(t)$?

Computer exercises

11 Calculate 1.01^{10} and $P(10)$ (as in Example 1) with a hand calculator. Compare with the accuracy a computer gives.

12 Carry out a computer simulation of the situation in Exercise 10, but with this difference: at the splitting times ($t = 1, 2, \ldots$) there is only a 0.75 chance of the splitting occurring and a 0.25 chance that it does not occur. If no split occurs, then the next opportunity for that bacterium to split is 1 hour later. Tabulate the population for $t = 1, 2, 3, \ldots, 15$ and see whether this tabulation can be fit with a function of the form $P(t) = P(0)r^t$.

BIBLIOGRAPHY

Keyfitz, Nathan: "Applied Mathematical Demography," John Wiley & Sons, Inc., New York, 1977. Chapter 1, on aspects of exponential growth, is a nice supplement to our Section 4.

5 MACRO AND MICRO POPULATION MODELS II— THE LESLIE MATRIX

How many people aged 65 to 70 will there be in 10 years? This is often more useful to know than how many people there will be altogether. If we want to know how much social security will have to be paid out in 10 years or how many schools, nursing homes, or obstetricians will be needed, we'll need a model which recognizes and projects age groups. There is no way that the very macro model of the last section can help us. The task of this section is to develop a slightly less macro model in which age groups are separately "tracked" into the future.

A peculiarity our model will have is that men are completely ignored: we do not take into account how many there are, nor do we attempt to predict how many there will be. The reason for this female chauvinism is partly demographic tradition and partly because it helps overcome some technical problems, which we shall not describe.

In our model we imagine the female population divided into age categories $[0, \Delta), [\Delta, 2\Delta), \ldots, [(n-1)\Delta, n\Delta)$. Here Δ, which is the width of each age interval, can be any convenient number, and n is a number sufficiently large that only a negligible number of women survive beyond $n\Delta$ years. In practice $\Delta = 5$ and $n = 20$ are often used. We could achieve more precision by taking 1-year age groups ($\Delta = 1$), but then we would need about 100 of these age groups ($n = 100$), which would generate a lot of data to handle, a 100×100 matrix with 10,000 entries.

We will use the variable t to measure time (in years) with $t = 0$ being the present. Our model will not be able to tell us the populations of the age groups for all times in the future, but only for a series of instants in the future spaced Δ years apart: $t = \Delta$, $t = 2\Delta$, etc. Thus we are going to take a series of snapshots of the future population. There exist other models which, in effect, give a movie of the future population, by giving predictions for all times in the future, but we will not deal with these continuous models here.

Let $F_i(t)$ denote the number of females at time t in the ith age group, i.e., with ages in the interval $[i\Delta, (i+1)\Delta)$. We define the column vector $\mathbf{F}(t)$ by

$$\mathbf{F}(t) = \begin{bmatrix} F_0(t) \\ F_1(t) \\ \vdots \\ F_{n-1}(t) \end{bmatrix}$$

and call this the *age distribution vector* for time t. $\mathbf{F}(0)$ is the current age distribution and known to us from census data. Our task is to predict $\mathbf{F}(\Delta)$, $\mathbf{F}(2\Delta)$, ...

A graphical representation of $\mathbf{F}(t)$ is often made in the form of a "population pyramid" (Figure 1), which is simply a bar graph. The bars represent the various age groups: the lowest age group is represented by the bottom bar, and the highest by the top bar. The length of a bar is proportional to the population of the age group it represents. Thus we can think of our snapshots of the population at times $t = 0, \Delta, 2\Delta, \ldots$ as being a series of population pyramids. The problem of making a population projection is to determine the pyramids for $t = \Delta, 2\Delta, \ldots$, given the pyramid for $t = 0$.

Obviously, to do this trick of prediction, it is necessary to have some information about birth and death rates for the various age groups. Therefore let d_i

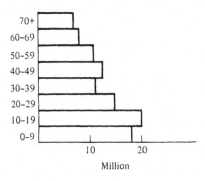

Figure 1 Population pyramid for American females (1969) with $\Delta = 10$.

be the death rate for the ith age group; specifically, d_i is the fraction of the ith age group which, on account of death,[†] will not be present in the $(i + 1)$st age group Δ years later. The fraction surviving is therefore $1 - d_i$, which we denote p_i. This survival rate p_i is assumed to be in effect for all future times covered by our projection. This means

$$F_{i+1}(t + \Delta) = (1 - d_i)F_i(t)$$
$$= p_i F_i(t) \qquad (1)$$

for $t = 0, \Delta, 2\Delta, \ldots$, up to the future-most time one intends to project for.

Let m_i denote the Δ-year maternity rate for the ith group. This means that, for any t value, the average woman in the ith age group at time t will, by having babies in between times t and $t + \Delta$, contribute m_i children to the lowest age group at time $t + 1$. The m_i are assumed constant in time. Thus the number of newborns (age 0) at time $t + \Delta$ is

$$F_0(t + \Delta) = \sum_{i=0}^{n-1} m_i F_i(t) \qquad (2)$$

Equations (1) and (2) can be visualized in terms of the population pyramids. Consider the $F_i(t)$ females in age group i at time t. Geometrically they are the "contents" of the ith bar in the pyramid for time t (see Figure 2a). What becomes of them as one time period (Δ years) passes? Some will die, and the survivors will be in the next age group $(i + 1)$ at time $t + \Delta$. Thus the fraction p_i of the ith bar at time t flows into the $i + 1$ bar at time $t + \Delta$. This is indicated by the arrows in the figure.

By following the arrows in reverse, we can trace all but one age group at $t + \Delta$ back to the age group it came from at t. The exception is the lowest age group at $t + \Delta$. It can't be traced back because these people weren't alive yet at time t. All members of this age group were born during the Δ-year period between t and $t + \Delta$. Therefore we must seek their origin as births to those females in the population who are capable of becoming mothers between t and $t + \Delta$.

In human populations, only certain age groups are capable of reproduction. Let α be the index of the lowest-age group capable of reproduction over the next Δ years and let β be the index of the highest-age group capable of reproduction over the next Δ years. For example, if $\Delta = 5$, then the lowest-age childbearing group is the [5, 10) group. (Over 5 years this group will mature into the [10, 15) group and, in the course of this growth, may produce some children.) The index of this group is 1; so $\alpha = 1$. The highest-age childbearing group is [40, 45); so $\beta = 8$. Thus $m_0 = 0, m_1 = m_\alpha > 0, m_2, \ldots, m_8 = m_\beta > 0$, and $m_9, \ldots = 0$.

In Figure 2b we have shaded certain bars to represent the childbearing age groups that contribute to the lowest-age group at $t + \Delta$. Since the average woman in the ith age group contributes m_i children, the total contribution from this age group is $m_i F_i(t)$. Adding these contributions for the various age groups gives the

[†] We do not take migration into account in our model.

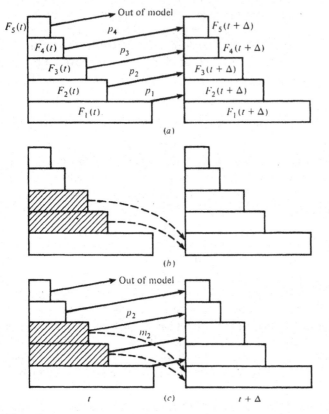

Figure 2

total size of the lowest-age group at $t + \Delta$. This is indicated by the dashed arrows in Figure 2b. We use dashed arrows instead of solid ones because in Figure 2a the solid arrows indicated an actual flow, or transfer, of people from one group to another, while here the arrows indicate a creation of new females.

Figure 2c shows both processes of mortality and maternity in the same scheme. Although this figure is logically identical to Equations (1) and (2) and is often considered superior to them for understanding the nature of population projection, numerical and theoretical work is usually done with the equations or a matrix reformulation of them, which we now describe.

Equations (1) and (2) can be written as a single matrix equation:

$$
\begin{bmatrix} F_0(t + \Delta) \\ F_1(t + \Delta) \\ \vdots \\ F_{n-1}(t + \Delta) \end{bmatrix} = \begin{bmatrix} m_0 & m_1 & m_2 & \cdots & m_{n-1} \\ p_0 & 0 & 0 & \cdots & 0 \\ 0 & p_1 & 0 & \cdots & 0 \\ \multicolumn{5}{c}{\dotfill} \\ 0 & 0 & 0 & \cdots & p_{n-2}\,0 \end{bmatrix} \begin{bmatrix} F_0(t) \\ F_1(t) \\ \vdots \\ F_{n-1}(t) \end{bmatrix} \quad (3)
$$

which is valid for $t = 0, \Delta, 2\Delta, \ldots$, up to the latest time one wishes to project to. The $n \times n$ matrix occurring in this matrix equation is called the Leslie matrix, and we will denote it by M. Equation (3) can then be abbreviated

$$F(t + \Delta) = MF(t) \qquad t = 0, 1, 2, \ldots \qquad (4)$$

Introducing this single matrix equation in place of the n equations in Equations (1) and (2) was a simple enough innovation by P. H. Leslie, but a very important one because it opens the way to the use of powerful matrix theory, as we will soon describe. The story goes that Leslie, who was completely self-educated in mathematics, spent some time during World War II in a remote part of the world, where his only scientific reading matter was P. A. M. Dirac's book on quantum mechanics. Leslie was impressed with how useful matrices were in quantum mechanics and concluded that they must be equally useful in demography!

Particular instances of Equation (4) for $t = 0$ and $t = \Delta$ are

$$F(\Delta) = MF(0) \qquad \text{and} \qquad F(2\Delta) = MF(\Delta)$$

Substituting the first of these into the second gives

$$F(2\Delta) = MMF(0) = M^2F(0)$$

Likewise, $F(3\Delta) = MM^2F(0) = M^3F(0)$. Continuing with this argument will show

$$F(k\Delta) = M^kF(0) \qquad k = 0, 1, \ldots \qquad (5)$$

Example 1

Imagine a population divided into three age groups. (This is too few to be realistic for a human population, but it makes a convenient numerical example which might be relevant to an animal population.) Initially ($t = 0$) the population of females is divided into three age groups, as in the pyramid below.

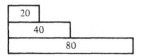

Suppose that, as one time unit passes, everyone in the oldest group dies and one-quarter of those in each of the other age groups dies. Suppose also that the age-specific maternity rates are $m_0 = 0$, $m_1 = 1$, and $m_2 = 2$. Find the age distribution vectors $F(\Delta)$ and $F(2\Delta)$ and represent them as population pyramids.

The information given about mortality implies $p_0 = \frac{3}{4}$ and $p_1 = \frac{3}{4}$. Therefore the Leslie matrix is

$$\begin{bmatrix} 0 & 1 & 2 \\ \frac{3}{4} & 0 & 0 \\ 0 & \frac{3}{4} & 0 \end{bmatrix}$$

and the initial vector is

$$\begin{bmatrix} 80 \\ 40 \\ 20 \end{bmatrix}$$

To find $F(\Delta)$, we use Equation (5) with $k = 1$:

$$F(\Delta) = \begin{bmatrix} F_0(\Delta) \\ F_1(\Delta) \\ F_2(\Delta) \end{bmatrix} = \begin{bmatrix} 0 & 1 & 2 \\ \frac{3}{4} & 0 & 0 \\ 0 & \frac{3}{4} & 0 \end{bmatrix} \begin{bmatrix} 80 \\ 40 \\ 20 \end{bmatrix} = \begin{bmatrix} 80 \\ 60 \\ 30 \end{bmatrix}$$

Observe that there is no change in the youngest age group. However the total population has increased by 30. Will the same be true as we project one additional period? In order to calculate $F(2\Delta)$, we can use either the equation $F(2\Delta) = MF(\Delta)$ or $F(2\Delta) = M^2F(0)$. The first of these is more convenient since we have already computed $F(\Delta)$, but we shall do the computation both ways:

1. $$F(2\Delta) = \begin{bmatrix} F_0(2\Delta) \\ F_1(2\Delta) \\ F_2(2\Delta) \end{bmatrix} = \begin{bmatrix} 0 & 1 & 2 \\ \frac{3}{4} & 0 & 0 \\ 0 & \frac{3}{4} & 0 \end{bmatrix} \begin{bmatrix} 80 \\ 60 \\ 30 \end{bmatrix} = \begin{bmatrix} 120 \\ 60 \\ 45 \end{bmatrix}$$

2. $$F(2\Delta) = \begin{bmatrix} F_0(2\Delta) \\ F_1(2\Delta) \\ F_2(2\Delta) \end{bmatrix} = \begin{bmatrix} 0 & 1 & 2 \\ \frac{3}{4} & 0 & 0 \\ 0 & \frac{3}{4} & 0 \end{bmatrix}^2 \begin{bmatrix} 80 \\ 40 \\ 20 \end{bmatrix}$$
$$= \begin{bmatrix} \frac{3}{4} & \frac{3}{2} & 0 \\ 0 & \frac{3}{4} & \frac{3}{2} \\ \frac{9}{16} & 0 & 0 \end{bmatrix} \begin{bmatrix} 80 \\ 40 \\ 20 \end{bmatrix} = \begin{bmatrix} 120 \\ 60 \\ 45 \end{bmatrix}$$

Here are the population pyramids for $t = 0$, Δ, and 2Δ.

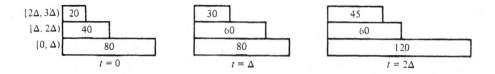

So far our Leslie-matrix model does not give us the same conclusion as the model of Section 4, namely that total population is an exponential function of time. It is true that Equation (5) bears a notational resemblance to Equation (5) of the last section, but the meanings of the terms in the two equations are different: M is a matrix while r is a number; P denotes total population while F is an age distribution vector. However, by applying some linear algebra to Equation (5), we can arrive at a conclusion analogous to exponential growth.

It is possible to show that, in most cases, there exist constant vectors E_0, $E_1, \ldots, E_{n-1}$ and numbers (possibly complex) $\lambda_0, \lambda_1, \ldots, \lambda_{n-1}$ so that

$$M^k F(0) = \lambda_0^k E_0 + \lambda_1^k E_1 + \cdots + \lambda_{n-1}^k E_{n-1} \qquad (6)$$

This is a bit more like Equation (5) of the last section in that the exponents k occur on numbers rather than on matrices. Furthermore, it is possible to show that, if k is large enough,[†] only the $\lambda_0^k E_0$ term on the right-hand side makes a significant contribution to the sum. [Technically, $M^k F(0)$ is asymptotically equivalent to $\lambda_0^k E_0$.] Thus we can write

$$\mathbf{F}(k\Delta) = M^k \mathbf{F}(0) \approx \lambda_0^k E_0$$

If we now wish to find $P(k\Delta)$, the total population at time $k\Delta$, we would add the components of $\mathbf{F}(k\Delta)$. Therefore

$$P(k\Delta) \approx s\lambda_0^k$$

where s is the sum of the components of E_0. By setting $k = 0$ in this equation, we observe that $s = P(0)$. If we set $k\Delta = t$ and $r = \lambda_0^{1/\Delta}$, then our equation becomes

$$P(t) = P(0)r^t \qquad (7)$$

which is identical to Equation (5) of the Malthusian model, described in the last section.

Exercises 6 and 7 show how one arrives at an equation like Equation (7) computationally if one starts with a particular Leslie matrix and initial age distribution.

Sidelight: The Story Behind the Constant-Rates Assumption

How true are the assumptions that the birth and death rates don't change? History gives conflicting evidence. In some times and places this assumption was ridiculously wrong. But in others it was not too bad an assumption, especially if one was not too fussy about exactness and willing to take Δ to be large enough that bad times (like epidemics and wars) got averaged out with better times.

Looking to the future and leaving aside sensational medical breakthroughs or catastrophes like war or new diseases, we have fair confidence that mortality rates will not change drastically in many countries. Fertility is harder to predict. In some countries, like the United States, it is about as low as it can go and has nowhere to go but up; but, in other countries, fertility rates are so high that they can only come down.

[†] For practical demographic purposes $k\Delta$ should be about 70 years.

The modern justification for population projections using constant rates is based on a combination of three main reasons:

1. Often the future is not drastically different from the present provided one doesn't try to project far. With a little luck, a 5-, 10-, or 15-year projection may be fairly accurate.
2. For many purposes an inaccurate projection is better than none at all.
3. Often the purpose of a projection is not to predict the future population, but to show that the continuation of present rates is impossible. For example, birth rates in many underdeveloped countries are quite high, whereas public health measures have lowered death rates to levels comparable with those in industrialized countries. When projections are made from these rates, one predicts staggering increases in population, which could not possibly be absorbed by the country in question. Limitations on food, resources, or just plain living space will intervene. Something has to give, and that something will have to be either a rise in death rates or a decline in birth rates. Since allowing death rates to rise is unpleasant, these projections usually are used to argue that governments had better sponsor birth control measures. In fact this argument has been fairly successful. It may well be that population projection is the single most influential mathematical model in use in the world today because of this political application.

EXERCISES

● **1** Assume $F(0)$ is $\begin{bmatrix} 800 \\ 1000 \\ 1200 \end{bmatrix}$, $m_0 = \frac{1}{2}$, $m_1 = 1$, and $m_2 = 0$ while $p_0 = \frac{3}{4}$ and $p_1 = \frac{1}{2}$.

(a) What is the Leslie matrix M?
(b) Find $F(\Delta)$.
(c) Find $F(2\Delta)$ in two different ways: first by $F(2\Delta) = MF(\Delta)$ and second by $F(2\Delta) = M^2F(0)$.
(d) Draw the population pyramids for $t = 0$, Δ, and 2Δ.

2 Do Exercise 1 under the assumption that survival rates are half as high ($p_0 = \frac{3}{8}$ and $p_1 = \frac{1}{4}$) and maternity rates are twice as high ($m_0 = 1$, $m_1 = 2$, and $m_2 = 0$). Use the same initial age distribution vector. Before computing the answers, can you guess how they will come out?

● **3** Do Exercise 1 with the same rates given in Exercise 1 but with an initial age distribution vector $F(0)$ twice as high in each age category as the one in Exercise 1. Before computing the answers, can you guess or predict how they will come out?

4 Suppose that survival and maternity rates change with time so that, instead of having one projection matrix M to deal with, we have a series: M_0 to get from $t = 0$ to $t = \Delta$, M_1 to get from $t = \Delta$ to $t = 2\Delta$, and, in general, M_i to get from $t = i\Delta$ to $t = (i + 1)\Delta$. What formula now connects $F(k\Delta)$ with $F(0)$ in place of Equation (5)?

● **5** (a) Show that, if $F(0)$ and $G(0)$ are age distribution vectors that differ only in their last component and M is a Leslie matrix, the particular form of M [see Equation (3)] implies that $MF(0) = MG(0)$. In nonmathematical language this says that the last age group plays no role in determining the population distribution at the next time step.

(b) Suppose we proceed from $MF(0) = MG(0)$ to $M^{-1}MF(0) = M^{-1}MG(0)$ and then to $F(0) = G(0)$. The latter is a contradiction because $F(0)$ and $G(0)$ are assumed to have different last components. Where is the flaw in the argument?

6 (a) For the Leslie matrix $M = \begin{bmatrix} \frac{3}{2} & 2 \\ \frac{1}{2} & 0 \end{bmatrix}$, show that $M\begin{bmatrix} 4 \\ 1 \end{bmatrix} = 2\begin{bmatrix} 4 \\ 1 \end{bmatrix}$ and $M\begin{bmatrix} -1 \\ 1 \end{bmatrix} = -\frac{1}{2}\begin{bmatrix} -1 \\ 1 \end{bmatrix}$.

(b) Let $\begin{bmatrix} x_0 \\ y_0 \end{bmatrix}$ be any initial population, broken into two age groups. Find a and b (as functions of x_0 and y_0) such that

$$\begin{bmatrix} x_0 \\ y_0 \end{bmatrix} = a\begin{bmatrix} 4 \\ 1 \end{bmatrix} + b\begin{bmatrix} -1 \\ 1 \end{bmatrix}$$

(c) Substitute the equation found in part (b) into the projection equation $\begin{bmatrix} x_n \\ y_n \end{bmatrix} = M^n\begin{bmatrix} x_0 \\ y_0 \end{bmatrix}$. [Here we are using x_n and y_n for the components of $F(n\Delta)$.] Then, taking part (a) into account, show

$$\begin{bmatrix} x_n \\ y_n \end{bmatrix} = a2^n\begin{bmatrix} 4 \\ 1 \end{bmatrix} + b(-\tfrac{1}{2})^n\begin{bmatrix} -1 \\ 1 \end{bmatrix}$$

(d) Finally, if the total population at time n is P_n, show that $P_n = ra2^n = P_0 2^n$.

7 This exercise is an alternate approach to deriving the equation $P_n = P_0 2^n$ of part (d) of Exercise 6.

(a) Show that, if there exist 2×2 matrices z_1 and z_2 such that:

1. $2z_1 - \frac{1}{2}z_2 = M$
2. $z_1 + z_2 = I$, the identity matrix
3. $z_1^2 = z_1$ and $z_2^2 = z_2$
4. $z_1 z_2 = z_2 z_1 = 0$, the zero matrix

then

$$M^2 = 2^2 z_1 + (-\tfrac{1}{2})^2 z_2$$

and, in general,

$$M^n = 2^n z_1 + (-\tfrac{1}{2})^n z_2$$

(b) If $\begin{bmatrix} x_0 \\ y_0 \end{bmatrix}$ is the initial age distribution and $\begin{bmatrix} x_n \\ y_n \end{bmatrix}$ is the age distribution after n time steps, then show that

$$\begin{bmatrix} x_n \\ y_n \end{bmatrix} = 2^n z_1\begin{bmatrix} x_0 \\ y_0 \end{bmatrix} + (-\tfrac{1}{2})^n z_2\begin{bmatrix} x_0 \\ y_0 \end{bmatrix}$$

(c) Find the z_1 and z_2 (numerically) that correspond to the M of Exercise 6. [*Hint*: Conditions 1 and 2 of part (a) give a system of eight equations for the entries of z_1 and z_2. After solving these equations, check whether conditions 3 and 4 hold.]

(d) If P_n is the total population after n time steps ($= x_n + y_n$),

$$P_n = (x_0 + y_0)2^n = P_0 2^n$$

8 (a) Show that, in a three-age-group model, where the rates are $m_0 = 0$, $m_1 = 0$, and $m_2 = 4$; $p_0 = \frac{1}{2}$ and $p_1 = \frac{1}{2}$; and the initial age distribution vector $F(0) = \begin{bmatrix} 1600 \\ 0 \\ 0 \end{bmatrix}$, then P_n, the total population after n time periods, is

$$P_n = \begin{bmatrix} 1600 \\ 800 \\ 400 \end{bmatrix} \quad \begin{array}{l} \text{if } n \text{ is divisible by 3} \\ \text{if } n \text{ leaves a remainder of 1 after division by 3} \\ \text{if } n \text{ leaves a remainder of 2 after divison by 3} \end{array}$$

(b) Show that this is incompatible with the concept of exponential growth. Specifically, there is no constant r such that $P_n = P_0 r^n$.

9 In our projection method, where the time step and the size of the age category are both the same (Δ years), all the members of an age category at time t will be together in one age category at time $t + \Delta$ (but not the same one they were in at time t). Show by actual example that this will not be true if the width of the age category is not the same as the number of years elapsing between the two population pyramids.

Computer exercises

10 (a) Write a computer program to carry out a population projection, as described in this section. Design your program to handle the data in the table below ($\Delta = 5$ and 17 age groups).

(b) At each of the projection times (1969, 1974, 1979, etc.) find the total population. Does it appear to be increasing exponentially?

1964 Population estimates and vital rates (U. S. females)

Age group	Number F_i	Maternity rate m_i	Survival rate p_i
[0, 5)	10,136,000	0	0.99661
[5, 10)	10,006,000	0.00103	0.99834
[10, 15)	9,065,000	0.08779	0.99791
[15, 20)	8,045,000	0.34873	0.99682
[20, 25)	6,546,000	0.47607	0.99605
[25, 30)	5,614,000	0.33769	0.99472
[30, 35)	5,632,000	0.18333	0.99229
[35, 40)	6,193,000	0.07605	0.98866
[40, 45)	6,345,000	0.01744	0.98304
[45, 50)	5,796,000	0.00096	0.97416
[50, 55)	5,336,000	0	0.96222
[55, 60)	4,642,000	0	0.94430
[60, 65)	4,002,000	0	0.91410
[65, 70)	3,332,000	0	0.86938
[70, 75)	2,809,000	0	0.80060
[75, 80)	2,036,000	0	0.68817
[80, 85)	1,179,000	0	

BIBLIOGRAPHY

Keller, Edward L.: "Population Projection," UMAP Module 345, available from COMAP, 271 Lincoln St., Lexington, MA 02173. More advanced analysis of the Leslie matrix.

Tuchinsky, Philip M.: "Management of a Buffalo Herd," UMAP Module 207, available from COMAP, 271 Lincoln St., Lexington, MA 02173. Application of the Leslie matrix with real data about buffalo.

6 MACRO AND MICRO POPULATION MODELS III— FAMILY-PLANNING MODELS

How many children will there be in your family in 10 years? Today many couples try to practice family planning. In effect, they are hoping to be able to answer this question with fair certainty. In this section we will discover that, under certain circumstances, this question cannot be answered with much certainty. For example, a couple who wish to have no children may find that they will not achieve this goal with current contraceptive techniques (excluding abortion).

Our model takes as its basic unit a single couple, living in a sexual union, and not an entire population as in the previous sections. Thus we can't use the ideas of the previous sections, based upon average rates of childbearing and dying, because they are only reasonable when applied to a large group of people. Instead, we use estimates of biological fertility and information about the use or nonuse of contraception. Our predictions will be probabilistic: we will compute the probabilities of various outcomes, average values, and deviations from the averages.

To begin with, we assume that there is, for each couple, a constant p, called the *fecundability*[†] by demographers, which is the probability of conception in any given month for that couple. When we assert the existence of such a constant probability, we are, of course, making a modeling assumption—the assumption that the probability doesn't change for different months. As a matter of fact, this is false: there seem to be more conceptions in the winter. However, we will make the assumption anyway because it is convenient and not too far from the truth.

The value of p depends on many factors: the age and biological state of the couple, their frequency of intercourse, and their use or nonuse of contraception are the main factors involved. For an average, healthy, young couple using no contraception, p may easily be as high as 0.3, while, for an older, still fertile couple using contraception, p may be as little as 0.001.

The commonly held notion that the use of contraception will always avoid conception is very far from the truth. Every known contraceptive has a failure rate.[‡] However, some methods are better than others, as we shall soon see.

The mathematical approach to the measurement of the effectiveness of a contraceptive begins with a refinement of the concept of fecundability. We define the natural fecundability of a couple as the probability of conceiving in any given month provided that no contraceptive measures are taken. We denote this by f. For most couples in the reproductive years, who are not completely infertile, f seems to range from about 0.1 to 0.3. A contraceptive can be thought of as nullifying a certain fraction, say e, of the natural fecundability. The remaining

---

[†] This tongue-twisting technical term is distinct from "fertility" and "fecundity," which have other meanings.

[‡] Here we are speaking of the pill, douche, condom, IUD, and diaphragm. We exclude surgical measures and abortion.

fraction would be the actual fecundability, denoted as before by p:

$$\begin{pmatrix} \text{Actual} \\ \text{fecundability} \end{pmatrix} = \begin{pmatrix} \text{natural} \\ \text{fecundability} \end{pmatrix} - \begin{pmatrix} \text{amount} \\ \text{nullified} \end{pmatrix}$$

$$p = f - ef$$

$$= (1 - e)f \tag{1}$$

The coefficient $1 - e$ may be thought of as the probability the contraceptive method fails in any given month.

The value of e is a number between 0 and 1 (although in the popular press it is often expressed as a percent, e.g., 0.7 is given as "70 percent effective"). If $e = 1$, we have a perfect contraceptive, currently unavailable. Table 1 gives the efficiencies of some methods in common use.

It is important to note that these figures are based on the experience of real people using these methods. Consequently, the failures implied by these figures are only partly due to inherent technical limitations of the methods, but are also due to careless or incorrect use, to some extent. A person who took special care to use a method properly might improve upon these efficiencies.

The e values in Table 1 show that there is no certainty about these family-planning methods. But how serious is this uncertainty really? After all, life is full of uncertainties, and we often do quite well by basing our planning on average outcomes. To explore the uncertainty further, we first obtain a formula for the average waiting time till conception. Following this we examine deviations from the average.

Observe that, if the probability of conceiving in a month is p, then the probability of not conceiving in a given month is $1 - p$. By the multiplication rule for independent binomial trials, we discover that the probability of not conceiving for $k - 1$ consecutive months and then conceiving in the kth month is $(1 - p)^{k-1}p$. Therefore, if we let w denote the random variable "waiting time till conception," then

$$\text{Prob}\{w = k\} = (1 - p)^{k-1}p \tag{2}$$

Figure 1 shows a histogram and table of values for the case $p = 0.1$.

The average value of w, also called the *mean* and denoted $\bar{w}$, is obtained by adding the values of k from 1 to ∞, with each value weighted by the probability that it turns out to be the waiting time. That is,

$$\bar{w} = 1p + 2(1 - p)p + 3(1 - p)^2p + \cdots \tag{3}$$

A trick makes this series easy to sum. Multiply both sides by $1 - p$ to obtain a new equation and then subtract this from Equation (3). After dividing by p, we get

$$\bar{w} = 1 + (1 - p) + (1 - p)^2 + \cdots$$

a geometric series, whose sum we obtain by the standard formula

$$\bar{w} = \frac{1}{p}$$

Table 1

Contraceptive method	$e^{\dagger}$	$\overline{w} = \dfrac{1}{p} = \dfrac{1}{f(1-e)}$	
		$f = 0.1$	$f = 0.3$
Douche	0.84	63 mos.	21 mos.
Rhythm	0.87	77	26
Foam	0.88	83	28
Diaphragm	0.91	111	37
Condom	0.93	143	48
IUD	0.96	250	83
Pill	0.98	500	167

† Estimated from 1970 National Fertility Survey.

In Table 1, we apply this formula, together with Equation (1), to calculate the months of average waiting time for various combinations of contraceptive method and fecundability.

The figures in Table 1 describe what would happen to an average couple. But what about a real flesh and blood couple? Can they assume that these figures will be approximately true for them? A study of Figure 1 shows that, for $p = 0.1$, a substantial fraction of couples seem to have a waiting time substantially less than the mean of 10 months. For example, 10 percent of these couples wait only 1 month. This suggests that the mean waiting time is not a very useful number for

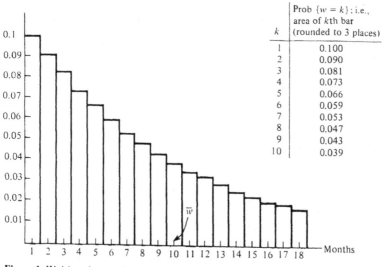

k	Prob $\{w = k\}$; i.e., area of kth bar (rounded to 3 places)
1	0.100
2	0.090
3	0.081
4	0.073
5	0.066
6	0.059
7	0.053
8	0.047
9	0.043
10	0.039

Figure 1 Waiting time probabilities based on $p = 0.1$.

a couple to have in mind, unless they are also aware that substantial deviations from the mean occur very often.

Let us explore this for an arbitrary value of p and in a slightly different way. To find the fraction of couples whose waiting time k is less than or equal to m months, we need to add terms given by formula (2) for $k = 1, 2, \ldots, m$. In terms of the histogram, this corresponds to adding the areas of the first m bars. Examination of Equation (2) shows that we are adding the first m terms of a geometric series, whose ratio is $1 - p$ and whose first term is p. If we call the sum S_m, then the familiar formula for the sum of the first m terms of a geometric series gives

$$S_m = \frac{p - p(1 - p)^m}{1 - (1 - p)} = 1 - (1 - p)^m \tag{4}$$

This formula was derived for integer values of m, the waiting time. In what follows it seems reasonable to admit that a waiting time could involve a fractional number of months. The proper way to do this is simply to use Equation (4), even for non-integer values of m.

Example 1

Find what fraction of women with $p = 0.1$ have a waiting time less than or equal to one-sixth of the mean waiting time for such women.

Solution Since the mean waiting time is $1/0.1 = 10$ months and one-sixth of this is 1.67, we need $S_{1.67}$:

$$S_{1.67} = 1 - (1 - 0.1)^{1.67}$$
$$= 1 - (0.9)^{1.67}$$
$$= 0.161$$

Since 0.161 is about $\frac{1}{6}$, we conclude that, for $p = 0.1$, about 1 in 6 women will have a waiting time only one-sixth of the average or less. It turns out that, even for arbitrary p in the range 0.001 to 0.3, this rule of thumb is approximately true.

Rule of Thumb

For p between 0.001 and 0.3, about one-sixth of all women with fecundability p will have a waiting time only one-sixth as long as the average or less.

Justification We proceed as in the previous example, but with $\frac{1}{6}(1/p)$ in place of 1.67:

$$S_{1/6p} = 1 - (1 - p)^{1/6p}$$

Figure 2 shows a table of values and graph of this as a function of p. Although we have only calculated the particular values shown in the table and represented by heavy dots on the graph, we can justify connecting them to form a mono-tonically increasing curve, as drawn (see Exercise 4). Although the function

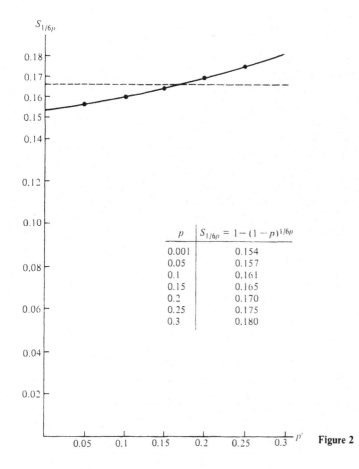

p	$S_{1/6p} = 1-(1-p)^{1/6p}$
0.001	0.154
0.05	0.157
0.1	0.161
0.15	0.165
0.2	0.170
0.25	0.175
0.3	0.180

Figure 2

$S_{1/6p}$ is not constant, the variation is not great, and we will not be far wrong if we estimate that $S_{1/6p} = \frac{1}{6}$ for p between 0.001 and 0.3. This concludes the justification for the rule of thumb.

Example 2

Consider a couple with a low natural fecundability of $f = 0.1$, who use the pill to avoid having children. Table 1 indicates that on the average such a couple will be safe for 500 months. This is over 41 years, quite enough to see them out of their fertile years. However, the rule of thumb shows that about 1 in 6 such couples will conceive before 83.33 months (about 7 years) elapse.

People often regard the rule of thumb as being contrary to common sense because it predicts so much contraceptive failure. Keep in mind that the model deals with average Americans. Certain groups such as college-educated persons or persons with high incomes might have a different experience.

In addition, common sense is a poor way to evaluate this model, because most of us do not have reliable information about the contraceptive practices of other people. By contrast, the e values of Table 1, which lie at the heart of our model, have been scientifically gathered and evaluated. It is true that oversimplified modeling assumptions may have introduced errors. For more elaborate models, see the book by Mindel C. Sheps and Jane A. Menken, cited in the Bibliography.

There are numerous morals one might draw from this model, depending on one's values and ethical perspectives:

1. Contraception doesn't work perfectly, so why bother.
2. We need abortion as a backup measure when contraception fails.
3. We need research to find more perfect contraceptives.
4. It is desirable to use two contraceptives at once.
5. We need to be able to adapt ourselves to unplanned pregnancies and regard them as happy accidents.

EXERCISES

● **1** Find the mean waiting times for women with $f = 0.2$ for the various contraceptive methods.

2 Suppose a couple uses two methods of contraception, with efficiencies e_1 and e_2 respectively. Assume that the event that one method fails is independent of the event that the other fails.
    (a) What is the efficiency of the combined approach as a function of e_1 and e_2?
    (b) Calculate the combined efficiencies for the following combinations of methods:
        1. Condom plus diaphragm
        2. Diaphragm plus douche
        3. Rhythm plus foam
        4. Condom plus pill
    (c) Calculate mean waiting times for each of the combined methods of part (b).

● **3** (a) Find what proportion of women with $p = 0.1$ have a waiting time less than the mean for $p = 0.1$.
    (b) For arbitrary p, find the formula which gives the proportion of women whose waiting times are less than the mean for that fecundability p.
    (c) Make a graph of the function found in part (b).

4 To show $1 - (1 - p)^{1/6p}$ is monotone increasing, it suffices to show $(1 - p)^{1/6p}$ is monotone decreasing, and this is equivalent to showing that $\log (1 - p)^{1/6p}$ is monotone decreasing.
    (a) Simplify $\log (1 - p)^{1/6p}$ and find its derivative.
    (b) Show that, if this derivative is 0 at some value of p, say p_0, then $-\log (1 - p_0) = p_0/(1 - p_0)$.
    (c) Find Taylor series for the right and left side of the equation in part (b) and, by comparing them term by term, show that the equality is impossible except for $p_0 = 0$.
    (d) As a consequence of part (c) $\log (1 - p)^{1/6p}$ doesn't "change direction" between $p = 0$ and $p = 0.3$. It is either monotone increasing (if the derivative is always positive for $p > 0$) or monotone decreasing (if the derivative is always negative for $p > 0$). Consequently $1 - (1 - p)^{1/6p}$ is either monotone increasing or decreasing. How would you decide which?

● **5** The median waiting time for fecundability p is defined to be that number M such that 50 percent of all women with fecundability p experience a waiting time of $\leq M$ (i.e., $S_M = 0.5$).
    (a) Find M for $p = 0.1$.
    (b) For arbitrary p, show that $M = -\log 2/\log (1 - p)$.

6 For the values of p of interest to us, log $(1 - p)$ can be accurately approximated by the Taylor polynomial of degree 1 at the point 0 [i.e., the constant and linear terms of the Taylor series at 0 for the function log $(1 - p)$].

(a) Find this approximation to log $(1 - p)$.

(b) What do we obtain as a formula for M [see part (b) of Exercise 5] if we use this approximation?

(c) Using the formula you found in part (b), plot a graph of the median as a function of p. On the same axes plot the mean as a function of p and compare.

● **7** In obtaining the average waiting time, we added all possible waits k from 1 to ∞ (each wait weighted by its probability).

(a) Explain why it is unreasonable to take k values approaching ∞.

(b) If we are dealing with a 20-year-old woman, whose fertility years will extend only for another 25 years, calculate the probability, as a function of f, that she doesn't conceive at all. (Use the multiplication rule applied to the 300 monthly experiments.)

(c) Calculate the values of the function found in part (b) for the p values which result from the various contraceptive methods, combined with a natural fecundability of $f = 0.2$.

8 The variance, denoted σ^2, of the waiting-time distribution is defined as $\sum_{k=1}^{\infty} (k - \mu)^2 p_k$, where μ is the mean $1/p$ and p_k is the probability of waiting k months. The standard deviation, denoted σ, is defined as the positive square root of the variance. It gives an indication of whether the histogram is narrow (bars close to the mean when σ small) or spread out (some bars far from the mean when σ large). Show algebraically that

$$\sigma^2 = \sum_1^{\infty} k^2 p_k - \mu^2$$

(*Hint*: Expand each term of the series defining σ^2 and collect terms into three series.)

9 Use the result of Exercise 8 to show that, for our waiting-time distribution,

$$\sigma^2 = \sum_{k=1}^{\infty} k^2 p(1 - p)^{k-1} - \frac{1}{p^2}$$

$$= 1^2 p + 2^2 p(1 - p) + 3^2 p(1 - p)^2 + \cdots - \frac{1}{p^2}$$

10 Use the result of Exercise 9 and the identity $n^2 = (n - 1)^2 + n + (n - 1)$ to show

$$\sigma^2 = p[1 + 2(1 - p) + 3(1 - p)^2 + 4(1 - p)^3 + \cdots]$$
$$+ p[1(1 - p) + 2(1 - p)^2 + 3(1 - p)^3 + \cdots]$$
$$+ p[1^2(1 - p) + 2^2(1 - p)^2 + 3^2(1 - p)^3 + \cdots]$$

11 (a) Use the method shown in the text to sum the series for $\overline{w}$ and apply it to find the sums of the first two series in Exercise 10.

(b) Show that

$$\sigma^2 = \frac{1}{p} + \frac{1 - p}{p} + (1 - p)\left(\sigma^2 + \frac{1}{p^2}\right) - \frac{1}{p^2}$$

(c) Solve the equation in part (b) to obtain

$$\sigma^2 = \frac{1 - p}{p^2}$$

12 Use the result of part (c) of Exercise 11 to do the following:

(a) Plot a graph of σ as a function of p.

(b) Find the maximum and minimum of σ for $p \in [0.001, 0.3]$ and find whether σ is monotone increasing or decreasing as a function of p.

(c) Show that, for any p, σ is at least 83 percent of the mean.

(*d*) If we regard σ as a measure of the uncertainty concerning the length of the waiting time and if f is fixed, show that, as the efficiency of contraception goes up, so does the uncertainty. Is this reasonable? Do you think "uncertainty" is a good word to apply to σ?

13 In principle it should be possible to create a picture of the entire moon by placing close-up shots side by side. Likewise, one might suppose that micro models for population growth could tell you anything a macro model could.

(*a*) Is there any type of information which we can obtain from the Malthusian model of Section 4 which we cannot obtain from the Leslie-matrix model in Section 5?

(*b*) Is there any type of information which we can obtain from the Malthusian model of Section 4 which we cannot obtain from our family-planning model of this section?

14 In the left-hand column below we have listed various predictions concerning next year's births. These projections may be thought of as arising from mathematical models occupying different positions on the macro-micro scale. In the right column we have listed various "consumers" of mathematical models, who might find some of these pieces of information useful. For each item in the left-hand column, list which groups in the right-hand column would find the information useful. Be prepared to explain your answers.

1. Total births in a country	(*a*) Baby-food manufacturers
2. Births per woman in a country	(*b*) Local school boards
3. Age specific birth rates for a country	(*c*) The gossip columnist of a local news-
4. A complete list of communities and the births expected in each community	paper
	(*d*) Manufacturers of birth control devices
5. A complete list of each family in a community and the births expected in each family	(*e*) The American Obsterical Society

Computer exercises

15 Write a computer program that accepts as input a value of f and the name of one of the devices listed in Table 1 and that prints out the expected waiting time till conception in years and months (e.g., "5 years and 7 months" if the waiting time is 67 months).

16 Write a computer simulation to determine the waiting time till conception of a couple with $p = 0.2$. [Use a random-number generator to pick a number in the interval $(0, 1)$. If the number drawn is in $(0, 0.2)$, this signifies conception.] Now repeat the simulation for this couple 100 times and compute the average waiting time over all trials. Compare the result with the theoretical value obtained from $w = 1/p$.

BIBLIOGRAPHY

Meyer, Rochelle W.: "Everything You Always Wanted to Know about the Mathematics of Sex and Family Planning But Were Afraid to Ask," *MATYC J.*, vol. 12, no. 1, pp. 7–12, 1978. A reproductive-cycle approach to the material of Section 6, including abortion, effect of breast-feeding, etc.

Sheps, Mindel C., and Jane A. Menken: "Mathematical Models of Conception and Birth," University of Chicago Press, Chicago, 1973. A comprehensive text on micro models of fertility.

7 DESCRIPTIVE AND PRESCRIPTIVE MODELS— INVENTORY POLICY

Abstract The difference between a descriptive model, which tells how something works, and a prescriptive model, which tells the ideal way for it to work, is illustrated. The particular example concerns how and when a firm should replenish its inventory. The trade-off between order cost and carrying cost is studied.

Prerequisites Calculus (minimizing a function of one variable).

Mathematical models can be divided into two categories: descriptive and prescriptive. A descriptive model is one which describes or predicts how something actually does work or how it will work. A prescriptive model is one which is meant to help us choose the best way for something to work. Alternative names which are sometimes used for prescriptive models are *normative* or *optimization* models.

For example, in the summer of 1979 the space station called Skylab became "captured" by the earth's gravitational field in such a way that it came crashing down to earth. Before it crashed, there was intense speculation throughout the world about when and where it would hit and what damage it would do. Suppose for the moment that we had had no control over Skylab's descent. Then the only contribution science and mathematics could have made to relieving public anxiety would be to make a prediction of when and where the impact point would be. A model that made this prediction would be a descriptive model. But the actual state of affairs was that we did retain some slight control over Skylab's orientation in space. Furthermore, small changes in orientation could have a major influence on the time and place of earth impact. Scientists were able to find an orientation that made it very likely for the crashdown to occur over water or unpopulated land. The mathematical model used for this decision-making problem was a prescriptive model. Its function was to help choose the best policy (the best orientation). What actually happened was that Skylab disintegrated into chunks which fell into the Indian Ocean and remote desert regions of Australia. No human or property damage was reported. This successful outcome was a tribute to the skill of the space scientists involved, but it also rested heavily on some good mathematical models.

The differences between descriptive and prescriptive models do not lie primarily in the mathematics. The main difference is in what the model is used for. A prescriptive model is a tool for human decision making, while a descriptive model just tells us "what makes it tick." Often a descriptive model can be turned into a prescriptive one, as we shall see in this section.

We shall illustrate these ideas in more detail with a model of inventory policy. Suppose you are the manager of a large retail store, which sells 20 soccer balls each day. How often and in what quantity should you order new balls from the factory? At one extreme, you can order 20 new balls from the factory each day. At the other extreme, you could order a whole years supply of $(365)(20) = 7300$ balls at once and have only one delivery per year.

Getting a delivery each day is expensive because there are *ordering costs* for each delivery. For example, the manufacturer (or wholesale distributor) generally charges a flat rate, say $100, each time a delivery is made, regardless of the size of the delivery. Of course, you will also have to pay for the balls and possibly a

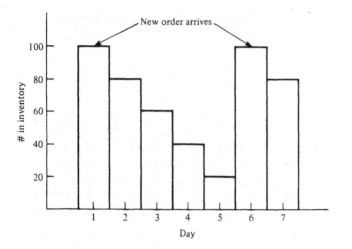

Figure 1

handling charge for each ball delivered, but these costs are not included in ordering costs. (For one thing, these costs come out the same over the course of a year—7300 times the cost for one ball—whether you get the balls in large or small lots.) Ordering costs also include salaries of personnel involved in drawing up the purchase order, following it up, paying the bill, etc. In general, ordering costs are always the same, regardless of the size of an order.

If we wish to minimize the ordering costs for a year, we should have a small number of large orders, for example, one order of 7300 balls. But this runs up another kind of expense called *carrying cost*. These carrying costs may include the rental of storage space, fire insurance on inventories, and interest on money invested in inventory. Carrying costs are always in proportion to the number of items in inventory and the time they spend in inventory and, therefore, are expressed as dollars per item per day. For example, if the carrying cost is $0.10 per ball per day and we have 100 balls in inventory for 365 days, the carrying cost is $3650. In reality, carrying costs are harder to compute than this because we don't have a constant number of balls in inventory; the number declines between one delivery and the next as the balls are sold (see Figure 1). Here is a numerical example that shows how we cope with this variability in inventory.

Example 1

Assume the following:

$$\text{Rate balls are sold} = 20 \text{ per day}$$

$$\text{Carrying cost} = \$0.05 \text{ per ball per day}$$

$$\text{Ordering cost} = \$100 \text{ per order}$$

We work out the yearly cost if we order in lots of 100. Since we sell 20 per day, we need a delivery every fifth day. We assume that deliveries come at the beginning of the day. In any given day the number in inventory declines during the day as balls are sold. But we assess the storage cost on the basis of the number present at the start of the day. Now we can make the following table that shows the situation on a day-by-day basis.

Table 1

Day	Delivery	Ordering cost	Number in inventory	Carrying cost for this day
1	Yes	$100	100	$5
2	No	0	80	4
3	No	0	60	3
4	No	0	40	2
5	No	0	20	1
6	Yes	100	100	5
⋮	⋮	⋮	⋮	⋮

Day 1 through day 5 constitutes one delivery cycle. The costs for one cycle are obtained from the table:

$$\text{Carrying cost} = 5 + 4 + 3 + 2 + 1 = 15$$
$$\text{Ordering cost} = 100$$
$$\text{Total cost} = 115$$

In a year of 365 days there will be $365/5 = 73$ cycles; so the total yearly cost is $73(\$115) = \$8395$.

We now describe the calculation of the previous example algebraically. We replace the specific numbers with variables as follows:

$$\text{Rate items are sold} = r \text{ per day}$$
$$\text{Storage cost} = s \text{ per item per day}$$
$$\text{Ordering cost} = k \text{ per order}$$
$$\text{Number in each order} = x$$

The x items in an order will last x/r days. So if we get a delivery on day 1, then the next delivery is on day $x/r + 1$ and so on. Here's a table that describes the situation.

Table 2

Day	Delivery	Ordering cost	Number in inventory	Carrying cost for this day
1	Yes	k	x	sx
2	No	0	$x - r$	$s(x - r)$
3	No	0	$x - 2r$	$s(x - 2r)$
$\vdots$	$\vdots$	$\vdots$	$\vdots$	$\vdots$
x/r	No	0	r	sr
$x/r + 1$	Yes	k	x	sx

Figure 2 shows how the carrying cost changes during the days of a cycle. To get the cycle carrying cost, we need to add the heights of the bars in Figure 2. Since each bar has width 1, this is the same as adding their areas. We can get an approximate answer by approximating the tops of the bars with some handy function, which we then integrate. In this case, a straight line suggests itself. We shall leave this for Exercise 10 and pursue another approach.

The carrying costs for the various days of one cycle form an arithmetic progression where the first term is sr and the difference between consecutive terms is also sr. There are x/r terms in the progression. The formula for the sum of an arithmetic progression, then, gives the total:

$$\text{Carrying cost} = \frac{x}{r} \frac{sr + sx}{2}$$

for one cycle.

A handy way to remember this formula is to think of the cycle carrying cost as the product of the number of days in the cycle x/r and the average daily carrying cost over all days of the cycle. Because our daily carrying costs are an arithmetic

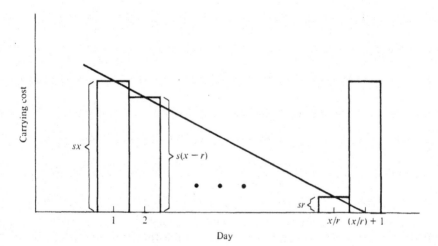

Figure 2

progression, the average carrying cost is obtained by averaging just the first and last carrying costs, $(sr + sx)/2$. (If we did not have an arithmetic progression, some other formula or method would have to be found to compute the average carrying cost.)

Next we add the ordering cost to get, for one cycle,

$$\text{Total cost} = k + \frac{(sx/r)(r + x)}{2}$$

The number of cycles in a year is $365/(x/r) = 365r/x$. So the yearly cost, which we denote $y(x)$, is

$$y(x) = 365\left(rkx^{-1} + \frac{sr}{2} + \frac{sx}{2}\right) \tag{1}$$

We can use this formula to compute the cost per day of any plan we might like to test, with any given values of r, s, k, and x.

Example 2

As in Example 1, let

$$r = 20$$

$$s = \$0.05$$

$$k = \$100$$

Using formula (1), we compare the two ordering policies $x = 50$ and $x = 500$:

$$y(50) = 365\left[\frac{(20)(100)}{50} + \frac{(20)(0.05)}{2} + \frac{(0.05)(50)}{2}\right] = \$15,238.75$$

$$y(500) = 365\left[\frac{(20)(100)}{500} + \frac{(20)(0.05)}{2} + \frac{(0.05)(500)}{2}\right] = \$6205.00$$

Formula (1) is just a descriptive model so long as we use it merely to predict the cost involved for a given choice of x. But we can also use this formula to pick out the value of x that gives the lowest cost. When we do this, we are using the formula as a prescriptive model. To find the optimal x, differentiate y and set the derivative equal to zero:

$$y'(x) = 365\left(-rkx^{-2} + \frac{s}{2}\right) = 0$$

$$x^{-2} = \frac{s}{2rk}$$

$$x = \sqrt{\frac{2rk}{s}} \tag{2}$$

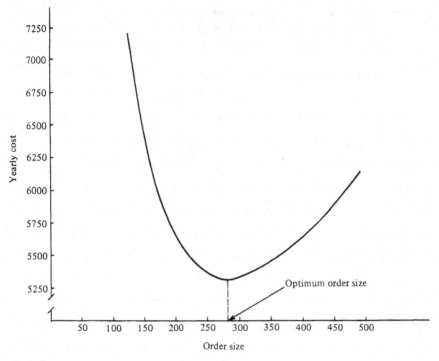

Figure 3

But is this critical point a relative maximum or a relative minimum? We could use a second derivative test to find out, but a simpler method is at hand. We observe that, for positive values of r, k, and s (the only kinds which are realistic), $y \to \infty$ if either $x \to \infty$ or $x \to 0$. Since the graph goes to ∞ at both ends and has only one critical point in the middle, this point must be a relative minimum. Figure 3 shows the graph of $y(x)$ for the r, k, and s values of Example 2.

Example 3

When $r = 20$, $s = 0.05$, and $k = 100$, as in Examples 1 and 2, the optimal value of the delivery size is

$$x = \sqrt{\frac{2(20)(100)}{0.05}} = 282.84$$

Since balls come only in whole numbers, the result cannot be taken literally. To find the integer value with the lowest cost we round 282.84 up and then down and compare these x values:

$$y(282) = \$5344.40$$

$$y(283) = \$5344.37 \quad \text{the minimum}$$

There is another practical consideration we may have to think about in problems of this sort: the maximum or minimum size of a delivery. For example, if a delivery can't exceed 200 balls, then the solution just found is useless. In that case, the interval of permissible delivery sizes is [20, 200] and $y(x)$ has no critical point on this interval. Then the maximum will occur at one of the endpoints. A similar problem occurs if the optimum value calculated from Equation (2) is less than the minimum delivery size.

EXERCISES

● 1(a) Determine the optimum order size if $r = 40$, $s = 0.10$, and $k = 100$. Compare to Example 3.

(b) Explain why the optimum order quantity stays the same if daily carrying costs and daily sales both double.

2 Suppose you have a choice of three suppliers for soccer balls. The first charges $100 per delivery but won't deliver more than 200 at a time; the second charges $100 also but won't deliver less than 300; the third will deliver any quantity but charges $200. Which supplier should you use? In all cases $r = 20$ and $s = \$0.05$.

● 3 Using the r, s, and k values of Example 3, find the optimal order quantity if it must be an integer multiple of 25 balls.

4 The model constructed in this section could be described as being unrealistic, except when x is an integer multiple of r. Explain. Do you agree with the criticism?

5 Suppose the optimum order quantity computed by Equation (2) turns out to be less than r. Is such an answer acceptable? Explain. If the answer is not acceptable, how would you obtain an acceptable one?

● 6 Suppose your supplier charges $100 per delivery plus $0.06 per ball in the order. How would you revise formula (1)? What change does this cause in formula (2)?

7 Suppose you get both basketballs and soccer balls from your supplier. To save money, you would like to get them in the same delivery. The daily sales of soccer balls and basketballs are, respectively, $r_s = 20$ and $r_b = 10$. The carrying costs are the same, $s = 0.05$, for both balls. Regardless of what the delivery consists of, k is $100. Let $x =$ the number of soccer balls per delivery and $y =$ the number of basketballs. Build a model that determines the x and y values that give minimum yearly cost. Is this an improvement over getting separate deliveries for each type of ball?

8 In the soccer ball model of Example 1, suppose you have room in the store for 20 balls. So each morning 20 new balls are transferred from a warehouse to the store. On the morning of a delivery, 20 balls from the delivery are put in the store and the remainder in the warehouse. Suppose the carrying cost of $0.05 per ball applies only to the balls in the warehouse and there is no carrying cost for the balls in the store. k and r are as in Example 1.

(a) Make a table, like Table 1 in Example 1, for an order size of 100, but based on the new assumption about carrying cost.

(b) Find the cost for one cycle (of 5 days).

(c) Carry this out algebraically, as in Table 2.

(d) Find a formula, analogous to Equation (1), based on the new assumption about carrying cost.

● 9 Suppose the manager of the store likes to think in terms of days between deliveries instead of the size of a delivery. Can you build a model in which the variable, call it z, is the number of days between deliveries? Use your model to find a formula for the value of z that gives the lowest yearly cost.

10 Find the sum of the areas of the bars in Figure 2 by integration. First find the equation of a straight line that fits the tops of the bars. Then find the area under it between appropriate limits. Compare the formula you obtain to Equation (1).

Computer exercise

11 Suppose the store is closed Sunday. Thus the r value applies only to the other 6 days. However, the carrying cost applies to Sundays as well.

(a) Using the data of Example 1, make a table like Table 1, by hand, covering 2 weeks.

(b) Based on your experience in part (a), write a computer program to figure out the yearly cost of any given policy (order size) given to the program as input.

(c) Is there any way you could use your program as a prescriptive model, i.e., a model that yields the optimum order quantity?

BIBLIOGRAPHY

Bender, Edward A.: "An Introduction to Mathematical Modeling," Wiley Interscience, New York, 1978. Section 4.1 discusses inventory optimization from a point of view "opposite" to ours, that of the factory instead of the store.

Levin, Richard I., and Charles A. Kirkpatrick: "Quantitative Approaches to Management," 4th ed., McGraw-Hill Book Company, New York, 1978. Chapters 7 and 8 cover inventory models with a minimum of mathematics and a lot of emphasis on the practical aspects of the subject.

CHAPTER TWO

THE RELATION OF MODELS TO DATA

1 SOURCES OF ERROR—UNDERGROUND EXPLORATION OF THE EARTH

Abstract Following a scheme of John von Neumann and Herman Goldstine, we discuss four categories of errors in mathematical models. Each type is illustrated with reference to a model for exploring the compositions of the earth's crust through *controlled source seismology* (setting off explosions and listening for the echoes).

Prerequisites Geometry and trigonometry. Power series.

Most mathematical models contain errors. In fact, if the average grocery clerk made as many errors in a day as a mathematical modeler does, the clerk would be replaced by someone with a better grasp of arithmetic. What is more, the errors a modeler makes are often deliberate! This state of affairs is largely due to the fact that mathematical modeling is normally a lot more difficult than adding up a grocery bill. Perfection is often unobtainable or too expensive.

This does not mean that mathematical modelers ought to be casual or unconcerned about errors. On the contrary, good modeling ought to include some formal or informal error estimation. At the very least, one should have an idea of where the errors in a model might be. In this section, we propose to give a survey by example of four categories of error. These categories were proposed by von Neumann and Goldstine in a landmark paper, and they serve as a rough but

**Table 1 Layering of earth's crust
—central Wisconsin**

Layer	Depth (kilometers)
1	1.4
2	6.1
3	30

useful map for the subject of error estimation. The four categories represent four sources of error:

1. Modeling assumptions
2. Observations
3. Approximations of uncomputable processes or functions, by computable processes or functions
4. Rounding off of the results of arithmetic calculations

We shall illustrate these kinds of errors in a single example, in which each type of error occurs.

If we were to dig down into the earth's crust, we would encounter a series of layers of different kinds of material. For example, under central Wisconsin, the crust has three layers, as described in Table 1.

Identifying the composition of these layers and their depths is very important to geologists, but, as you can imagine, digging holes is a tiring and impractical way of doing it. For this reason, geologists have developed the method called *controlled source seismology*. Typically one sets off a small explosion at one location, called the *source*, and records the reverberations at various receiving sites (Figure 1). It is a little like the way a doctor taps a finger on your chest and back and listens for the echoes with a stethoscope. What the doctor hears tells a little about what is inside.

When we set off our explosion, the energy of this disturbance is transmitted in all directions at once, in the same way that sound spreads in all directions from its source. At any given time, the leading edge of the disturbance is called a *wavefront* (dashed semicircle in Figure 1). The various directions of spreading are called *wavepaths* (arrows in Figure 1).

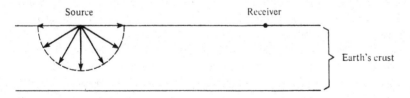

Figure 1

Table 2 Seismic velocities in various materials

Material	Kilometers/second
Soil and sand	0.11 to 1.85
Sandstone	1.4 to 4.3
Shale and slate	2.3 to 4.7
Limestone (hard)	2.8 to 6.4
Granite	4.8 to 5.6
Basalt	5.1 to 5.6
Dunite	7.4 to 8.6

Among the wavepaths there is one which travels along the surface, in a straight line (more or less) to the receiving station. The time it takes to make this trip depends on the distance D from source to receiver and on the velocity with which it can travel through the top layer of the earth's crust. This, in turn, depends on the kind of material this layer is made of. Table 2 shows characteristic velocities of transmission for some of the most common materials.

In practice, this way of thinking is turned on its side: we measure the time T between the explosion and our detection of it at the receiving station. We also measure D. Then we compute the velocity.

$$v = \frac{D}{T} \tag{1}$$

We locate v in a table, such as Table 2, and see what kinds of material it corresponds to. (Many of the velocity ranges in Table 2 overlap and thereby prevent us from making definite determinations. For example, a measured velocity of 3 kilometers/second could indicate sandstone, or shale and slate, or limestone.)

Our main interest, however, is to measure the depth of the first layer. We assume that the boundary between this layer and the one below it is parallel to the surface. When the wavefront reaches this boundary, some of the energy is bounced back up into the upper layer. Figure 2 shows two different wavepaths being bounced back. The law that governs this bounce-back is the same that applies to billiard balls bouncing off a cushion or light rays bouncing off a mirror: the angle of incidence equals the angle of reflection.

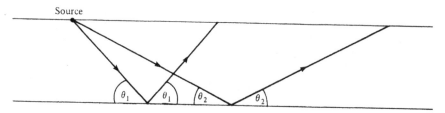

Figure 2

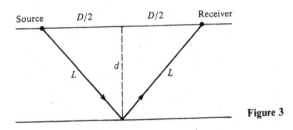

Figure 3

One of the wavepaths from the source will strike the boundary at a point where it gets bounced back to the receiver (Figure 3).

Let T' denote the time required for this echo to arrive. Then,

$$2L = vT'$$

$$= \frac{D}{T} T'$$

Applying the pythagorean theorem gives

$$d = \sqrt{L^2 - \frac{D^2}{4}}$$

$$= \sqrt{\frac{D^2(T'/T)^2}{4} - \frac{D^2}{4}}$$

$$= \left(\frac{D}{2}\right)\sqrt{\left(\frac{T'}{T}\right)^2 - 1} \tag{2}$$

By measuring T, T', and D and plugging into this formula, we can compute the depth d without digging a hole.

If we accept Equation (2) without further thought, we may fall prey to an error of category 1. An important assumption needed to derive this equation is that the boundary between layers is parallel to the surface. If this is not true, Equation (2) is not true. If the angle the boundary makes with the direction of the surface is α (Figure 4), then it turns out that the correct formula (see Exercise 8 for the derivation) is

$$d = D \frac{-\tan \alpha + (1/\cos \alpha)\sqrt{(T'/T)^2 - \cos^2 \alpha}}{2} \tag{3}$$

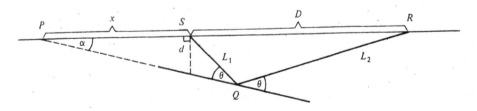

Figure 4

Table 3

α (degrees)	d by Equation (2)	True d by Equation (3)	Difference	Percent error[†]
1	1.031	0.997	0.034	3.41
5	1.031	0.874	0.157	18.0
10	1.031	0.752	0.279	37.1
30	1.031	0.504	0.527	105

[†] Computed by dividing difference by true d value.

Table 3 compares formulas (2) and (3) for the case where $D = 4$, $T'/T = 1.125$, and α takes on various values. Notice that even for $\alpha = 1°$, where the boundary is nearly parallel to the surface, the difference between the true value of d and the value given by formula (2) is over 3 percent. For larger values of α the errors are very substantial.

We now turn to category 2, errors in observations. Suppose that the boundary between layers is indeed parallel to the surface, so that Equation (2) is correct. If our time keeping is somewhat inaccurate, then the values of T and T' will be wrong. Suppose, for example, that the measured values of D, T, and T' are 4 kilometers, 1 second, and 1. 2 seconds, but our clock makes errors as great as 1 percent. Then the true value of T could be as great as 1.01 or as low as 0.99, and the true value of T' could be as great as 1.21 or as low as 1.19. Table 4 shows the resulting values of d, computed by Equation (2), for some of these combinations. Although our computation would be 1.33 kilometers, the true value could be as high as 1.41 (about 6 percent higher) or as low as 1.27 (about 5 percent lower).

We turn now to category 3 and, in doing so, suppose that there are no errors of the first two categories. In particular this means Equation (2) and the values $D = 4$, $T = 1$, $T' = 1.2$ are exactly correct. There is still a possibility for error arising through the calculation of the square root. Except for lucky cases where an exact square root exists, we have to settle for approximate values.

There are a number of ways to get approximate square roots. One way is to use logarithms, but this only shifts the difficulty to finding good approximations

Table 4

T	T'	$d = (D/2)\sqrt{(T'/T)^2 - 1}$ (kilometers)
1.01	1.21	1.32
1.01	1.19	1.27
0.99	1.21	1.41
0.99	1.19	1.33
1.00[†]	1.20[†]	1.33

[†] Actually measured values.

for logs. Another way, called the Newton-Raphson method, is to produce a series of approximations $r_0, r_1, r_2, \ldots \to \sqrt{a}$ as follows:

Step 1: Pick any positive guess at the answer, r_0.
Steps 2, 3, ...: Continue producing new approximations from the previous approximation by repeated use of the formula $r_{n+1} = \frac{1}{2}(r_n + a/r_n)$.

By carrying this out often enough, you can get as much accuracy as you want (see Exercise 11).

Similar problems arise when we wish to compute trigonometric functions, as would be necessary in formula (3). One thing we can do (and this is the strategy used to construct trigonometric tables by computers and calculators) is to take a finite number of terms of the Taylor series for $\sin x$:

$$\sin x = x - \frac{x^3}{3!} + \frac{x^5}{5!} + \frac{x^7}{7!} + \cdots \tag{4}$$

How many terms shall we take? It can be shown that, if we specify a desired degree of accuracy, we can determine the number of terms that will give that accuracy (see Exercise 15). This is typical. Usually category-3 error can be made completely negligible. The systematic study of how to do this is called *numerical analysis* and is beyond the scope of this book, but see Exercises 13 through 18 for a brief taste.

Other potential examples of type-3 error (which do not occur in the case at hand) are: solving a differential equation by the Cauchy polygon method, obtaining an area under a curve by the trapezoid rule, and estimating roots of various polynomials by the Newton-Raphson method.

In actual practice, many mathematical modeling calculations are carried out by computers, which have built-in procedures for computing square roots, trigonometric functions, and so on. The accuracy of these built-in functions, even in the least-powerful machines such as hand calculators, is enough for most purposes. For this reason, we nearly always accept the accuracy given by the machines and think no more of it. But it is good to keep in mind that there are cases where we might need more accuracy or where we might need some idea of the magnitude of the errors being made. Indeed, the availability of computers has stimulated the study of numerical analysis rather than making it obsolete.

Finally, let us suppose that we assume the model is correct (no error of category 1), the errors of observation (category 2) are inconsequential, and the errors of approximations by computable processes (category 3) can also be ignored. There is still another source of error due to the fact that in practice only a fixed and finite number of digits can usually be carried along in our arithmetic. Numbers with more digits need to be rounded off.

To illustrate this most simply, we suppose that our computations are carried out in scientific notation. That means each number is written in the form

$$a \times 10^b$$

where b is an integer and a is a decimal with its first nonzero digit occurring right after the decimal point. For example 100 becomes 0.1×10^3, 0.043 becomes 0.43×10^{-1}, and 0.7392 becomes (remains) 0.7392×10^0. In any sort of computation, there is a limit to how many decimal places we are willing or able to carry along in the number symbolized by a. (There is a similar limit for b, but, in practice, this is rarely exceeded, so we'll ignore it.) For example, a typical contemporary electronic computer might carry 11 places in the number a.

What happens if we multiply two numbers in scientific notation together? That is, suppose we want to compute $(a_1 \times 10^{b_1}) \times (a_2 \times 10^{b_2}) = a_1 a_2 \times 10^{b_1 + b_2}$. For example, suppose we are carrying three places in a and need to multiply 0.432×10^2 and 0.321×10^4. The correct answer is 0.138672×10^6. Except that, when we round to three places and get 0.139×10^6, we incur some error. This is a category-4 error.

The same sort of thing can happen with addition. Again keeping three places, if we add $0.782 \times 10^0 + 0.431 \times 10^0$, we obtain 1.213 or 0.1213×10^1, which gets rounded off to 0.121×10^1. This is another example of category-4 error. Can this sort of error occur with subtraction? Division?

Many numerical computations, even something as apparently simple as calculating a sine of an angle [see Equation (4)], require large numbers of repetitions of the four basic operations of arithmetic. Since one may incur a roundoff error at each stage, there is a possibility that the errors will accumulate, even though they are separately insignificant. For example, reference to Equation (4) shows that to calculate $\sin x$ one needs to compute terms of the form $x^n/n!$ for possibly very high values of n. In the table below, we give an example of how the errors in x^n can accumulate. In the third column, we compute the powers of 1.2300001, using eight-place accuracy (each time we multiply we round off to eight digits). In the second column we use three-place accuracy, starting with 1.23 (which is the three-place version of 1.2300001). The discrepancies due to the

Table 5

	$(1.2300001)^n$		
n	Three-digit accuracy	Eight-digit accuracy	Discrepancy, column 3 − column 1
1	1.23	1.2300001	0.0000001
2	1.51	1.5129002	0.0029002
3	1.86	1.8608674	0.0008674
4	2.29	2.2888671	−0.0011329
5	2.82	2.8153068	−0.0046932
6	3.47	3.4628276	−0.0071724
7	4.27	4.2592783	−0.0107217
8	5.25	5.2389127	−0.0110873
9	6.46	6.4438631	−0.0161369
10	7.95	7.9259523	−0.0240477

differences in the number of digits are in column 4, and they show a definite snowballing effect. The table does not show a column with the exact powers of 1.2300001. (How many digits would this require?) But eight-digit accuracy is much closer to exactness than three-digit accuracy.

What can be done about these sources of error? Category-1 errors are the toughest to avoid. There are few handy hints or rules of thumb to help make sure one's models are based on correct (or not-too-incorrect) assumptions. The only guideline (a pretty obvious one) is that, if you are building a model in an area where you are not expert, then you should get the best advice available.

Category-2 errors can occasionally be somewhat overcome by taking many measurements and averaging them. The next section gives details about this.

Category-3 and category-4 errors can be minimized, even though in any model of significant complexity it is unusual to be able to avoid them completely. The methods of minimizing these errors form a subject called numerical analysis. If you are thinking of a career in a branch of applied mathematics or computer science, you should definitely study this subject.

Sidelight: Lobster Locomotion

Once upon a time two pure mathematicians (who must remain nameless to protect their reputations) were honeymooning on Prince Edward Island, Canada. Hoping to see the lobster boats come in with their catch, they drove to one of the fishing villages in the late afternoon, but were surprised to be told by a salty native, "The season isn't here anymore. It's moved to Cambelltown, around the coast a piece." The inquisitive mathematicians discovered that the season moved from town to town around the coast of the island and took 1 year to make the complete cycle. Our mathematical couple was aware that many animal species had migratory habits, and they decided to calculate the migration rate of the lobster. With a road map and ruler, they ascertained that Prince Edward Island is roughly 50 miles wide at its widest point and 112.5 miles long. Using a rectangle as a model for the shape of the island, the intrepid calculators fixed the approximate perimeter at 325 miles. "Aha," they said, "if the lobsters cover 325 miles in 365 days that means they move at the rate of almost 1 mile each day (0.89 mile per day to be exact)!"

Our heroes were landlubbers and accustomed to viewing lobsters crawling lazily over one another in fish-store windows; so the calculated rate seemed high. But the distractions of their vacation overcame their curiosity, and they went on to other adventures. Some weeks later they learned that the "season" was moved by local authorities to prevent overfishing, and it had nothing to do with the migration of lobsters. For all anyone knows, lobsters are as lazy in the briny deep as they are in fish stores or, indeed, on dinner plates.

What kind of error did the two mathematicians make?

Sidelight: Human Mortality

Occasionally errors are deliberately introduced into a mathematical model as the price paid for other advantages. Here is an example from the early history of the insurance industry.

If x denotes age, let $l(x)$ denote the fraction of people who will survive till age x or beyond. This function, called the *survivorship function*, naturally varies from time to time and place to place. Figure 5 shows a typical shape for $l(x)$. This function is rarely given as a formula; usually it is tabulated for the various values of x.

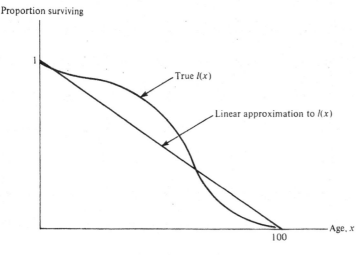

Figure 5

The reason is that formulas which fit the table of values are rather ugly. Abraham De Moivre, one of the founders of the actuarial profession, used a linear approximation to $l(x)$ when he required a formula, and he did this in full knowledge that it was not completely accurate.

To understand his motivation, one must be aware of two facts: first, the job of an actuary requires carrying out a large number of calculations and, second, aids to calculation such as adding machines or computers hadn't been invented. De Moivre must have decided that the financial loss to the company from inaccurate predictions of mortality were compensated by the savings in not having to hire as many calculators.

Which category of error is involved in this approximation?

EXERCISES

● **1** (*a*) Suppose that, in an actual experiment, the time required for the direct wavepath to travel the 0.9 kilometer between source and receiver is 0.6 second. Consult Table 2 and state what the possibilities are for the composition of the top layer.

(*b*) In the same experiment, the reflected wavepath required 1 second. What is the depth of the top layer? Assume the lower boundary of this layer is parallel to the surface.

(*c*) If the lower boundary of the top layer makes an angle of 5° with the surface, what do you estimate the depth to be?

2 (*a*) Verify the calculations in Table 3.

(*b*) Compute what the table entries would be for $\alpha = 20°$.

3 (*a*) Verify the calculations in Table 4.

(*b*) Make a similar table, based on the assumption that the measurement inaccuracy could be as much as 5 percent.

● **4** There is yet another way in which an echo of the explosion can reach the receiver. A ray that hits the lower boundary of the first layer at a critical angle of $\theta = \arcsin(v/v')$, where v is the velocity of transmission in the top layer and v' is the velocity of transmission in the bottom layer (and where we assume $v' \geq v$), will be bent and travel exactly along the boundary of the two layers. At each point of its travel along this boundary it sends a ray upward, making this same angle θ with the boundary. As shown in Figure 6, one of these will arrive at the receiver. Find a formula for the time required in terms of d, D, v, v', and θ. Can you solve for v' in terms of the other quantities?

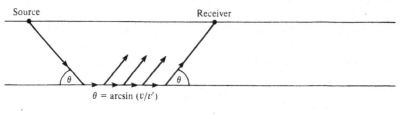

$$\theta = \arcsin(v/v')$$

Figure 6

5 Suppose you detect two shocks at the receiving station and know that one is the direct wavepath and the other is the reflected one. How do you know which is which?

6 Suppose you detect three shocks at the receiving station—the reflected one, the direct one, and the one described in Exercise 4. Can you tell which the first, second, and third shock is? If $v' > v$, is it possible that the refracted wavepath (described in Exercise 4) could arrive first?

● **7** Suppose a billiard ball were to travel from a source at point (a, b) in the xy plane to a receiver at (c, b), after bouncing off $(x, 0)$ on the x axis.

(*a*) Find a formula for the distance traveled.

(*b*) What value of x makes this distance a minimum?

(*c*) Show that, for this value of x, the angle the ball's path makes with the vertical direction is the same before and after the bounce. (This is how billiard balls, light rays, and the wavepaths of our explosion actually behave.)

8 In Figure 4, use the law of sines in triangle PQS to find an expression for L_1/x. Then find an expression for $L_2/(x + D)$. Use the law of cosines to prove $D^2 = L_1^2 + L_2^2 - 4L_1L_2 \sin^2 \theta$. Combine this with the other relationships you found and show $4x^2 \sin^2 \alpha + 4xD \sin^2 \alpha + D^2 - (L_1 + L_2)^2 = 0$. Solve this equation for x and deduce Equation (3).

9 Which category of error is involved in the lobster story at the end of this section?

10 In reference to the model of mortality, show that:

(a) The number of persons dying between age x and $x + 1$ is $l(x) - l(x + 1)$.

(b) If each of these persons is credited with $x + \frac{1}{2}$ years of life (this averages out those who die closer to age x with those dying closer to age $x + 1$), show that the average life span, denoted $\mathring{e}_0$, is

$$\mathring{e}_0 = \sum_{x=0}^{w} (x + \tfrac{1}{2})[l(x) - l(x + 1)]$$

$$= l(0) + l(1) + \cdots + l(w) + \tfrac{1}{2}$$

where w is the highest possible age, say 100 for the sake of practicality.

(c) If you approximate $l(x)$ by the linear function $l(x) = ax + b$, show that substitution in the formula of (b) gives

$$\mathring{e}_0 = a(1 + 2 + \cdots + 100) + 101b + \tfrac{1}{2}$$

$$= (50)(101)a + 101b + \tfrac{1}{2}$$

$$= 5050a + 101b + \tfrac{1}{2}$$

(d) How many arithmetic operations are required to carry out the calculation of $\mathring{e}_0$ by the method of part (b)? How many operations are needed when using the final formula of part (c)?

11 In the Newton-Raphson method given in the text for finding $\sqrt{a}$, show that:

(a) If $r_n < \sqrt{a}$, then $r_{n+1} > \sqrt{a}$.

(b) If $r_n > \sqrt{a}$, then $\sqrt{a} < r_{n+1} < r_n$.

(c) If we let $f(x) = \frac{1}{2}(x + a/x)$ so that $r_{n+1} = f(r_n)$, show that $0 < f'(x) < \frac{1}{2}$ provided $x > \sqrt{a}$.

(d) According to the law of the mean, $f(x) - \sqrt{a} = f(x) - f(\sqrt{a}) = f'(b)(x - \sqrt{a})$ for some b between x and $\sqrt{a}$. Use this and parts (a), (b), and (c) to show that, for $n \geq 1$, $0 < r_{n+1} - \sqrt{a} < \frac{1}{2}(r_n - \sqrt{a})$.

(e) Use part (d) to show that $0 < r_{n+1} - a < (\frac{1}{2})^n (r_1 - \sqrt{a})$. [Since $(\frac{1}{2})^n \to 0$, this proves $r_n \to \sqrt{a}$.]

● **12** (a) Could you try to calculate a square root by first computing a Taylor series for $\sqrt{x}$ around the point $x = 0$? If so, do it. If not, explain why.

(b) What about setting $y = x - 1$ and getting a Taylor series for $\sqrt{x} = \sqrt{y + 1}$ in powers of y? (Don't forget to check the radius of convergence.)

13 (a) Write out the Taylor series expanded around $x = 0$ for e^{-x} to 10 terms. Use this to calculate $e^{-5.5}$. In your calculations round off to five digits after every operation; e.g., 41.94036458 becomes 41.940.

(b) Now compute $e^{-5.5}$ to five-place accuracy using the formula $e^{-x} = 1/e^x$ and replacing e^x by the number obtained from adding the first 10 terms of its Taylor series (rounding to five places after every operation).

(c) The true value of $e^{-5.5}$ (to five places) is 0.00409. Which of your calculations comes closer?

14 Repeat the comparative calculations of Exercise 12 for $x = -20$ and $x = -0.1$. Do you see any pattern?

● **15** (a) If two decimals a and b do not differ until the kth digit after the decimal point, then what is the maximum difference between a and b?

(b) If two decimals differ by less than 0.000001, then for how many places after the decimal point must they be identical? What if the difference is less than 0.000009?

(c) If we estimate $f(x)$ by taking n terms of the Taylor series for f expanded around $x = a$, then the error is exactly given by the formula

$$E = \frac{f^{n+1}(\bar{x})(x - a)^{n+1}}{(n + 1)!}$$

where f^{n+1} means $(n + 1)$st derivative and $\bar{x}$ is some point in the open interval with endpoints a and x. In practice the exact location of $\bar{x}$ cannot be found; so we have to settle for an upper bound for the magnitude of E (i.e., $|E|$), based on the worst location for $\bar{x}$. Show that, when we estimate $\sin x$ with n terms of its Taylor series around 0, $|E| \leq x^{n+1}/(n+1)!$.

(d) For $x = \pi/4$, what value of n is large enough so that n terms of the Taylor series give an estimate for $\sin x$ which agrees with the true value in six decimal places?

16 (a) Use the quadratic formula to calculate the roots of the quadratic equation $0.001x^2 + x + 0.001 = 0$ with five-place accuracy (at every stage of the calculation numbers are rounded to five places). Compare your answer to the answers obtained from a seven-place calculation. (The seven-place answers are $x_1 = -999.9990$ and $x_2 = 0.0010005$.)

(b) Show that an alternate set of formulas for the two roots of $ax^2 + bx + c = 0$ is

$$x_1 = \frac{2c}{-b - \sqrt{b^2 - 4ac}} \qquad x_2 = \frac{2c}{-b + \sqrt{b^2 - 4ac}}$$

Hint: There is no need to derive these. Merely show that they satisfy the equation or equal the expressions in the more customary quadratic formula.

(c) Use the formulas of part (b) to compute the roots of the equation in (a). Round off to five places at every stage. Do the formulas of part (b) work better than the customary ones?

(d) Suppose we "mix and match" to get the following set of formulas:

$$x_1 = \frac{-b + \sqrt{b^2 - 4ac}}{2a} \qquad x_2 = \frac{2c}{-b + \sqrt{b^2 - 4ac}}$$

Are these results more accurate than those of part (a) or (c)? Can you mix and match the formulas yet another way to get better accuracy?

● **17** Is it always true that the product of two n-digit numbers is a $2n$-digit number? Work out some examples by hand for small values of n.

18 Suppose a_1 and a_2 are numbers, which have been rounded off to a'_1 and a'_2, and the errors are ε_1 and ε_2. That is, $a'_1 = a_1 + \varepsilon_1$ and $a'_2 = a_2 + \varepsilon_2$. Suppose we desire to multiply a_1 and a_2 but attempt to do this by computing the product $a'_1 a'_2$.

(a) Find the formula for the error in terms of a_1, a_2, ε_1, and ε_2. (Assume that $a'_1 a'_2$ is calculated exactly, not rounded off after being multiplied.)

(b) Illustrate your formula with examples.

(c) Find the formula for the error if we take $(a'_1)^3$ as the estimate of a_1^3. [Again, assume $(a'_1)^3$ is calculated exactly.] Use the binomial formula to generalize to $(a'_1)^n$.

Computer exercises

19 Compute $\sqrt{5}$ using the Newton-Raphson method explained in the text. Start with 5 itself as the first guess (r_0). Compute as many iterations as are needed to get an answer accurate in the third place after the decimal point. (Use the built-in square root function of your computer as a comparison to check accuracy). Now do the same, but with four-place accuracy, five-place accuracy, etc. Can you devise a formula or rule of thumb about how many iterations are required to get a certain number of decimal places of accuracy? Does your rule hold up for calculating $\sqrt{50}$? How about $\sqrt{500}$?

20 On your computer, calculate $1/N$ for $N = 10$, $N = 100$, $N = 1000$, etc. After each calculation, print the answer. Where is the first power of 10 whose reciprocal is printed out incorrectly?

BIBLIOGRAPHY

Forsythe, George E.: Pitfalls in Computation, or Why a Math Book Isn't Enough, *Amer. Math. Monthly*, November, 1970. A nice introduction to numerical analysis (errors of categories 3 and 4).

Montgomery, R. G.: "Listening to the Earth: Controlled Source Seismology," UMAP Module 292, available from COMAP, 271 Lincoln St., Lexington, MA 02173.

Neumann, John, von, and Herman Goldstine: Numerical Inverting of Matrices of High Order, *Bull. Amer. Math. Soc.*, vol. 53, pp. 1021–1099, 1947.

2 ADJUSTING DATA I (THE EASY WAY)—THE MEAN AND THE MAXIMUM LIKELIHOOD

Abstract When many measurements of something give different results, we often average the results. Common sense suggests that the average will be a truer measure than the individual measurements, which can be distorted by quirks in the measuring process. In certain circumstances, mathematical theory justifies the commonsense approach. To investigate this, we introduce the normal curve.

Prerequisites Histograms, elementary calculus, and the multiplication rule for independent events.

It is one thing to show a man he is in error, and another to put him in possession of truth.

John Locke

Most mathematical models require some input data; that is, certain measurements and counts made in the real world must be supplied in order for the model to crank out predictions or conclusions. In a few cases, these numbers or measurements can be obtained with perfect accuracy. For example, if we need to know how many senators there are in the U.S. Senate, we can surely find this out with certainty. But more commonly we are unable to be so certain.

Table 1 shows the results of weighing the same standard weight 100 times on the same highly accurate scale. These experiments were conducted at the National Bureau of Standards over a 2-year period as part of its ongoing program of calibrating measuring instruments. The weight is nominally a 10-gram weight (in fact it's serial number is NB 10), but, despite rigorous attempts at keeping the cirumstances of the weighings as identical and as ideal as possible,[†] there are variations in the weights obtained. The first three weights are

$$9.999591$$
$$9.999600$$
$$9.999594$$

These numbers have the first four digits in common. To avoid repeating all the digits, we replace each of these numbers by the number of micrograms by which it falls short of 10 grams. (1 gram = 1,000,000 micrograms.) Thus the three numbers above are represented by the following "shortfalls" measured in micrograms.

$$409$$
$$400$$
$$406$$

All the 100 measurements are shown in Table 1.

[†] Air pressure, temperature, and other variables were kept the same.

Table 1 One-hundred measurements on the weight of NB 10†

Meas. no.	Result	Meas. no.	Result	Meas. no.	Result	Meas. no.	Result
1	409	26	397	51	404	76	404
2	400	27	407	52	406	77	401
3	406	28	401	53	407	78	404
4	399	29	399	54	405	79	408
5	402	30	401	55	411	80	406
6	406	31	403	56	410	81	408
7	401	32	400	57	410	82	406
8	403	33	410	58	410	83	401
9	401	34	401	59	401	84	412
10	403	35	407	60	402	85	393
11	398	36	423	61	404	86	437
12	403	37	406	62	405	87	418
13	407	38	406	63	392	88	415
4	402	39	402	64	407	89	404
15	401	40	405	65	406	90	401
16	399	41	405	66	404	91	401
17	400	42	409	67	403	92	407
18	401	43	399	68	408	93	412
19	405	44	402	69	404	94	375
20	402	45	407	70	407	95	409
21	408	46	406	71	412	96	406
22	399	47	413	72	406	97	398
23	399	48	409	73	409	98	406
24	402	49	404	74	400	99	403
25	399	50	402	75	408	100	404

†Made by Almer and Jones at the National Bureau of Standards. Each measurement was some number of micrograms below 10 grams. These numbers are shown in the column headed "Result."

Figure 1 shows these measurements displayed in a histogram: the area of each bar represents the fraction of all measurements falling in the interval represented by the base of the bar. The area can also be thought of as the probability that a randomly chosen measurement falls within the interval represented by the base of the bar.

Where do these deviations come from? The object NB 10 itself is surely not likely to vary by any detectable amount. The measuring instrument is a more likely source of these chance errors. Although this instrument is one of the most accurate ever built, it is composed of various wheels and other moving parts, whose alignments may change slightly from one weighing to the next and give rise to errors.

The highest and lowest numbers in Table 1 are 437 and 375. The difference between these numbers is 437 − 375 = 62 micrograms, and this serves as one measure of the error in the measuring process. Since the object we are measuring

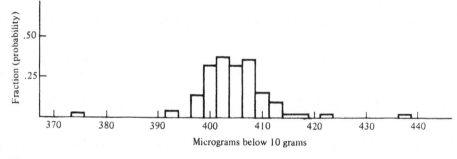

Figure 1

weighs approximately 10,000,000 micrograms, an error of 62 is very small, a percent error of (62/10,000,000) × 100 percent or 0.00062 percent). Who cares about errors that small? Certainly if we are buying cheese at the supermarket, we don't insist that the error be that small. The earliest verifications of Newton's laws were only made to an accuracy of 4 percent, which is probably not as good as the scale used to measure cheese at your supermarket. On the other hand, recent experiments in particle physics have depended on measurements whose errors were less than 0.00062 percent.

If the problem of accurate measurement is difficult in the physical sciences, it can be downright baffling in the social sciences. Take the apparently simple question of how many people lived in New York State in 1975. If you look this up in a book, you might well find a number, but this has to be taken with many grains of salt. The first grain of salt is for the fact that the question is meaningless (if you want to be fussy) because 1975 is a long period of time, during which people are dying, being born, and migrating in and out of the state. The question becomes meaningful if we specify a definite instant, but this makes the practical problem of doing the count impossible. After all, we can't freeze the world and its activities for the benefit of the census takers. Leaving all this aside, we need another grain of salt for the fact that no census of any sort was taken in 1975. Censuses were taken in 1970 and 1980, and we might use these figures to arrive at a figure for 1975 (for example, by interpolation), but this is surely not completely reliable. A final hefty grain of salt is needed to help swallow the sad truth that even in a census year the census may miss people. By its own admission the Bureau of Census undercounts various groups in the population by anywhere from 1 to 16 percent. This should not be considered a criticism of the Bureau of the Census, which is a dedicated, highly professional agency. However, the problems it faces are very great.

Counting people presents particular problems: people lie, hide from census takers, have faulty memories, or may be missed by the interviewer. One might suppose that counting something other than people would be easier. Not so! In 1935, the United States reported gold imports of 64,485 thousands of pounds sterling from the United Kingdom. But the United Kingdom claims to have supplied the United States with 85,883 thousands of pounds sterling!

Table 2   1949 agricultural production[†]

Crop	Census estimate	Department of agriculture estimate	Percent difference
Corn (for grain)	2,778,190 bushels	3,114,726	− 12.1
Sorghums (for grain and seed)	140,835 bushels	152,630	− 8.4
Wheat	1,006,559 bushels	1,141,188	− 13.4
Barley	220,963 bushels	236,737	− 7.1
Oats	1,136,642 bushels	1,329,473	− 17.0
Rice	40,244 bags	40,747	− 1.2
Soybeans (for beans)	212,440 bushels	230,897	− 8.7
Alfalfa	32,254 tons	38,645	− 9.6
Cotton	15,419 bales	16,128	− 4.6
Irish potatoes	219,802 hundredweight	246,939	− 12.3
Tobacco	1,769,769 pounds	1,972,541	− 11.5

[†] Adapted from table 15 of O. Morgenstern, "On the Accuracy of Economic Observations," Princeton University Press, Princeton, N.J., 1963.

Table 2 shows discrepancies in measuring agricultural production when two different agencies do the counting.

Before discussing how to cope with discrepancies in measurements, we should mention a useful distinction between systematic and random errors. A random error is one which is unlikely to be operating the same way in different measurements. For example, vibrations due to outside road traffic may be present during one weighing and not another. Or, a small air current in the room, perhaps set off by movements of the human operator, might influence the scale slightly. But since the direction and force of such vibrations and air currents would probably vary unpredictably from one weighing to another, they would contribute to the random error and not the systematic error.

On the other hand, if there is an air conditioner blowing at the scale, with the same force each time the weighing is carried out, this would give rise to systematic error. A systematic error is one which effects the results in about the same way each time. Systematic error is also called *bias*. Another example of bias would be to have the scale on a surface which is not absolutely level. Bias may also occur if some of the internal parts of the scale were not made with precisely the right weight or shape.

Unfortunately, systematic error is often not easy to detect. Except in obvious cases, it is normally only discovered if the results of two or more independent and distinct methods of observation (e.g., two different scales) are compared with each other. This is normally the domain of the experimental scientist or data collector and not of the mathematical modeler. Therefore, in this section we proceed on the assumption that the errors we are dealing with are random errors.

A common and well-known procedure for dealing with random errors is to take the mean (average) of the various measurements available. This appeals to common sense. The purpose of this section is to find out whether theory agrees

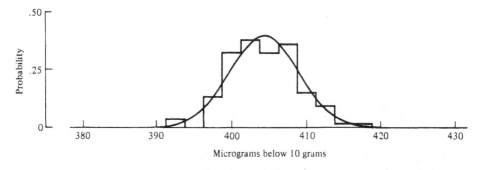

Figure 2

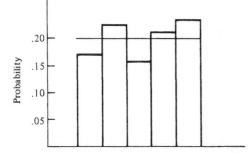

Figure 3 Fitting a histogram with a uniform density.

with common sense. The answer we will obtain is "sometimes." For example, averaging is a good idea if the errors in measurement can be closely fit by a normal curve. Figure 2 shows a so-called normal curve fitted to the histogram for the weights of NB 10.[†] The fit appears pretty good. The equation of this particular normal curve is

$$y = \frac{1}{4.4\sqrt{2\pi}} \exp\left[-\frac{1}{2} \left(\frac{x - 404.3}{4.4} \right)^2 \right] \tag{1}$$

Although normal curves are very frequently used, we can imagine fitting other sorts of curves to histograms. Figure 3 shows a histogram for which a straight line segment provides a pretty good fit. Figure 4 shows a histogram for which a curve with equation of the form $y = x^{a-1} e^{-x/b}/b^a(a!)$ provides a good fit. These smooth curves are examples of probability density curves. Figure 2 is called a *normal probability density*, Figure 3 a *rectangular* or *uniform density*, and Figure 4 a *gamma density*.

[†] The histogram used here differs from the one in Figure 1 because three measurements (423, 437, and 375) have been thrown out as "flukes" or, as statisticians put it, "outliers." There are statistical procedures for identifying such outliers.

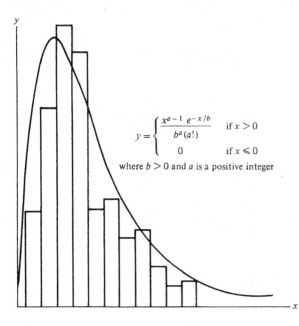

$$y = \begin{cases} \dfrac{x^{a-1}\, e^{-x/b}}{b^a\,(a!)} & \text{if } x > 0 \\ 0 & \text{if } x \leqslant 0 \end{cases}$$

where $b > 0$ and a is a positive integer

Figure 4 Fitting a histogram with a gamma density.

Definition

Any function $y = f(x)$ for which $f(x) \geq 0$ and

$$\int_{-\infty}^{\infty} f(x)\, dx = 1$$

is called a *probability density function*. The graph of $y = f(x)$ is called a *probability density curve*.

The reason we require the area under the curve to be 1 is that areas have a special interpretation as probabilities:

The area under a probability density curve between $x = a$ and $x = b$ represents the probability that x (a particular measurement) will have a value between a and b.

In the following examples we will let $P(a, b)$ stand for the probability that a measurement comes out between a and b.

Example 1

Suppose a set of measurements of the weight x of a speck of dust is fit (Figure 5) by the uniform probability density function

$$y = \begin{cases} \frac{1}{10} & \text{if } 5 \leq x \leq 15 \\ 0 & \text{for all other } x \end{cases}$$

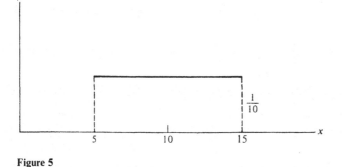

Figure 5

What is the probability of getting a measurement between 5 and 8? If two measurements are taken, what is the probability that the first falls between 5 and 8 and the second between 8 and 10?

The first question is routine.

$$P(5, 8) = \int_5^8 \tfrac{1}{10} \, dx = .3$$

For the second question, we assume that the two separate measurings are independent so that we can make use of the multiplication rule. This rule asserts that, if we carry out a series of independent measurements, the probability of a specified sequence of outcomes is the product of the probabilities of the separate outcomes. This means that in our problem the probability of a sequence in which the first measurement is between 5 and 8 and the second is between 8 and 10 will be

$$P(5, 8)P(8, 10) = (.3)(.2) = .06$$

Example 2

If Equation (1) describes the process of weighing NB 10 on a certain scale, what is the probability that the next two weighings of NB 10 will result in values of 415 ± 1 and 400 ± 1, in that order?

A reading of 415 ± 1 means the instrument could not be read with exact accuracy, and so the measurement could be as low as 414 or as high as 416. The probability of such a reading is

$$P(414, 416) = \int_{414}^{416} \frac{1}{4.4\sqrt{2\pi}} \exp\left[-\frac{1}{2}\left(\frac{x - 404.3}{4.4} \right)^2 \right]$$

This integral is not easily evaluated. There are tables that will help. However, we would like to have the answer given by a formula suitable for applying optimization methods from elementary calculus. Therefore, we approximate the area under the curve with the area of a suitable rectangle. We will use the rectangle whose base

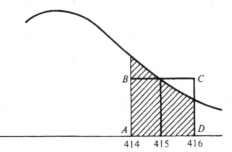

Figure 6 Rectangle *ABCD* approximates shaded area.

is the interval [414, 416] which we are integrating over, and whose height is the ordinate of the normal curve at the midpoint of the interval, $x = 415$ (Figure 6):

$$P(414, 416) = 2 \frac{1}{4.4\sqrt{2\pi}} \exp\left[-\frac{1}{2}\left(\frac{415 - 404.3}{4.4}\right)^2\right]$$

$$= .0094$$

Likewise

$$P(399, 401) = 2 \frac{1}{4.4\sqrt{2\pi}} \exp\left[-\frac{1}{2}\left(\frac{400 - 404.3}{4.4}\right)^2\right]$$

$$= .1125$$

Therefore, the probability of getting the outcomes 415 ± 1 and 400 ± 1 in sequence is $P(414, 416)P(399, 401) = .0011$.

Normal probability densities Because so many measurement processes involve normal probability density curves, we need some special information about them. A curve with the equation

$$y = \frac{1}{\sigma\sqrt{2\pi}} \exp\left[-\frac{1}{2}\left(\frac{x - \mu}{\sigma}\right)^2\right]$$

where μ is a real number and $\sigma > 0$, is called a *normal curve*.[†] In our examples the variable x will represent the outcome of the measurement process (weight, blood pressure, etc.). It is sometimes thought that y represents the probability of obtaining the value x. This is wrong and can lead to much confusion. The height y has no direct interpretation as a probability. Probabilities are represented as areas under the curve.

Figure 7 shows some normal curves. You will notice that each has an axis of symmetry. It can be shown (Exercise 5) that this axis is the line $x = \mu$. Therefore, one significance of the parameter μ in the equation of the normal curve is that it

---

[†] μ and σ are called, respectively, the *mean* and *standard deviation* of the particular normal curve. These terminologies and the reasons for them need not concern us.

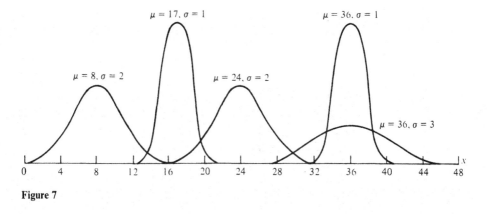

Figure 7

determines where the axis of symmetry of the curve is. Changing μ causes the curve to move horizontally by an amount equal to the size of the change.

The right-left location of the curve is the only graphical significance μ has. The shape of the curve, whether its hump is tall and skinny or short and spread out, is determined by the size of σ. The smaller σ is, the taller and skinnier the curve is.

A more specific interpretation of σ is contained in the following rules of thumb:

1. Approximately 68 percent of the area under a normal curve is contained between the lines $x = \mu - \sigma$ and $x = \mu + \sigma$.
2. Approximately 95 percent of the area under a normal curve is contained between the lines $x = \mu - 2\sigma$ and $x = \mu + 2\sigma$.

Since areas under a normal curve can be interpreted as probabilities, these rules of thumb can be restated

1′. The probability is approximately .68 that a value of x (a randomly taken measurement) will be within σ units of μ.
2′. The probability is approximately .95 that a value of x will be within 2σ units of μ.

We now turn to the question of how normal curves are connected to measurement processes. We have seen an example of this in Figure 2. But this is a highly misleading example, because one rarely has the luxury of making 100 measurements. To explore this further, it is helpful to have a more precise idea of what we mean by a measuring process.

By a measuring process we mean a particular instrument or set of instruments and a particular set of procedures for using these instruments and "reading" the final measurement. For example, it may consist of a bathroom scale and the procedure of placing the item to be weighed exactly in the center of the scale. Alternatively the measuring process might consist of a more accurate scale, a

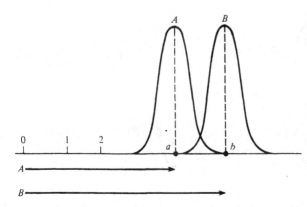

Figure 8 Measuring two arrows with the same measuring process. The measuring scale is shown on the horizontal axis. The normal curve *A* describes the various measurements of arrow *A*, while the normal curve *B* describes the various measurements of arrow *B*. When measuring *A*, the most likely values are those near the number *a*, which is the true length of arrow *A*. When measuring arrow *B*, the most likely values are those near *b*, the true length of arrow *B*.

carpenter's level (to make sure the platform of the scale is exactly level), instructions about centering the object, and a magnifying glass to help in reading the scale.

We do not consider the object being measured to be part of the measurement process. In other words, a measurement process is something that can be applied to various objects. When we apply a measuring instrument to a particular item to be measured, there is one particular probability density function that describes the probabilities of various intervals of values that might be obtained. (But, unless we have a large number of measurements of the object, we won't know which function.) If we change the object being measured, there will very likely be a new probability density function (Figure 8).

When faced with a small number of measurements of a particular object, it is hard to know which probability density curve to use to fit them. The practical approach is to narrow down to some familiar family of probability densities. One might, for example, assume that the densities that correspond to the items one has to measure would be found among the normal curves (Figure 7) or, perhaps, the subset of normal curves that have $\sigma = 3$ and differ only in their μ value (i.e., that have the same degree of "flatness" and differ only in right-left location along the axis; see Figure 9). Or one might select the set of uniform densities (Figure 10).

In this section we will make the following assumptions:

1. The probability density that describes the measurement process is normal.
2. For any item to be measured, the true value coincides with μ, the location of the axis of symmetry of the probability density function describing the measurement process. Roughly speaking, this asserts that one is just as likely to obtain underestimates of the truth as overestimates.

It must be emphasized that these are assumptions about our measurement process, not mathematical certainties. It is quite remarkable how often these assumptions are made without the least bit of critical thought. It has been remarked (by G. Lippmann, a French physicist) that "everybody believes in the [normal approximation]: the experimenters because they think it is a mathematical theorem,

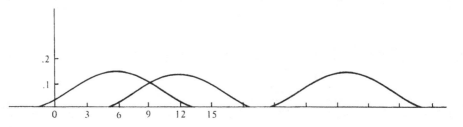

Figure 9 Three normal curves having $\sigma = 3$. (There are infinitely many normal curves having $\sigma = 3$, one for each choice of μ.)

the mathematicians because they think it is an experimental fact." The quotation is very apt. There are theorems in statistics which suggest (but do not prove) that many measurements processes would be described by normal curves. Experience confirms this, in the sense that many measurement processes have been carefully investigated and shown to be fit closely by normal curves. Nevertheless, there are measurement processes which are best described by other kinds of probability density functions than the normal. As an example, there is an old procedure in astronomy in which visual estimates are made of the ratio of brightnesses of two stars. Here the measuring instrument is the human eye (aided by a telescope). The human nervous system, for example the eye, is a very peculiar measuring instrument. As a consequence, it turns out that, if $r_1, r_2, \ldots, r_n$ are the ratios obtained in various repetitions of the measurement process, then the probability density of the r's is not normal at all. Furthermore, the best way of obtaining an estimate of the true ratio x is not by taking the mean of the r_i but, instead, using the formula $x = (r_1 r_2 \cdots r_n)^{1/n}$.

Henceforth, we assume that theory or past experience with a particular measurement process has convinced us that assumptions 1 and 2 are reasonable. Imagine we are now faced with a particular item to measure, have taken n measurements, and obtained the numbers $x_1, x_2, \ldots, x_n$. How do we estimate the true value? In view of assumption 2, this is equivalent to finding out which normal curve governs the measurement process, more precisely, what the μ value of this curve is.

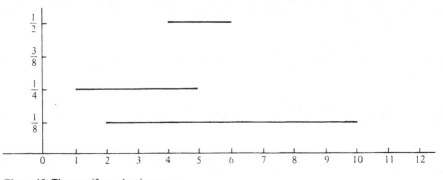

Figure 10 Three uniform-density curves.

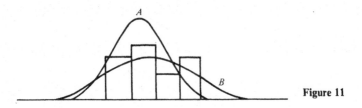

Figure 11

If one has lots of measurements, one can often get a good estimate by eye. For example, for the histogram of Figure 2 most people could, even without seeing the best-fitting normal curve, guess that its μ value would be somewhere near 404. But it is rare to be able to do as many measurements as were done on NB 10 by the National Bureau of Standards. It is more common to have a small handful. For example, suppose we found seven values for a weight, as shown in Figure 11. Which normal curve fits this data best, the one labeled A or the one labeled B?

Since each member of the family of normal curves is determined by the values of the parameters μ and σ, we can think of this as choosing the optimum pair of values for these parameters. Actually, we don't really need the value of σ. It turns out that we can find the best μ without knowing the best σ. Finding the best μ turns out to be a standard 1-variable max-min problem. (If we also want to know the best σ, we need to do a max-min problem in the two variables μ and σ.) But before we get to the mechanics, we need to have something precise to maximize. We have spoken of getting the best fit of a normal curve, but how do we measure fit?

The principle we use to answer this question is called the *method of maximum likelihood*.

We select that normal curve which would give the highest probability (maximum likelihood) of yielding the observations that were actually obtained.

To illustrate the logic of this, suppose that, at your college, the economics department is very stingy with high grades, while almost everyone gets high grades in sociology. If a student claims a grade-point average which is usually high and you want to guess that student's major, the principle of maximum likelihood says that you should guess sociology. Here are some examples related to observation error and the normal curve.

Example 3

The diameter of a circular disc is measured and found to be somewhere between 71 and 72 millimeters. Which of the following normal curves is more likely to be the one governing the measurement process?

1. $\mu = 71$ and $\sigma = 1$.
2. $\mu = 71.5$ and $\sigma = 2$.

The method of maximum likelihood says that, for each of these two cases, we must calculate the likelihood, denoted $L(\mu, \sigma)$, that the normal curve with those parameters gives a value between 71 and 72.

$$L(71, 1) = \frac{1}{\sqrt{2\pi}} \int_{71}^{72} \exp\left[-\frac{1}{2}\left(\frac{x-71}{1}\right)^2\right] dx$$

$$L(71.5, 2) = \frac{1}{2\sqrt{2\pi}} \int_{71}^{72} \exp\left[-\frac{1}{2}\left(\frac{x-71.5}{2}\right)^2\right] dx$$

Once again we are faced with tough integrals. Using the method of approximating with a rectangle, we can work the problem if we have tables for the exponential function or a calculator with an e^x key.

$$L(71, 1) = \frac{1}{\sqrt{2\pi}} \exp\left[-\frac{1}{2}\left(\frac{71.5-71}{1}\right)^2\right](1)$$

$$= 0.352$$

$$L(71.5, 2) = \frac{1}{2\sqrt{2\pi}} \exp\left[-\frac{1}{2}\left(\frac{71.5-71.5}{2}\right)^2\right](1)$$

$$= 0.199$$

Since $L(71.5, 2) < L(71,1)$, our best guess, according to the method of maximum likelihood, is that the better-fitting curve of the two being compared, is the one with $\mu = 71$ and $\sigma = 1$.

The previous example presents a number of features worthy of comment:

1. Only one measurement was taken. This is unusual in practice and was done merely to provide a simple example. Our next example is more realistic.
2. The result of the single measurement taken was not a definite number but a range (71 to 72). This is quite realistic since it is hard to read an instrument precisely. If one must report a definite value, one should also report the range of uncertainty in the reading of the instrument. In the previous example, we might report a value of 70.5 with an uncertainty of ± 0.5. In general, a measurement of $x \pm \Delta$ means that the measurement appeared to be somewhere between $x - \Delta$ and $x + \Delta$.
3. We had only two normal curves to choose between. This is highly unrealistic: why not admit all the other infinitely many normal curves into the competition? The method for doing this is shown after Example 4.

Example 4

Suppose that the disc of Example 3 is measured twice, the values obtained are 71 and 72, and in each case the value of Δ (the uncertainty in the reading) is 0.5.

Calculate which of the normal curves, 1 or 2, has the highest likelihood of yielding these measurements.

In case 1 the probability of obtaining a value of 71 ± 0.5 is approximately

$$\frac{1}{\sqrt{2\pi}} \exp\left[-\frac{1}{2}\left(\frac{71-71}{1}\right)^2\right](1) = 0.399$$

and the probability of obtaining 72 ± 0.5 is approximately

$$\frac{1}{\sqrt{2\pi}} \exp\left[-\frac{1}{2}\left(\frac{72-71}{1}\right)^2\right](1) = 0.242$$

What we need now is a way to estimate the probability of getting *both* these measurements. We will make the assumption that these measurements are independent events so that we can apply the multiplication rule: the probability of getting 71 ± 0.5 and 72 ± 0.5 in two measurements is the product of the separate probabilities. Thus

$$L(71, 1) = (0.399)(0.242) = 0.097$$

In the same way, we can determine that

$L(71.5, 2)$

$$= \frac{1}{2\sqrt{2\pi}} \exp\left[-\frac{1}{2}\left(\frac{71-71.5}{2}\right)^2\right](1) \frac{1}{2\sqrt{2\pi}} \exp\left[-\frac{1}{2}\left(\frac{72-71.5}{2}\right)^2\right](1)$$

$$= (0.193)(0.193)$$
$$= 0.037$$

Since $L(71, 1) > L(71.5, 2)$, the method of maximum likelihood suggests that $\mu = 71$ and $\sigma = 1$ is a better guess than $\mu = 71.5$ and $\sigma = 2$.

What we have learned so far may be summarized: if measurements $x_1, x_2, \ldots, x_n$ have been obtained, each with a reading error of $\pm \Delta$, in a measurement process conforming to assumptions 1 and 2, the likelihood of obtaining these measurements from the normal curve with parameters μ and σ is

$$L(\mu, \sigma) = \frac{1}{\sigma\sqrt{2\pi}} \exp\left[-\frac{1}{2}\left(\frac{x_1-\mu}{\sigma}\right)^2\right]2\Delta \cdots \frac{1}{\sigma\sqrt{2\pi}} \exp\left[-\frac{1}{2}\left(\frac{x_n-\mu}{\sigma}\right)^2\right]2\Delta$$

$$= \left(\frac{2\Delta}{\sigma\sqrt{2\pi}}\right)^n \exp\left[-\frac{1}{2}\sum_{i=1}^{n}\left(\frac{x_i-\mu}{\sigma}\right)^2\right]$$

Let us temporarily treat σ as a constant and try to find the value of μ that maximizes $L(\mu, \sigma)$. Equivalently, we try to maximize $y = \log L(\mu, \sigma)$ because this will be a little simpler.

$$y = \log L(\mu, \sigma) = n \log \frac{2\Delta}{\sigma\sqrt{2\pi}} - \frac{1}{2} \sum_{i=1}^{n} \left(\frac{x_i - \mu}{\sigma} \right)^2$$

$$\frac{dy}{d\mu} = \frac{1}{\sigma} \sum_{i=1}^{n} (x_i - \mu) = 0$$

$$\left(\sum_{i=1}^{n} x_i \right) - n\mu = 0$$

$$\mu = \frac{\sum_{i=1}^{n} x_i}{n}$$

This is the familiar formula for the mean of the x_i. This value of μ will be a critical point for the function y. It is possible to show that it is actually a maximum. (Try this as an exercise.) Thus we arrive at the commonsense notion that the best guess for the true value, according to the method of maximum likelihood, is obtained by averaging the measurements.

What if the measurement process is not well described by a normal probability density function? Specifically, suppose we abandon assumption 1 but keep assumption 2. For some nonnormal probability densities the method of maximum likelihood still justifies taking the average as the estimate of the true value. (See Exercise 7 for an example.) But in other cases taking the average has no theoretical justification. The next example illustrates this.

Example 5

Suppose the probability density function which best fits a measurement process is a uniform probability density. Specifically, if the true value is μ, then any value between $\mu - k$ and $\mu + k$ can occur and all values are equally likely. If we set $\alpha =$ the smallest possible measurement ($\alpha = \mu - k$) and $\beta =$ the largest possible measurement ($\beta = \mu + k$), then we can write the equation of this probability density function $y = u(x)$, where

$$u(x) = \begin{cases} \dfrac{1}{\beta - \alpha} & \text{if } x \in [\alpha, \beta] \\ 0 & \text{for all other } x \end{cases}$$

See Figure 12. Verify that this is a probability density function by integrating from $-\infty$ to ∞.

A uniform density is determined by the two parameters α and β. Our task is to choose those α and β values that give the best fit to a set of observations $x_1, x_2, \ldots, x_n$, which have been read to an accuracy of Δ units. We will assume

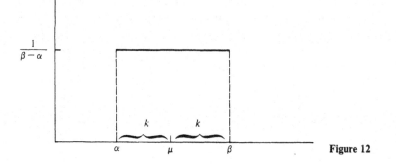

Figure 12

x_1 is the least of these values and x_n the largest. Following the principles of this section, we define $L(\alpha, \beta)$ as the following product of integrals.

$$L(\alpha, \beta) = \int_{x_1 - \Delta}^{x_1 + \Delta} u(x)\, dx \cdots \int_{x_n - \Delta}^{x_n + \Delta} u(x)\, dx \tag{2}$$

We shall show that the optimal α lies in $[x_1 - \Delta, x_1 + \Delta]$ and the optimal β lies in $[x_n - \Delta, x_n + \Delta]$.

The first thing we observe is that the product in Equation (2) may be zero if one of the intervals $[x_i - \Delta, x_i + \Delta]$ lies outside the interval $[\alpha, \beta]$ for, in that case, the integral over that interval is 0. [Remember, $u(x)$ is zero except from α to β. See Figure 13.] Since x_1 is the smallest of the measurements and x_n the largest, the observation we have just made implies $\alpha < x_1 + \Delta$ and $\beta > x_n - \Delta$. Is there any value in choosing $\alpha < x_1 - \Delta$? No, because decreasing α beyond $x_1 - \Delta$ lowers $1/(\beta - \alpha)$ and each factor of Equation (2) becomes smaller. Therefore the optimal α lies in $[x_1 - \Delta, x_1 + \Delta]$. Likewise the optimal β lies in $[x_n - \Delta, x_n + \Delta]$. If Δ is small enough to be considered negligible, then this nearly pinpoints α and β as

$$\alpha = x_1 \qquad \text{and} \qquad \beta = x_n \tag{3}$$

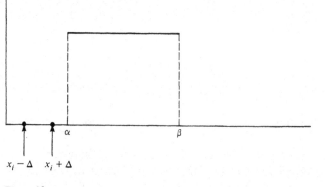

Figure 13

Since μ is halfway between α and β, $\mu = (\alpha + \beta)/2 = (x_1 + x_n)/2$. This says that the approximate best guess for the true value is obtained by averaging the largest and smallest measurements—and ignoring all the rest! This is very different from the idea of taking the average of all measurements, which is the correct procedure for normal curves. It is true that formulas (3) are only approximate, but even the exact formulas ignore the size of all the measurements except the largest and smallest.

Here is a summary of this section:

1. Measurement processes are rarely perfect and repeating a measurement will often give different values.
2. The outcomes of measuring a particular object with a particular measurement process are assumed to be described by a probability density curve. In most of the section we make the further assumption that the curve is normal.
3. We also assume that the axis of symmetry of the normal curve describing the measurements is located at the true value of the object being measured. Consequently, if we can find the best-fitting normal curve, we can estimate the true value of the object as the value at the foot (on the horizontal axis) of the axis of symmetry.
4. If we only have a few actual measurements to go on, it's hard to know which normal curve fits best. The method of maximum likelihood is a way of overcoming these difficulties.
5. The method of maximum likelihood implies that, if the measurement process is described by normal curves, then the best estimate of the true value is obtained by taking the mean of the measured values.
6. But this commonsense notion of taking the mean of the measured values is not necessarily right if the curve is not normal. Example 5 illustrates this.

Sidelight: Gauss and the Validity of the Mean

The theory of this section seems strongly tied to the normal curve. Indeed, Example 3 shows that, for at least one nonnormal probability density, it is not sensible to take the mean. Is it always wrong to take the mean of our measurements when the probability density is not normal? According to a theorem by Karl Friedrich Gauss there is a very large category of curves for which it is wrong. What Gauss proved was that, if our measurement process is described by a family of curves, which differ only by the location of the mean (i.e., each curve can be gotten from any other by a right or left translation), and the curves are not normal but are continuously differentiable for all real values, then, at least occasionally, it will be wrong to take the mean. That is, there will be at least one set of measurements for which the method of maximum likelihood does not specify the mean of the measurements as the best guess for the true value. (See, for example, the book by J. Aczel, listed in the Bibliography.)

EXERCISES

● **1** Suppose a measurement process applied to one particular object has a uniform probability density function with the equation

$$y = \begin{cases} \frac{1}{12} & \text{if } 0 \le x \le 12 \\ 0 & \text{for all other } x \end{cases}$$

(a) What is the probability of getting a measurement between 2 and 3? Between 2 and 4? Between 20 and 24? Between 10 and 14?

(b) If two measurements are taken, what is the probability of getting a number between 1 and 2 followed by a number between 2 and 3?

● **2** Suppose a measurement process applied to a certain object has a normal probability density function whose equation is

$$y = \frac{1}{\sqrt{2\pi}} \exp\left(-\tfrac{1}{2}x^2\right)$$

(a) What is the probability of getting a number between 1.4 and 1.5? Between -0.2 and 0.2? (*Hint:* Find the rectangles that approximate the areas you need.)

(b) If two measurements are taken, what is the probability of getting a number between -0.2 and 0.2 on the first and a number between 1.4 and 1.5 on the second?

3 Show that the normal curve takes on its maximum value at $x = \mu$. (*Hint:* You can do this as a calculus-type max-min problem, but you can avoid having to do that by appealing to general properties of the exponential function.) What is the maximum value?

4 Show by a change of variables that the area from x_1 to x_2 under a normal curve, with parameters μ and σ, is the same as the area under the so-called standard normal curve, i.e., the one with parameters 0 and 1, from $(x_1 - \mu)/\sigma$ to $(x_2 - \mu)/\sigma$.

5 Show that a normal curve is symmetrical about $x = \mu$. That is, if $x_1 < \mu < x_2$ and $\mu - x_1 = x_2 - \mu$, then $f(x_1) = f(x_2)$, where $y = f(x)$ is the equation of the normal curve.

6 Show that a normal curve with parameters μ and σ has points of inflection at $x = \mu \pm \sigma$.

● **7** Suppose a measurement process applied to something whose true value is μ is governed by the non-normal probability density $f(x) = (1/\mu)e^{-x/\mu}$. This means that the probability of obtaining a measurement between a and b is

$$\int_a^b \frac{1}{\mu} e^{-x/\mu}\, dx$$

Suppose measurements $x_1, x_2, \ldots, x_n$ have been taken with reading accuracy Δ. It is desired to use the method of maximum likelihood to find the value of μ which would give the highest probability of yielding those measurements.

(a) If $L(\mu)$ denotes the likelihood function, i.e., the probability of getting $x_1, x_2, \ldots, x_n$ if the true value is μ, find the formula for the exact value of $L(\mu)$.

(b) Find an approximate formula for $L(\mu)$, using the same method of approximating integrals that we used to obtain the formulas in Example 3.

(c) Find the value of μ which maximizes $L(\mu)$. [*Hint:* You may wish to maximize log $L(\mu)$ instead.]

(d) Does your result contradict the information given in the sidelight entitled "Gauss and the Validity of the Mean"?

8 Final course grades are often computed by taking the mean of all the test scores. Is this consistent with a hypothesis that errors in measuring a student's "true ability" by tests are governed by a uniform distribution? What justification is there for using the mean of all scores to determine the final grade?

● **9** Suppose a measurement process is governed by a normal curve. You have taken measurements $x_1, x_2, \ldots, x_n$ with reading accuracy Δ, and you wish to find the best-fitting normal curve. The theory of this section says pick one with $\mu = (x_1 + \cdots + x_n)/n$. But we didn't work out which value of σ to use. Can you do this? What good would it do to find σ?

10 Suppose you decide that the measurement of the length of an item is best described by the normal curve with $\mu = 5$ and $\sigma = 1$. What is the probability of getting a negative value? If you can't estimate the value of this probability, can you decide whether it is 0 or not? Is your answer realistic?

11 In Example 4 would your estimate of μ change if Δ, the measure of reading accuracy, changed from $\Delta = 0.5$ to $\Delta = 0.25$?

12 The mean of the normal probability density function with parameters μ and σ is defined as

$$\frac{1}{\sigma\sqrt{2\pi}} \int_{-\infty}^{\infty} x \exp\left[-\frac{1}{2}\left(\frac{x-\mu}{\sigma}\right)^2\right] dx$$

Show that this integral equals μ. (This integral is a generalization of a sum in which we multiply various x's by their probability of occurrence and add the results. Thus it is a generalization of an average and, therefore, is called the *mean of the density*. It is also called the *expectation*.)

13 This exercise deals with the computation of the exact formula for $L(\alpha, \beta)$ in Example 5 of the text. From the discussion in Example 5, we know that we must choose $\alpha \in [x_1 - \Delta, x_1 + \Delta]$ and $\beta \in [x_1 - \Delta, x_n + \Delta]$.

(a) Show that, with these restrictions on α and β,

$$L(\alpha, \beta) = \left(\frac{2\Delta}{\beta - \alpha}\right)^{n-2} \left(\frac{x_1 + \Delta - \alpha}{\beta - \alpha}\right)\left[\frac{\beta - (x_n - \Delta)}{\beta - \alpha}\right]$$

(See Figure 14.)

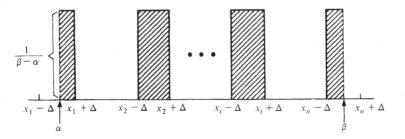

Figure 14 $L(\alpha, \beta)$ is the product of the n shaded areas. $n - 2$ of these (of which only two are shown, the middle two in the figure) have dimensions 2Δ by $1/(\beta - \alpha)$. The two end areas have other dimensions.

(b) Find the values of α and β that maximize $L(\alpha, \beta)$. In doing this, please recall that α and β are restricted to $x_1 - \Delta \le \alpha < x_1 + \Delta$ and $x_n - \Delta < \beta \le x_n + \Delta$. Since there are two boundary edges in this rectangular domain, the maximum might occur at a point on these boundary edges and where the partial derivatives are not 0. (Can you determine, without going through the horrible details, that your answers won't involve $x_2, x_3, \ldots, x_{n-1}$?)

14 In Table 2 the census estimate is lower in every single crop than the Department of Agriculture estimate. Why do you suppose this is? Could it be pure chance?

Computer exercise

15 Verify the "0.68 rule" that the area between $x = \mu - \sigma$ and $x = \mu + \sigma$ under a normal curve is 0.68. Use the normal curve with $\mu = 0$ and $\sigma = 5$. Approximate the area with rectangles: subdivide the interval into a number of subintervals and erect a rectangle on each subinterval to approximate the area corresponding to that subinterval, as in Figure 6. If the sum of your rectangles is not nearly 0.68, use a larger number of smaller subintervals to improve the accuracy of your approximation.

BIBLIOGRAPHY

Aczel, J.: "Lectures on Functional Equations and their Applications," Academic Press, New York, 1966. A discussion, beginning on page 106, of when the method of maximum likelihood prescribes taking the mean of a series of measurements. It is shown that under many circumstances it is necessary to have normally distributed measurements.

Freedman, D., R. Pisani, and R. Purves: "Statistics," W. W. Norton & Company, Inc., New York, 1978. A good discussion of the chance errors made by the National Bureau of Standards in weighing its 10-gram weight.

Hoel, P.: "Introduction to Mathematical Statistics," John Wiley & Sons, Inc., New York, 1947. A standard introductory book.

Morgenstern, O.: "On the Accuracy of Economic Observations," Princeton University Press, Princeton, N.J., 1963. A provocative, nontechnical book by a prominent economist.

Whittaker, E., and G. Robinson: "The Calculus of Observations," Blackie and Sons, Ltd., Glasgow, 1924. Some nice material on normal curves not found elsewhere appears in chapters 8 and 9.

3 ADJUSTING DATA II (THE HARD WAY)—MISCELLANEOUS METHODS AND EXAMPLES

Abstract When one doesn't have repeated measurements of a quantity, taking the mean is impossible. Not only that; it isn't even clear whether the measurement has any error in it. Sometimes one can see a pattern involving other related measurements, but one also needs a theory about what lies behind the numbers. This need for understanding how the numbers arise is the main theme of this section. The adjustment methods discussed are: grouping, moving averages, least squares, and interpolation.

Prerequisites Multiplication of matrices, systems of linear equations, and elementary calculus.

In the last section we justified taking the average of a number of measurements. But it is not always possible to have multiple measurements of the same thing. This is especially true for social science data which is obtained through survey or census, at considerable expense and trouble. Doing a survey three or four times is generally out of the question. But, if the measurement is part of a series of related measurements, we might look for a pattern. A measurement that doesn't conform to the pattern might be a candidate for adjustment.

For example, if every age group in a survey gives the President an approval rating of about 60 percent except for the 35–39 year age group, which rates him at over 70 percent, then this latter figure is possibly wrong. If we could take this measurement a few more times, we could average the results obtained for this age group. If this is impractical, then somehow the ratings of the other age groups might somehow guide us. But how? There are many methods which have been used for this sort of thing, and they go by various names: adjusting data, smoothing, filtering out noise, etc. We will discuss four examples which illustrate four different methods.

Our discussion will show that there is a large role for personal judgement in all this. We often need a hunch or theory about what lies behind the data in order to know how or whether to adjust. We have to go out on a limb, and that's what makes this the "hard way." Another way to put it is this: when we adjust data in a certain way, we are adopting a model that expresses our belief about the process that gave rise to the data.

Example 1 Grouping

The 1970 National Fertility Study asked married women how often they had sexual intercourse in a 4-week period. Part of the results are shown in Table 1 and Figure 1a. Some observers felt the results were too peculiar to be believed.

Why should 8 times be so much more common than 7 or 9? One could probably find out by doing some more fact gathering, but there is a limit to the amount of time and money available for taking surveys. In this instance, instead of doing more fact gathering, the researchers formulated a theory. The theory was that women often chose a likely estimate of twice per week and then multiplied by 4 to get a 4-week estimate of 8. For many of these women, according to the theory, the true frequency might well have been 7 or 9 (or another number).

It is hard to know whether this theory is true or not. One could formulate an alternate theory that people do fall into weekly schedules (perhaps without planning to) with regard to intercourse and have intercourse only on certain days of the week. If this is true, then the 4-week total would truly be 4 times the size of the 1-week total, and the 4-week total of 8 would naturally be very common. Numbers like 7 and 9 could only come about from a small group of people not on a schedule. According to this theory there is no reason to distrust the numbers in Table 1. They may be exactly correct!

Table 1[†]

Times in 4 weeks	Percent	Times in 4 weeks	Percent
0	6.0	11	0.5
1	3.3	12	9.3
2	6.7	13	0.3
3	5.5	14	1.2
4	11.7	15	3.9
5	6.0	16	2.3
6	8.7	17	0.2
7	2.5	18	0.8
8	12.8	19	0.1
9	1.7	20	4.0
10	8.6		

[†] Frequencies over 20 are omitted.

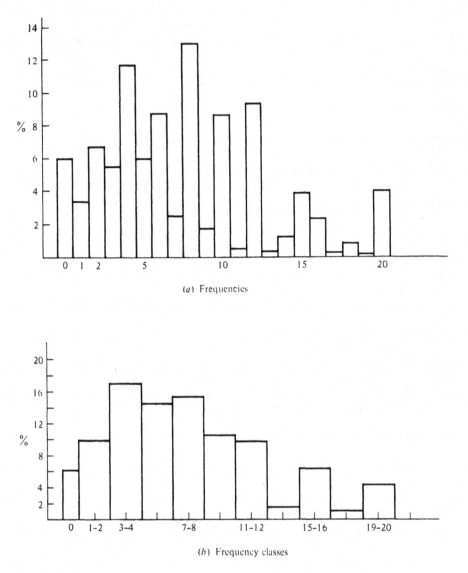

(a) Frequencies

(b) Frequency classes

Figure 1

Which theory should we believe? You are entitled to your own opinion. The researchers who collected this data inclined to the first theory and used it as a license to tinker with the data. Unfortunately, the theory doesn't tell us how to do this tinkering.

The strategy adopted by the sex researchers was *grouping*. The various possible frequencies of intercourse are grouped into categories, as shown in Table 2. By doing this, the high percentage of people having a frequency of 8 is merged with the low percentage whose frequency is 7. The graph which results from this (Figure 1*b*)

Table 2 Grouped data

Coital frequency classes (times in 4 weeks)	Percent[†] 1970
0	6.1
1–2	9.9
3–4	17.0
5–6	14.6
7–8	15.2
9–10	10.4
11–12	9.9
13–14	1.5
15–16	6.2
17–18	1.0
19–20	4.2

[†] Discrepancies between these percents and Table 1 are due to roundoff from original data.

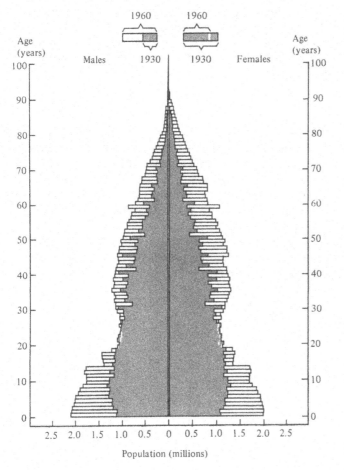

Figure 2 Population by single years of age and sex, United States, 1930 and 1960. (*Reprinted from "Introduction to Demography," by Mortimer Spiegelman, by permission of Harvard University Press.*)

looks less erratic than the original (Figure 1a) and may (if one believes the theory) give a more correct impression of intercourse frequencies.

There are many other occasions where grouping is used to smooth out data which are unbelievable because of their erratic jumps. For example, census takers have often discovered that many people report their ages incorrectly. In Figure 2 census figures from 1930 and 1960 for both men and women illustrate this. The 1930 figures (gray bars) show that, for both men and women, the totals for ages 30, 35, 38, 40, 42, 45, 50, and 60 are much larger than the totals for directly adjacent ages. Why? More specifically, why should there be so many more people aged 45 than aged 44 or 46? Does it mean that lots more people were born 45 years earlier than 44 or 46 years earlier? This can be checked against birth records and shown to be an unlikely explanation. The preferred explanation is that people don't report their ages accurately. They may be deliberately lying, or they may not remember exactly. The fact that the high ages end in the digit 5 or 0 or are even numbers (38 and 42) suggests that people may simply be rounding their ages to the nearest handy or pleasant number to remember. As you can see from Figure 2, this phenomenon hardly appears at all in the 1960 census.

Partly because of the "age heaping" just discussed, demographers often break down a population into 5-year age groups instead of 1-year age groups. Even this is not fooproof, as Example 4 shows.

Example 2 The Moving Average

Another method of tinkering with the data of Table 1 would be to decide arbitrarily that a certain portion of those answering "8," say a quarter of them, actually had intercourse 7 times, while another portion, say a quarter again, actually had intercourse 9 times. In this way we conclude that the true percentage having intercourse 8 times is only 6.4 percent, while the true percentages having intercourse 7 and 9 times are 5.7 and 4.9 percent, respectively.

We can express this algebraically as follows: let $x_i, i = 1, \ldots, 20$, be the percent who report intercourse i times and let x_i' be the percent who (according to our assumption) actually have intercourse i times. Then

$$x_1' = x_1$$
$$\vdots$$
$$x_7' = x_7 + 0.25x_8$$
$$x_8' = 0.5x_8$$
$$x_9' = x_9 + 0.25x_8$$
$$x_{10}' = x_{10}$$
$$\vdots$$
$$x_{20}' = x_{20}$$

We note that this is a system of linear equations and can be expressed in matrix form as

$$
\begin{bmatrix} x'_1 \\ \vdots \\ x'_{20} \end{bmatrix} =
\begin{bmatrix}
1 & & & & & & \\
 & \ddots & & & & & \\
 & & 1 & 0.25 & & & \\
 & & & 0.5 & & & \\
 & & & 0.25 & 1 & & \\
 & & & & & \ddots & \\
 & & & & & & 1
\end{bmatrix}
\begin{bmatrix} x_1 \\ \vdots \\ x_{20} \end{bmatrix}
\tag{1}
$$

Naturally the constant 0.25 and 0.5, which appear in the matrix, are somewhat arbitrary. Another arbitrary aspect is the fact that we only adjusted the frequencies 7, 8, and 9. Maybe all the reported frequencies are wrong and need adjusting. Here is a set of equations which adjusts all the frequencies:

$$
\begin{aligned}
x'_1 &= \tfrac{2}{3}x_1 + \tfrac{1}{3}x_2 \\
x'_2 &= \tfrac{1}{3}x_1 + \tfrac{1}{3}x_2 + \tfrac{1}{3}x_3 \\
&\ \vdots \\
x'_i &= \tfrac{1}{3}x_{i-1} + \tfrac{1}{3}x_i + \tfrac{1}{3}x_{i+1} \\
&\ \vdots \\
x'_n &= \tfrac{1}{3}x_{n-1} + \tfrac{2}{3}x_n
\end{aligned}
\tag{2}
$$

Figure 3 shows first a graph of the data in Table 1 and then a graph of the data which result from applying Equation (2). The main effect of the adjustment seems to be to smooth the graph somewhat. However there is still plenty of jaggedness and some "unbelievable" patterns in the data.

These ideas also find application in the field of stock-market analysis where stock prices show a day-by-day up-and-down fluctuation, which appears partly random. In order to remove this random noise and expose the underlying trend, we may compute what is called the *moving average* in the following way:

1. Select an integer N ($N \geq 2$) to use in step 2. N will be the number of values averaged at each stage.
2. If $x_1, x_2, \ldots, x_N, \ldots, x_m$ are the daily values of the price,

$$
\begin{aligned}
x'_N &= \frac{x_1 + x_2 + \cdots + x_N}{N} \\[2mm]
x'_{N+1} &= \frac{x_2 + x_3 + \cdots + x_{N+1}}{N} \\[2mm]
&\ \vdots \\
x'_m &= \frac{x_{m-N+1} + x_{m-N+2} + \cdots + x_m}{N}
\end{aligned}
\tag{3}
$$

Note that we did not produce adjusted values to replace $x_1, \ldots, x_N$. Here is an illustration of this procedure.

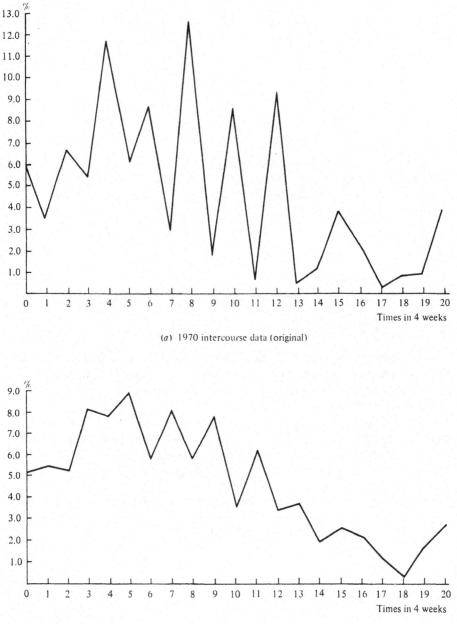

(a) 1970 intercourse data (original)

(b) 1970 intercourse data [adjusted by Eqs. (2)]

Figure 3

Suppose the New York Stock Exchange market index of stock prices has the following values during 12 consecutive trading days: 50, 51, 50, 52, 52, 54, 53, 52, 53, 54, 56, and 55. If we take a 3-day moving average of these numbers we get:

$$x_1' = x_1 = 50 \qquad\qquad x_7' = 53$$

$$x_2' = x_2 = 51 \qquad\qquad x_8' = 53$$

$$x_3' = \frac{x_1 + x_2 + x_3}{3} = 50.33 \qquad x_9' = 52.66$$

$$x_4' = \frac{x_2 + x_3 + x_4}{3} = 51 \qquad x_{10}' = 53$$

$$x_5' = \frac{x_3 + x_4 + x_5}{3} = 51.33 \qquad x_{11}' = 54.33$$

$$x_6' = \frac{x_4 + x_5 + x_6}{3} = 52.66 \qquad x_{12}' = 55$$

Figure 4 shows the two series of numbers plotted.

The argument in favor of taking a moving average of the stock market index is based on a theory of the stock market, which we will call the *underlying-trend theory* and which goes something like this: there is a trend to stock market prices, and this trend is based on underlying factors concerning the state of the economy, the political situation, and so on. The investor's job is to discover this trend, predict whether it will continue or turn around, and invest accordingly.

But in addition to this trend in the index, the theory maintains that there is a certain amount of random up-and-down motion of the index, which is super-imposed upon the trend, muddies the waters, and makes it hard to see the trend.

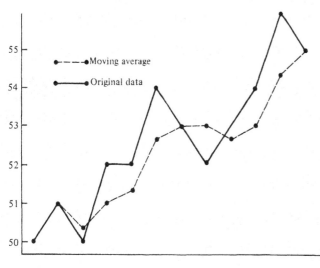

Figure 4

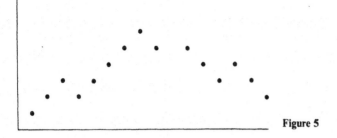

Figure 5

This random up-and-down motion is sometimes called *random noise*, by analogy to the static and other minor noises found in phonograph records alongside the music. Just as a fancy hi-fi set has electronic circuitry that can filter out some of this noise, the stock-market analyst can use a moving average to filter out some of the noise and lay bare the trend.

However, there is an alternative theory of stock prices, which can be used to interpret the market index and which questions the usefulness of the moving average. This is the so-called *pure-random-walk theory* of stock prices. For a simple example of a random walk, imagine a drunk walking aimlessly around, the twistings and turnings of the drunk's path being determined purely by chance. The kind of random walk some analysts claim to see at work in the stock market is one which can be described as follows:

1. The walk takes place on a two-dimensional coordinate system, e.g., a piece of graph paper.
2. A particular intersection point of the grid is designated as the starting point.
3. Each segment of the walk is a two-step process. First one moves one step horizontally to the right (the number of squares of graph paper which make one step has to be specified in advance). Second, one uses some chance device to determine how many squares up or down to move.
4. Step 3 may be repeated as many times as desired.

The positions arrived at after each completion of step 3 are called the *stages* of the random walk. In graphing the random walk, one may simply indicate the stages with dots on the graph paper, as in Figure 5, or one can plot the stages and also connect them, as in Figure 6.

Figure 6

Suppose the chance device in the random walk is a coin to be flipped. If the result is heads H, we move 1 square up. If the result is tails, we go 1 square down. Figures 5 and 6 show the result of such a random walk in which the coin gave the following sequence of outcomes:

H H T H H H H T T H T T H T T

Here is a different chance device that also yields a random walk: a spinner (Figure 7) whose sectors represent various up and down moves, as shown in the table. The unequal sizes of the sectors are arranged to give the probabilities shown in the table. Figure 8*a* and 8*b* shows two different random walks generated by this spinner.

Sector	Meaning	Probability of occurrence
+2	2 squares up	.05
+1	1 square up	.20
0	0 squares up or down	.50
−1	1 square down	.20
−2	2 squares down	.05

Now back to the stock market. The underlying-trend theorist accepts the idea that the graph of stock prices has some random up-and-down motion in it, but not the notion that it is a pure random walk. People who believe in trends might compare the market index to the walk of a slightly intoxicated person attempting to get from point *A* to point *B*. Being only slightly drunk, the person will succeed in getting to *B*, but at the cost of a few random deviations. When we observe the path, we see clearly where the person is going, and we can separate the trend of the motion (the person's intention) from the random component. Likewise, the underlying-trend theory goes, we can use the moving average to see the trend in the graph of the market index.

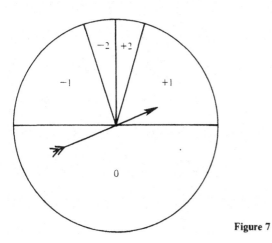

Figure 7

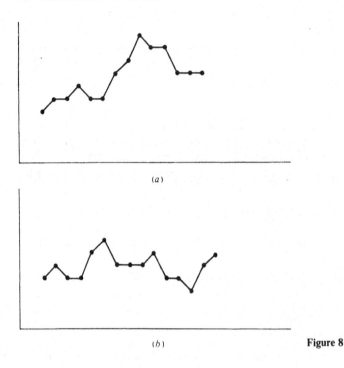

(a)

(b)   **Figure 8**

By contrast, the random-walk theory holds that most of the time the market is all random walk and no trend. Once in a while, events in the outside world push the market up or down. But between these relatively rare jolts, the market is purely random. A pure-random-walk theorist would be skeptical of the moving average. The theorist might even consider it a form of numerical hocus pocus, which produces the appearance of smooth trends where there really are none.

Which theory do you think makes more sense? Occasionally people doubt that a random walk could produce a graph that looks at all like an actual stock-market graph. You may judge for yourself from the two charts in Figure 9. One was produced as a pure random walk, with the use of the spinner. The other is a chart of the New York Stock Exchange market index for the period from March 3 to May 20, 1979. This index gives a sort of an average value of the shares of the firms whose stocks are traded on the exchange. We have rounded off the index values to the nearest quarter for convenience. Can you tell which chart is the real one and which is a pure random walk?

Example 3 Least Squares

Sometimes we have an idea about what kind of equation might describe the pattern we see in a table of values relating two variables x and y. In such a case, the least-squares method can be used to make adjustments to the numbers in the table.

For example, suppose a company wants to estimate what the demand would be for a certain new product and how this demand would depend on the price.

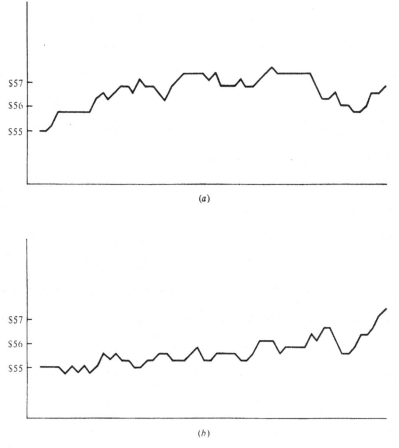

Figure 9

The company conducts a trial by selecting three cities whose populations are 100,000 and offering the product for sale at a slightly different price in each city. Suppose the data in the first two columns of Table 3 is collected. The company regards each of these numbers as a somewhat inaccurate estimate of what nationwide demands per 100,000 people would be. How should the individual cities' numbers be adjusted to form estimates of nationwide demands?

Table 3

p (dollars)	d (items per month)	Adjusted d (items per month)
9	1200	1169
10	1000	1052
11	975	956

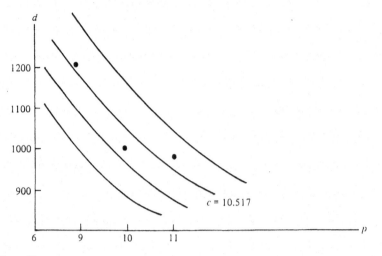

Figure 10

To begin with, we plot the points (p, d), as in Fig. 10. The key idea in the least-squares method is that the company has a prior notion of what kind of curve these points should approximately lie on, i.e., what kind of equation describes the relationship between price and demand. For example, it might believe that

$$d = \frac{c}{p} \tag{4}$$

is the right type of equation, where d stands for demand, p for price, and c is some appropriate constant. For each different choice of the constant c we get a different equation of type (4). Figure 10 shows graphs of some members of this family of equations—all hyperbolas. The problem is to find one of these hyperbolas that fits the data points of Figure 10 as closely as possible. (Ideally, we would like to find a hyperbola that passes exactly through the points, but this is usually not possible.) Algebraically, our problem is to find the best value of c to choose for Equation (4).

Let us skip ahead a bit and suppose we have found that

$$d = \frac{10{,}517}{p} \tag{5}$$

is the best-fitting curve of type (4). We can now use this to adjust the data in Table 3. We do this by using formula (5) and calculating the d values corresponding to the p values in the table. The resulting demand values will be regarded as the nation-wide demand values (see the last column of Table 3).

Now here is the least-squares method of selecting the best-fitting equation of type (4). For any curve of this family, we measure its fit to the data points by computing the vertical distances h_i from the data points to the curve. See Figure 11.

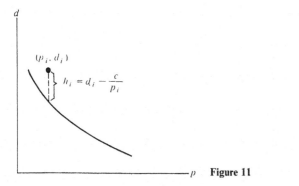

Figure 11

The least-squares distance from that curve to the data points is defined to be $\sqrt{f}$, where

$$f = \sum h_i^2$$

Our goal is to find the curve that gives the smallest f value.

Let us represent the prices in Table 3 as p_1, p_2, and p_3 and the associated demands as d_1, d_2, and d_3. Then the plotted points are (p_i, d_i) and $h_i = d_i - (c/p_i)$. Therefore

$$f = \sum_1^3 \left(d_i - \frac{c}{p_i}\right)^2$$

$$= \left(1200 - \frac{c}{9}\right)^2 + \left(1000 - \frac{c}{10}\right)^2 + \left(975 - \frac{c}{11}\right)^2 \qquad (6)$$

Notice that f is a function of c. Our task is to minimize f. (This also minimizes the least-squares distance $\sqrt{f}$.) This is a standard minimization problem in one variable (the d_i and p_i are constants), and so we apply the usual method—set $f'(c) = 0$ and so on:

$$f'(c) = -\frac{2}{9}\left(1200 - \frac{c}{9}\right) - \frac{2}{10}\left(1000 - \frac{c}{10}\right) - \frac{2}{11}\left(975 - \frac{c}{11}\right)$$

We set $f'(c) = 0$ and solve for c:

$$c\left(\frac{2}{9^2} + \frac{2}{10^2} + \frac{2}{11^2}\right) = \frac{2}{9}1200 + \frac{2}{10}1000 + \frac{2}{11}975$$

$$c = \frac{321.97}{0.03061}$$

$$= 10{,}517 \text{ (to the nearest integer)}$$

In the general case, where we have n points (p_i, d_i), formula (6) becomes a summation from 1 to n, and, using the same steps we just applied in the numerical example, we derive a formula for c.

$$f(c) = \sum_1^n \left(d_i - \frac{c}{p_i} \right)^2$$

$$f'(c) = \sum_1^n \left(-\frac{2}{p_i} \right) \left(d_i - \frac{c}{p_i} \right) = 0$$

$$\sum_1^n \frac{2c}{p_i^2} = \sum_1^n \frac{2d_i}{p_i}$$

$$c = \sum_1^n \frac{d_i}{p_i} \bigg/ \sum_1^n \frac{1}{p_i^2}$$

Needless to say, this formula only works when we are trying to fit a hyperbola of type (4) to data. If we believe that a straight line, parabola, or something else is called for, then the method changes somewhat. For some curves, the mathematics may involve partial derivatives, as in the following example.

The great chemist Mendeleev once obtained the following data on the solubility of sodium nitrate ($NaNO_3$) in water and how that varies with the tempera-

Temperature t_i	0	4	10	15	21	29	36	51	68
Solubility s_i	66.7	71.0	76.3	80.6	85.7	92.9	99.4	113.6	125.1

ture of the water. Mendeleev believed that there was a linear relationship of the form

$$s = mt + b \tag{7}$$

between temperature t and solubility s. To find m and b, we express the least-squares distance between a line of type (7) and the data points (t_i, s_i) in the table as follows:

$$f(m, b) = \sum_1^n (mt_i + b - s_i)^2$$

Here f is a function of two variables m and b, and so minimizing it requires setting partial derivatives equal to 0. Section 2 of Chapter 3 is devoted to this kind of problem. By the way, the line Mendeleev obtained is $s = 67.5 + 0.87t$.

Example 4 Interpolation

Interpolation means inserting one or more values between known values. There are many ways to do it; so there is no single technique or equation for interpolation.

For a simple example, suppose your sister measured 54 inches when she reached age 10, and 60 inches when she reached age 12. She wasn't measured at 11, but can

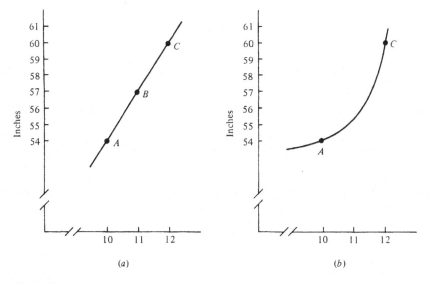

Figure 12

you estimate her height at age 11? Suppose we assume her growth was steady and uniform. Then, since 11 is midway between 10 and 12, we would estimate a height midway between 54 and 60 (i.e., 57).

A graphical view of this is shown in Figure 12a. Points A and C represent actual measured data. Point B is the interpolated point. Our assumption that her growth is steady and uniform turns out to be equivalent to the assumption that her height graph is a straight line. Other assumptions are possible. A curve like that in Figure 12b would reflect the assumption that her growth is faster in the second year of the 2-year interval.

A more complex example of interpolation occurred in connection with the 1960 census of the United States. In that census, many nonwhite males in their late 50s seemed to report their ages incorrectly.[†] Figure 13 shows some of the data reported. Census analysts felt that the 50–54 age totals and the 65–69 age totals were accurate but many of those who reported their age in the 55–59 age group were actually in the 60–64 age group. In effect, some portion of the second bar in Figure 13 needs to be chopped off and added to the third. But how is this adjustment to be done?

We begin by representing the age groups by their midpoints: $x_1 = 52.5$, $x_2 = 57.5$, $x_3 = 62.5$, and $x_4 = 67.5$. We denote the corresponding population totals by y_1, y_2, y_3, and y_4. The adjusted figures we will produce will be called y_1', y_2', y_3', and y_4'. The ordered pairs (x_i, y_i) can be plotted as in Figure 14.

[†] See M. Spiegelman, cited in the Bibliography, for a discussion of why people lie about their age and more on what can be done about it.

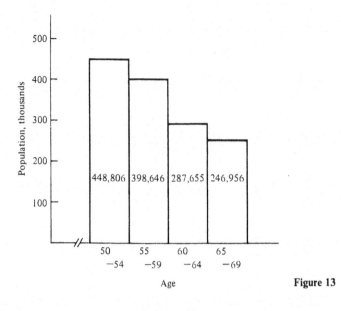

Figure 13

The method of adjustment chosen by the Bureau of the Census was based on the following requirements:

1. The four points (x_i, y_i') should satisfy a second-degree polynomial of the form $y = ax^2 + bx + c$. In other words,

$$y_i' = ax_i^2 + bx_i + c \qquad \text{for } i = 1, 2, 3, \text{ and } 4$$

Geometrically this means the points (x_i, y_i') lie on a parabola.

2. Since y_1 and y_4 are correct values, $y_1' = y_1$ and $y_4' = y_4$. In other words the curve should pass through the points (x_1, y_1) and (x_4, y_4) exactly. See Figure 14.

3. The new (adjusted) values y_2' and y_3' should have the same sum as y_2 and y_3 since we are merely moving some people from one of these age groups to another.

The straightforward way to find the new values of y_2' and y_3' is to write down the equations implied by these three conditions and solve them. We outline this below. It is numerically very tedious. Exercise 13 shows a more elegant approach, in which clever algebra reduces the arithmetic drastically.

Condition 2 implies

$$y_1 = ax_1^2 + bx_1 + c$$
$$y_4 = ax_4^2 + bx_4 + c$$

Condition 3 implies

$$ax_2^2 + bx_2 + c + ax_3^2 + bx_3 + c = y_2 + y_3$$

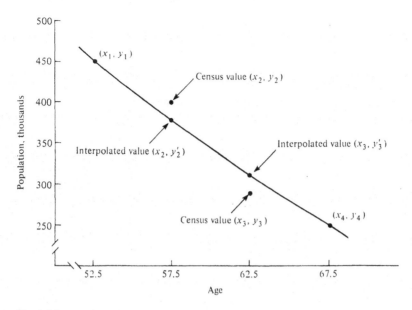

Figure 14

Since the x_i and y_i are constants, these equations are a system of three linear equations in a, b, and c. When we substitute the x_i and y_i values shown on Figure 13, we obtain the system

$$7212.50a + 120b + 2c = 686,301$$
$$2756.25a + 52.5b + c = 448,806$$
$$4556.25a + 67.5b + c = 246,956$$

The solution to this system is

$$a = 94.6100$$
$$b = -24,809.8667$$
$$c = 1,490,555.1875$$

Therefore, the second-degree polynomial we use to interpolate is

$$y = 94.6100x^2 - 24,809.8667x + 1,490,555.1875$$

When we substitute the values $x_2 = 57.5$ and $x_3 = 62.5$ successively for x in this polynomial, we obtain the new population totals for the age groups: $y_2' = 376,792$ and $y_3' = 309,509$.

Could other adjustment methods have been used for these data? Grouping would have been unacceptable because the census traditionally (and for good reason) publishes data by 5-year age groups. Least squares would not have been satisfactory because it would undoubtedly have called for adjustments in y_1 and y_4. What do you think about some kind of moving average? This example shows how special situations often call forth special adjustment methods.

There is an important theme playing under the mathematics of this section: the role of belief. You either believe the data, or a certain theory about them, or you don't. This is common in the social sciences. But the physical sciences are not completely immune to this subjective element. One of the most spectacular examples of unbelievable data occurred in connection with Albert Einstein's theory of relativity.

A crucial part of the theory concerned the velocity of light–specifically, whether the velocity of a light ray which an observer would measure would be different, depending on whether the observer is moving toward or away from the oncoming light ray. Common sense in the nineteenth century said there would be a difference; Einstein said no. Obviously, such a question should be settled by experiment. An experiment was performed in 1887 by Albert Michelson and Edward Morley, who came to the astounding conclusion that the velocity of light did not depend on the observer's own motion. But Einstein began moving toward this same conclusion for theoretical reasons and before he was aware of the Michelson-Morley experiment. According to his own words, "the Michelson-Morley experiment had a negligible effect on the discovery of relativity."

The story gets better. In 1902 a reexamination of the data suggested that Michelson and Morley had interpreted them incorrectly and there really was a difference in velocities. There followed a long period of experimentation by D. C. Miller, stretching from 1902 to 1926, during which improved versions of the Michelson-Morley experiment were performed thousands of times. When the evidence was all in, it showed a small but definite difference in velocity, a conclusion that flew in the face of Einstein's theory of relativity. If this evidence was to be believed, it would be necessary to abandon relativity. What to do?

It is important to be aware that Miller was not some obscure physicist of small talent or reputation; he was a respected experimenter who, at one point, was president of the American Physical Society. Furthermore, no one could point to any mistakes in his experimental procedures. Consequently, it wasn't easy to ignore his experiments. Easy or not, most physicists did exactly that. They ignored Miller's data and hoped that there was some unknown source of error in the experiment that would someday be discovered.

The reason they were able to reject Miller's data was that there were other kinds of experiments that confirmed relativity. Thus, physicists had to decide which experiments to believe. In the early days of relativity experiments, only a few experiments had been done and physicists were, to some extent, playing their hunches in deciding which to believe. Now that many more experiments have been performed—some of which are simpler than Miller's to carry out and which give more clear-cut results—the weight of evidence is strongly in favor of relativity, and Miller's experiments are all but forgotten.

This phenomenon is common in many fields of study: when new models or theories are first proposed, their value isn't clear. Experts who are brave enough to express an opinion sometimes make up their minds partly on the basis of subjective judgement.

EXERCISES

● **1** (a) What is the matrix of Equation (2) in the case $n = 5$?

(b) Write Equation (3) in a matrix form analogous to Equation (1). Use $n = 3$ and $m = 5$.

2 (a) If x' is a vector, which results from a vector x by an adjustment process represented by matrix M being applied (as in Example 2), and x'' results from the process being applied to x', then write a matrix and vector equation relating x'' to x.

(b) In the spirit of part (a) suppose we apply the adjustment process k times and thereby produce x^k. How is x^k related to x?

● **3** Is there an initial set of data x which is unaffected by the adjustment process of (2) being applied? Try this for the case $n = 4$ first. Explain what connection this question has with the concepts of eigenvector and eigenvalue.

4 In our adjustments of intercourse frequency in Example 1, in each case we have $\sum x_i' = \sum x_i = 1$. What is the significance of this? Why would it be unreasonable to change the 0.5 in Equation (1) to 0? Is this kind of consideration relevant to the moving-average process applied to stock prices?

5 In the chart below we give population figures for ages 15 through 24 reported by the 1930 census. It has been suggested that males may have been reporting their ages as 21 even when they had not yet attained that age, the reason being that at 21 males reach the so-called "age of majority." At this age males enjoy certain social advantages, such as being able to marry without parental consent. In 1930 the age of majority for females was 18, and it has been suggested that the numbers of females reporting their age as 18 is unusually high.

Native white population of native parentage, 1930

Age	Males	Females	Ratio (males to females)
15	704,157	688,760	1.022
16	718,684	709,811	1.013
17	701,581	681,343	1.030
18	692,100	704,958	0.982
19	664,085	667,024	0.996
20	630,916	667,874	0.945
21	663,246	654,273	1.014
22	629,853	650,476	0.968
23	613,504	628,388	0.976
24	601,319	615,659	0.977

(a) Do you agree that these ages are being overreported? What evidence do you have?

(b) Here are three ways you could modify the data or the presentation of the data. Which seems best to you? Say why you think your choice is the best.

1. A moving average in which Equation (2) is used
2. Grouping in 2-year age groups
3. Grouping in 5-year age groups

(c) Is there a fourth alternative you like better than any of the previous three?

6 What relevance does the ratio of males to females (the last column of the table) have in determining whether the ages of majority are overreported in the chart of Exercise 5. (*Remark*: It may help to know that it is a biological fact that the sex ratio of males to females among newborns is about 1.05 and doesn't fluctuate from year to year.)

● **7** In the price-demand model of Example 3, suppose it costs \$6 to produce the product. Suppose also that half of the difference between this \$6 and the selling price is profit for the retailer and half is profit for the manufacturer. What selling price gives the manufacturer the greatest total profit? (Total profit is the sum of the profits on all items sold.) What selling price is best for the retailer? In answering these questions, assume that $d = 10{,}517/p$ is the correct price-demand equation. Also, assume that any price (not just the three in Table 3) is possible.

8 In the least-squares method, the condition that $f'(c) = 0$ only guarantees that the value of c satisfying this equation is either a relative maximum or a relative minimum. Can you see why it is not a relative maximum?

● **9** In the price-demand model of Example 3, assume that the best-fitting curve is in the family of curves with equations $d = c/p^2$, c being a positive constant.

(a) Can you find the optimum value of c for the data of Table 3?

(b) Can you find a formula for the optimum c in the general case, where the data consists of the prices $p_1, \ldots, p_n$ and the corresponding demands $d_1, \ldots, d_n$?

10 In Example 4, involving the adjustment of census figures, suppose there were five ages $x_1, \ldots, x_5$ and corresponding population totals $y_1, \ldots, y_5$. Again, suppose that y_1 and y_5 are correct and the total $y_2 + y_3 + y_4$ needs to be redistributed among the three intermediate ages.

(a) Could the same method be used? If not, what changes would be needed?

(b) Use the method or your modification of it on the data below.

Age group	Population (in millions)
45–49	1.0
50–54	0.8
55–59	0.9
60–64	0.6
65–69	0.5

11 (a) Using the interpolation method corresponding to Figure 12a, what height would you estimate at $10\frac{2}{3}$ years of age?

(b) Using this interpolation method, give a formula that estimates the girl's height at x years of age.

● **12** The parabola in Figure 14 is visually nearly indistinguishable from a straight line. This suggests the question: what would happen if, in place of assumption 1, we assumed that the four points (x_i, y_i') satisfied a first-degree polynomial of the form $y = mx + b$? Try to work it out this way.

13 If $y_1', y_2', y_3', y_4', \ldots$ is a sequence of numbers, the sequence of *first differences* is obtained by subtracting each number from the one following it: $y_2' - y_1', y_3' - y_2', y_4' - y_3', \ldots$. If we carry out this process again on the new sequence, we get the sequence of second differences.

(a) If $(x_1, y_1'), (x_1 + d, y_2'), (x_1 + 2d, y_3'), \ldots$ are points satisfying a second-degree polynomial

$$y' = ax^2 + bx + c$$

show that the terms of the sequence of second differences are all the same.

(b) In the census interpolation problem described in Example 4, derive the following formulas [as consequences of the three conditions listed there and of part (a)].

$$y'_2 = \tfrac{1}{6}(y_1 + 3y_2 + 3y_3 - y_4)$$

$$y'_3 = \tfrac{1}{6}(-y_1 + 3y_2 + 3y_3 + y_4)$$

Computer exercises

14 Suppose the adjustment process represented by Exercise 2 is carried out repeatedly: it first produces vector x' from x, then x'' from x', and so on. Try this for a particular choice of x and large value of n and see what happens to the data. Try plotting the coordinates x_i angainst i on a graph. The data should begin to fall on a straight line. Try other choices of x and n.

15 Work Exercise 14 but with this change in Equation (2). Replace the first equation with $x'_1 = x_1$ and $x'_n = x_n$. Do you still get a straight line? Can you find a simple formula or procedure to find this line without lots of computation?

16 Write a computer program to compute c according to the least-squares procedure in Example 3.

17 Write a computer program to carry out the census interpolation of Example 4. Input to the program consists of the four population totals, y_1, y_2, y_3, and y_4, and the midpoints of the age intervals x_1, x_2, x_3, and x_4. Output should be the values y'_1, y'_2, y'_3, and y'_4. Solving the equations could be done by gaussian elimination or Cramer's rule. If your answers don't match ours, try having the program use double precision.

BIBLIOGRAPHY

Deming, W. E.: "Statistical Adjustment of Data," Dover Publications, Inc., New York, 1964. A standard work on the least-squares approach.

Klein, L. R.: "An Introduction to Econometrics," Prentice-Hall, Englewood Cliffs, N.J., 1962, A standard work on the empirical determination of supply and demand curves.

Malkiel, B. W.: "A Random Walk Down Wall Street," Norton, New York, 1973. A nontechnical book for the layman.

Polanyi, Michael: "Personal Knowledge," University of Chicago Press, Chicago, 1962. The story of the Michelson-Morley experiment is told here.

Spiegelman, M.: "Introduction to Demography," Harvard University Press, Boston, 1970. Chapter 3 has a nice survey of errors in census data. Our Example 4 comes from the last paragraph of section 3.5.6. But note that the age group which is too big is incorrectly identified.

Westoff, C. F.: Coital Frequency and Contraception, *Family Planning Perspectives*, vol. 6, no. 3, Summer, 1974.

4 TESTING MODELS—WEIGHTED-VOTING GAMES

Abstract This section shows that weighted-voting systems are often unfair. It is difficult to verify this conclusion by checking the results of the votings. Even if we could determine when a voter had less power than he "ought to have," there could be other reasons besides the voting system that account for this. Thus we must decide whether the model is convincing without reference to empirical observations.

Prerequisites Simple counting arguments, factorial notation, and limits.

Perhaps the moral of the previous sections is that, since there is so much room for error in modeling, it is a good idea to test the predictions of the model against reality. That's good advice, but sometimes it is hard to follow. For example, in 1963 N. Keyfitz formulated a mathematical model that allowed him to conclude that about 69 billion people had lived on the earth during its entire history up till 1960. Naturally, testing this is impossible because most of these folks are no longer around to be counted.

Sometimes testing is possible but undesirable. Much of the strategic planning that lies behind the United States' nuclear deterrent system is based on mathematical models. We do partial tests of these models with war games, but a true test would require a war. Obviously, we'd rather do without the experiment.

Finally, there are cases where testing would be possible and not catastrophic, but impractical. For example, about 750,000 insect species have been discovered and cataloged. Mathematical models have been used to estimate that there are millions more waiting to be discovered. We could try to verify this by finding and cataloging the rest of the world's insects. But carrying out this huge task would probably mean that every biologist, naturalist, and amateur butterfly chaser in the world would have to devote full time to it for many years.

Now let's look at an example in more detail.

In ordinary elections where each person has one vote, it is obvious that all the voters have equal power. However, there exist weighted-voting systems where the voters may not all have the same number of votes. Our aim here is to estimate the fairness of unfairness of such systems.

Example 1

In 1958 the Nassau County (New York) Board of Supervisors consisted of six supervisors from the five towns making up Nassau County (one town got two representatives). Because these towns had very different sizes (see Table 1), it would clearly not be fair to give each supervisor one vote. Instead, the number of votes assigned to a supervisor was approximately in proportion to the population of

Table 1 Nassau County weighted voting

Municipality	Population 1954	Number of votes 1958	Population 1960	Number of votes 1964
S_1 Hempstead 1[†]	618,065	9	728,625	31
S_2 Hempstead 2		9		31
S_3 North Hempstead	184,060	7	213,225	21
S_4 Oyster Bay	164,716	3	285,545	28
S_5 Glen Cove	19,296	1	22,752	2
S_6 Long Beach	17,999	1	25,654	2
Total	1,004,136	30	1,275,801	115

[†] Hempstead got two representatives.

town the supervisor represented. The question is: if the votes are assigned this way, will the actual voting power come out to be in proportion to the number of votes assigned to a supervisor? Oddly enough, the answer may be no!

The question proposed in the previous example is easier to study with a simpler example. Suppose we have three voters A, B, and C, where A and B each have one vote and C has 3. As in all the examples in this section, we suppose that a simple majority determines the winning side. Clearly A and B have no power at all even though each has one-fifth of the total votes. This example shows that the fraction of total votes held by a voter may not be a good measure of the voter's power.

How then should we measure voting power? The method we are going to present here was devised by John F. Banzhaf III, a mathematically inclined lawyer. Not only is this system mathematically persuasive, but it has been accepted by numerous courts in the United States, which have been called upon to render judgements in cases stemming from the Supreme Court's "one person–one vote" decision of 1962.

But before we present Banzhaf's system, let us consider the difficulties if we tried to measure voting power by direct observation of the deliberations of the Nassau County Board of Supervisors. There are many factors beside number of votes which influence how much power a member of this board would have. A member with only a few votes may have great power because that member can compel others to follow on account of personality, intelligence, political connections, greater experience in politics, etc. How can we assess the impact of the numerous factors which determine power? We can't. Certainly there is no way of doing controlled experiments in which we equalize personality, intelligence, political connections, etc., among the supervisors.

Luckily, finding the actual power of an actual supervisor is not the problem at issue. What we want to know is how much power does the *voting system* give a supervisor (relative to the other supervisors)?

To answer this question, let's consider a particular vote in which the six supervisors line up as shown in the first vote of Table 2. In this table Y means "yes," N means "no," and the number of votes for the various supervisors are in parentheses. The set of voters who vote yes is called the *yes coalition*, and the remainder make up the *no coalition*. A pair of coalitions is called a *line-up*. In the first vote of Table 2 the yes position gets 16 votes and wins. However, if S_3 changes from yes to no, the noes win. In this sense, S_3 has power, and S_3's vote is crucial. By contrast, S_2, S_4, S_5, and S_6 have no power *in this line-up* because, if any of their votes change, it will have no effect on which position wins. But this applies only to the first line-up of yes versus no votes in Table 2. With a different line-up, as in

Table 2

	$S_1(9)$	$S_2(9)$	$S_3(7)$	$S_4(3)$	$S_5(1)$	$S_6(1)$
First vote	Y	N	Y	N	N	N
Second vote	Y	N	N	N	N	N

the second row of Table 2 for example, the list of supervisors who have power is different.

Can you invent a line-up of yeses and noes where S_5 has power? The surprising result is that you can't, and therefore it is really as though S_5 has no votes at all. S_5 is an example of a "dummy."

Definition

A voter is said to *crucial* for a given line-up of yes-and-no votes provided that, when that voter's vote changes, the outcome changes.

Definition

A voter in a weighted-voting situation is called a *dummy* if there is no line-up of yes-and-no votes where the outcome would be different if that voter's vote changed.

In 1958 Oyster Bay, Glen Cove, and Long Beach are all dummies, while in 1964 North Hempstead, Glen Cove, and Long Beach are all dummies. Can you prove these assertions?

These ideas need to be slightly modified when the total number of votes is even and there is the possibility of tie votes. In real-life politics there are various ways of breaking ties: a new election may be held, a third party may be given a tie-breaking vote, etc. We shall not specify any particular method, but we shall make this assumption: in a tied election, the tie-breaking method is just as likely to make the yes position the winner as it is likely to make the no position the winner.

We shall illustrate how this affects our concept of a crucial voter with the example of four voters A_1, A_2, A_3, and A_4, having respectively 1, 1, 2, and 2 votes. In the first yes-no line-up shown in Table 3, the yes position wins (4 to 2). If A_1 changes to vote no, the vote will be tied. Does this make A_1 crucial? Not necessarily. It depends on the results of the tie breaking. If, after tie breaking, yes becomes the winner again, then A_1's vote change had no real effect. On the other hand, if no becomes the winner after tie breaking, then A_1's vote change did change the winner. In such a case we say that A_1 was semicrucial. Obviously, if a voter is semicrucial in a coalition, that voter has half the power to change things as in a coalition where that voter is crucial.

Line-up 2 of Table 3 shows another case where A_1 is semicrucial, namely where A_1 can convert a tie to a win by a vote change.

Table 3

	$A_1(1)$	$A_2(1)$	$A_3(2)$	$A_4(2)$
First line-up	Y	Y	Y	N
Second line-up	Y	N	N	Y

Definition

A voter is semicrucial in a yes-no line-up if that voter's vote change either converts a win to a tie or a tie to a win.

The concept of a dummy can be defined in the same way as before: a voter who never has the capability of changing the outcome.

Here is our definition of the voting strength conferred by a given set of weighted votes.

Definition

Suppose there are n voters $A_1, \ldots, A_n$, whose voting weights are $w(A_1), \ldots, w(A_n)$. There are 2^n possible line-ups of the voters into yes and no groups. In each line-up, we make a note of those voters whose votes are crucial. Let $c(A_i)$ be the number of yes-no line-ups in which A_i is crucial, and $s(A_i)$ the number where A_i is semicrucial.

$$\beta_i = \frac{c(A_i) + \tfrac{1}{2}s(A_i)}{2^n}$$

is called the *Banzhaf power index* of voter A_i.

We may interpret β_i as the probability that A_i can change the winner in a randomly chosen vote. As the next example shows, it is possible that $\sum_{i=1}^n \beta_i \neq 1$. It is sometimes convenient to redefine the power indices so that they add to 1. This way each voter's index can be interpreted as a fraction of the total power. This is done by setting $\beta_i' = \beta_i / \sum_{i=1}^n \beta_i$ and calling β_i' the *normalized Banzhaf power index*.[†]

Example 2

Table 4 gives a detailed illustration of this for a system of four voters, whose voting weights are 2, 3, 1, and 2, respectively. There are 16 different Yes-No line-ups, and, for each one, we list the winning position and those voters who are crucial and semicrucial. Table 5 summarizes the information and gives β_1, β_2, β_3, and β_4. Notice that these power indices are not in proportion to the number of votes a voter has. For example, A_2 has 3 times as many votes as A_3 but has 5 times as big a power index.

As you can see from this example, calculating power indices by this method of tabulating all possible yes-no line-ups is time-consuming. In the general case of n voters, there are 2^n different coalitions, and, so, except for very small values of n, hand tabulation is out of the question. Most of the time there are no theoretical shortcuts to help out; so one is left with computer tabulations.

[†] Some authors call β_i' the Banzhaf power index and do not have a separate term for β_i.

Table 4

A_1 (2)	A_2 (3)	A_3 (1)	A_4 (2)	Winning position	Crucial voters	Semicrucial voters
Y	Y	Y	Y	Y	None	None
Y	Y	Y	N	Y	A_2	A_1
Y	Y	N	Y	Y	None	A_2
Y	Y	N	N	Y	A_1, A_2	None
Y	N	Y	Y	Y	A_1, A_4	A_3
Y	N	Y	N	N	A_2, A_4	None
Y	N	N	Y	Tie	None	A_1, A_2, A_3, A_4
Y	N	N	N	N	A_2	A_4
N	Y	Y	Y	Y	A_2	A_4
N	Y	Y	N	Tie	None	A_1, A_2, A_3, A_4
N	Y	N	Y	Y	A_2, A_4	None
N	Y	N	N	N	A_1, A_4	A_3
N	N	Y	Y	N	A_1, A_2	None
N	N	Y	N	N	None	A_2
N	N	N	Y	N	A_2	A_1
N	N	N	N	N	None	None

But there is one case where there is a large number of voters and where theory can help. This is the case of elections in which all voters have the same number of votes, for example, a statewide election for U.S. Senator. It is not hard to see, even without calculation, that in such an election no voter has more power than another. But we have a different question in mind. Suppose we wish to compare the power of a voter in a small state with the power of a voter in a large state in their respective senatorial elections. It stands to reason that β_i, the probability of being crucial in a small state, like Connecticut with about 3 million people, should be higher than β_i in a large state, such as Pennsylvania with about 12 million. But are the (unnormalized) power indices in the same ratio as the populations, namely 4 to 1 in the case of Connecticut and Pennsylvania? The following theorem will show us that the answer is no!

Table 5

Voter	Number of crucial line-ups	Number of semicrucial line-ups	Banzhaf power index β	Normalized Banzhaf power index β'
A_1	4	4	$\frac{6}{16}$	$\frac{1}{4}$
A_2	8	4	$\frac{10}{16}$	$\frac{5}{12}$
A_3	0	4	$\frac{2}{16}$	$\frac{1}{12}$
A_4	4	4	$\frac{6}{16}$	$\frac{1}{4}$

Theorem 1

In an election with n voters, all with one vote, the (unnormalized) Banzhaf power index for each voter, denoted $\beta(n)$, is such that, for large n,

$$\beta(n) \sim \frac{2}{\sqrt{2\pi n}}$$

More precisely, $\beta(n)/(2/\sqrt{2\pi n}) \to 1$ as $n \to \infty$.

Proof The proof is in two parts, depending on whether n is even or odd. We present only the even case and leave the odd case for the exercises. Therefore let $n = 2\rho$, where ρ is an integer.

The first thing to observe is that, in the even case, there can't be a line-up in which there is a crucial voter. (Do you see why?) Consequently, to calculate $\beta(n)$, all we have to do is to count, for a particular voter named A, the number of line-ups in which that voter is semicrucial. These line-ups can be divided into the following four categories:

1. A is part of a winning coalition of $\rho + 1$ yes voters; the no coalition will have $\rho - 1$ members (Table 6a).
2. A is part of a winning coalition of $\rho + 1$ no voters; the yes coalition will have $\rho - 1$ members (Table 6b).
3. A is a member of a tied coalition of ρ yes voters (Table 6c).
4. A is a member of a tied coalition of ρ no voters (Table 6d).

How many different line-ups of type 1 are there? We can think of this as being the problem of counting the number of ways of filling in the blanks with voters in Table 6. No two of these ways will be counted as different if they merely differ in the order in which the voters occupy the blanks (e.g., the yes coalition ABC is the same

Table 6[†]

	Yes voters	No voters
(a)	A $\underbrace{-\ -\ \cdots\ -}_{\rho}$ $\ \ \ \ \ \ $ 1	$\underbrace{-\ -\ \cdots\ -}_{\rho - 1}$
(b)	$\underbrace{-\ -\ \cdots\ -}_{\rho - 1}$	A $\underbrace{-\ -\ \cdots\ -}_{\rho}$ $\ \ \ \ \ \ $ 1
(c)	A $\underbrace{-\ -\ \cdots\ -}_{\rho - 1}$ $\ \ \ \ \ \ $ 1	$\underbrace{-\ -\ \cdots\ -}_{\rho}$
(d)	$\underbrace{-\ -\ \cdots\ -}_{\rho}$	A $\underbrace{-\ -\ \cdots\ -}_{\rho - 1}$ $\ \ \ \ A$

[†] Dashes represent voters.

as the coalition *CBA*). Notice that once we fill in the blanks in one of the coalitions in the line-up, this determines the composition of the other coalition. Thus the number we want for type 1 is the number of ways of choosing ρ items from among $n - 1$. (This is the number of ways of filling the blanks in the yes coalition.) Namely

$$\binom{n-1}{\rho} = \frac{(n-1)!}{p!(n-1-\rho)!}$$

$$= \frac{(n-1)!}{\rho!(\rho-1)!} = \frac{1}{2}\frac{n(n-1)!}{\rho\rho!(\rho-1)!}$$

$$= \frac{1}{2}\frac{n!}{(\rho!)^2}$$

$$= \frac{1}{2}\frac{n!}{[(n/2)!]^2}$$

With similar reasoning we can determine that this is also the number of type 2, 3, and 4 line-ups. Consequently,

$$\beta(n) = 4\frac{1}{2}\frac{\binom{n-1}{\rho}}{2^n} = \frac{1}{2^n}\frac{n!}{[(n/2)!]^2} \tag{1}$$

This is an awkward formula, and so we trot out an old workhorse, Stirling's formula, which says that as $x \to \infty$, $x!$ is "asymptotically equal to" $\sqrt{2\pi x}\,x^x e^{-x}$, a formula we abbreviate by $s(x)$. What this means precisely is

$$\lim_{x \to \infty} \frac{x!}{s(x)} = 1 \tag{2}$$

Now we apply these approximations to $\beta(n)$ as follows. From Equation (1) we get, by simple algebra:

$$\frac{2^n \beta(n)[s(n/2)]^2}{s(n)} = \frac{n!/s(n)}{[(n/2)!/s(n/2)]^2} \tag{3}$$

By Equation (2) the right side of Equation (3) $\to 1$ as $n \to \infty$. Therefore the left side does too. This fact may be slightly rewritten as

$$\frac{\beta(n)}{s(n)/\{2^n[s(n/2)]^2\}} \to 1 \qquad \text{as } n \to \infty$$

But

$$\frac{s(n)}{2^n[s(n/2)]^2} = \frac{1}{2^n}\frac{\sqrt{2\pi n}\,n^n e^{-n}}{(2\pi n/2)(n/2)^n e^{-n}}$$

$$= \frac{2}{\sqrt{2\pi n}}$$

Thus

$$\frac{\beta(n)}{2/\sqrt{2\pi n}} \to 1 \qquad \text{as } n \to \infty \qquad (4)$$

A quick and dirty approach to the foregoing calculation is just to substitute $s(n)$ and $s(n/2)$ for $n!$ and $(n/2)!$ in (1) and then carry out the algebra, leading to

$$\beta(n) = \frac{2}{\sqrt{2\pi n}}$$

Here the equal sign has to be understood as only approximate.

Using this theorem, we can compare the Banzhaf power indices of a voter in Pennsylvania with one in Connecticut for senatorial elections as follows. For Connecticut, $n = 3{,}000{,}000$, a large enough value that we can replace the "$\to$" in Equation (4) with "$=$." Thus

$$\frac{\beta(\text{Connecticut})}{2/\sqrt{2\pi(3{,}000{,}000)}} = 1$$

Likewise

$$\frac{\beta(\text{Pennsylvania})}{2/\sqrt{2\pi(12{,}000{,}000)}} = 1$$

Therefore

$$\frac{\beta(\text{Connecticut})}{\beta(\text{Pennsylvania})} = \sqrt{\frac{12{,}000{,}000}{3{,}000{,}000}} = 2$$

Although Pennsylvania has 4 times the population, a Pennsylvanian doesn't have one-quarter the power as a voter in Connecticut. Does this seem strange to you?

This argument can be generalized.

Theorem 2

If state S_1 has voting population n_1 and state S_2 has voting population n_2, then the ratio of Banzhaf indices $\beta(S_1)$ and $\beta(S_2)$ is

$$\frac{\beta(S_1)}{\beta(S_2)} = \sqrt{\frac{n_2}{n_1}}$$

Proof Omitted. Try it as an exercise. It follows the pattern of the Connecticut-Pennsylvania calculation.

Let us summarize two specific conclusions made by our voting power model.

1. In Nassau County, between 1958 and 1964, three town supervisors (Glen Cove, Oyster Bay, and Long Beach) were operating under a handicap imposed by the weighted-voting systems in use.
2. An individual citizen of Connecticut has twice the power of a citizen of Pennsylvania in influencing the election for the Senate.

Is proposition 1 empirically testable? Suppose we examine all the votes taken from 1958 to 1964. Is there any way in which the data could tell us whether Long Beach operates under a handicap? One thing which comes to mind is to see if Long Beach is on the losing side of votes too often. But how much is too often? Even if we could agree on what is too much, isn't it possible to find other reasons, having nothing to do with the voting system, which might explain Long Beach being on the losing side of the vote "too often"?

It may be that the natural interests of all the other towns always coincide and are against the interests of Long Beach. In such a case, even if all towns had equal votes, Long Beach would never be on the winning side. But this should not be blamed on the voting system for it would happen if all towns had the same number of votes.

Or consider the contrary case where Long Beach is surprisingly often on the winning side. Do we conclude that Long Beach is not handicapped by the voting system, or do we have to first consider the possibility that it has the handicap but overcomes it in some way (e.g., maybe the Long Beach Supervisor is a very persuasive debater).

The factors mentioned above (political interests of Long Beach and personality of the Long Beach Supervisor) cannot be "controlled" in scientific experiments. This means that it is difficult or impossible to determine whether the voting system imposes a handicap on Long Beach by examining the votes taken by the board.

What one is left with is this question: is the logic of the model convincing? Since this logic is expressed in the language of mathematics, it makes things difficult for judges, who are the people who must rule on the constitutionality of weighted-voting schemes. Some courts have been convinced by the Banzhaf model described here and have used it as a guide in their rulings. For example, the New York State Court of Appeals Decision, *Iannucci v. Board of Supervisors* (1967), mandated the Banzhaf index as *the* test of fairness of apportionment in county legislatures that use weighted voting. On the other hand, in *Whitcomb v. Chavis*, a case argued before the U.S. Supreme Court in 1970, the Banzhaf index was decisively rejected.

EXERCISES

● 1 In the Nassau County example with the 1958 weights, show that, in a line-up where S_1 and S_2 vote the same way, they are the only possible crucial voters.

2 Use Exercise 1 to help prove that in 1958 S_4, S_5, and S_6 are all dummies in Example 1.

● **3** Suppose we replace the majority-rule assumption by the following: in order for the yes side to win it must obtain two-thirds or more of the total votes. Otherwise, the no side wins. There are no ties.

(*a*) Show that, using this assumption and the 1958 weights in Example 1, there is a line-up in which Oyster Bay is crucial.

(*b*) In the same way, show that neither Glen Cove nor Long Beach is a dummy.

(*c*) Calculate the Banzhaf power indices of S_1, S_2, S_3, and S_4 under this rule.

4 Suppose the table below shows the voting system of the Division of Mathematical Sciences at a large university. The division has four departments, whose faculty sizes are shown. Each department has one representative on the divisional steering committee and the number of votes of that representative is shown in column 3. Suppose each representative agrees to vote exactly in accordance with the wishes of the majority in the department. Show that it is possible for a majority of the 64 professors in the division to oppose a measure but for it to receive a majority of the weighted votes when the representatives vote on it.

Department	Size	Number of votes
Mathematics	29	6
Computer science	21	4
Statistics	9	2
Operations research	5	1

● **5** Can you invent a weighted-voting game in which each line-up has exactly one crucial voter?

6 Calculate the Banzhaf power indices for three voters, each with one vote, under the assumption that any or all voters may abstain in any particular vote. (Thus each voter has three possible votes, yes, no, and abstain. Consequently there are 3^3 line-ups.)

● **7** Show that, if a voter is crucial in a line-up, then that voter must be on the winning side. Prove that, if *A* is not a dummy and *B* has more votes than *A*, then *B* is not a dummy either.

8 Show that, in any voting game, there is no line-up in which every voter is crucial. Is the same statement true if we replace the word "crucial" by "semicrucial"?

9 Suppose that D_1 and D_2 are dummies in some weighted-voting game and D_2 agrees always to vote the same as D_1. Show that this bloc of two voters is a dummy bloc, i.e., that it can never affect the outcome. (*Hint*: Suppose we have a line-up L_1 where the D_1-D_2 block could change the outcome by switching sides and creating a new line-up L_2. Now suppose we modify L_2 by switching D_1 back to his original position. Compare this new line-up L_3 to L_1 in terms of which side wins. Then compare it to L_2.)

10 Complete the proof of Theorem 1 by carrying out the analysis for odd values of *n* as follows:

(*a*) Show that there are no line-ups where *A* is semicrucial.

(*b*) Show that the line-ups where *A* is crucial fall into two categories: first, *A* votes yes along with $(n-1)/2$ others and, second, *A* votes no along with $(n-1)/2$ others.

(*c*) Count how many line-ups fall into each of the two categories mentioned in (*b*).

● **11** Suppose we are dealing with a voting game in which the no position wins automatically in case of a tie. Now there is no such thing as a semicrucial voter. How would you redefine the Banzhaf index? Calculate the revised Banzhaf index for each voter in the voting game of Table 4.

12 Many times coalitions don't just spring into existence, but grow one voter at a time, starting from a single one. For a winning coalition that grows this way, the voter whose joining makes the coalition a winner (puts it "over the top") is said to be the *pivot*. For any given ordering (permutation) of the *n* voters, we can determine who the pivot would be if a coalition were to grow according to that ordering. Now suppose we fix attention on a fixed voter, say voter *i*, and determine the number of orderings in which voter *i* is the pivot. If we divide by the total number of orderings possible for *n* voters, we obtain the Shapley-Shubik power index for voter *i*, denoted ϕ_i. Calculate this index for each player of the voting game described in Table 4.

13 Let ϕ_i denote the Shapley-Shubik index (see Exercise 12).

(a) Show that $\sum \phi_i = 1$.

(b) Show that $\phi_i = \sum [(s - 1)!(n - s)!/n!]$, where the sum is taken over all winning coalitions S for which $S - \{i\}$ is losing and $s = |S|$.

Computer exercises

14 Write a computer program that determines all crucial and semicrucial voters for a given line-up in a weighted-voting game. Inputs to the program are:

The number of players
The number of votes each player has
The line-up (how each player voted)

15 Starting with the program of Exercise 14, enlarge it so it creates and prints out tabulations like those in Tables 4 and 5. The input to the program is the number of voters and the number of votes each voter has. Apply your program to the four-voter game described at the top of Table 4, and check your results against the results of Tables 4 and 5.

16 Write a computer program to verify the validity of Stirling's formula. For $n = 1, 2, 3, \ldots, 10$ calculate $n!, s(n)$, and $s(n)/n!$. Does it appear that $s(n)/n!$ approaches 1 as x increases? Does it appear that $s(n) - n!$ approaches 0 as n increases?

BIBLIOGRAPHY

Banzhaf, J. F., III: Weighted Voting Doesn't Work: A Mathematical Analysis, *Rutgers Law Rev.*, vol. 13, pp. 317–343, 1965.

Brams, Steven: "Game Theory and Politics," The Free Press, New York, 1975.

—— "Paradoxes in Politics," The Free Press, New York, 1976.

Grofman, B.: Fair Apportionment and the Banzhaf Index, *Amer. Math. Monthly*, vol. 88, no. 1, January 1981.

—— and H. Scarrow: Iannucci and Its Aftermath: Game Theory and Weighted Voting in the State of New York, in Steven Brams (ed.), "Game Theory," Physica Verlag, Vienna, 1979.

Keyfitz, N.: How Many People Have Lived on the Earth? *Demography*, vol. 3, p. 581, 1966.

Lucas, W. F.: "Measuring Power in Weighted Voting Systems," Technical Report 227, Department of Operations Research, Cornell University, Ithaca, N.Y., 1974.

5 PUTTING THE FACTS IN ORDER—MENDELEEV AND THE PERIODIC TABLE

Abstract The world is full of data, many of them bewildering. Generally, data are not useful unless we have a pattern or theory that they fit into. In Mendeleev's time, the facts about the chemical elements were thought to be confusing. People who tried to fit a system to the facts were considered foolishly optimistic. However, Mendeleev chose to believe the facts and to believe that they were telling an important story, and his faith was dramatically rewarded.

Prerequisites Graphing functions. This section is a mathematical anecdote with no technicalities.

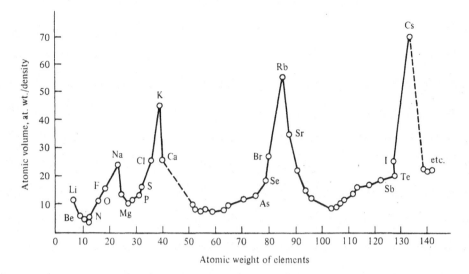

Figure 1 Atomic volumes for the first portion of the table of elements. (*After Lothar Meyer, but using modern values.*)

Around 1868 the Russian chemist Mendeleev was preparing a textbook on chemistry and looking for a systematic way of describing the relationships among the 63 elements known at the time. Often the elements were listed in order of their atomic weights. Chemists thought it was peculiar that when such a list was made, elements which were similar in chemical behavior did not cluster together in the list. For example, according to Figure 1, chlorine (Cl), Bromine (Br), and Iodine (I) have nearly the same atomic volume, but their atomic weights are far apart. There are other ways in which these three elements are similar, and chemists had long since regarded them as members of the so-called halogen family. It seemed very confusing to chemists that listing elements by atomic weight would break up the family. The halogens were not the only family that was broken up in such a list. The question was whether these family relationships could be deduced from some mathematical scheme based upon atomic weights.

One idea which popped up a few times was to abandom the idea of a list in favor of some kind of two-dimensional scheme. For example, the French chemist Chancourtois arranged the elements along a spiral that wound its way around and up the surface of a two-dimensional cylinder. The elements were put on the spiral in order of atomic weight, but the spacing between the elements was arranged to try to have family members appear in vertical stacks on the cylinder. See Figure 2.

Chancourtois named this spiral the *telluric helix*, but even with this fancy name chemists of the time paid no attention. Perhaps it was just as well for, when chemists did pay attention to efforts of this kind, they heaped scorn on them instead. For example, in 1886 John Newlands created a two-dimensional table which was a crude form of the one Mendeleev eventually became famous for. But his paper on the subject was refused publication, and Newlands was thoroughly ridiculed.

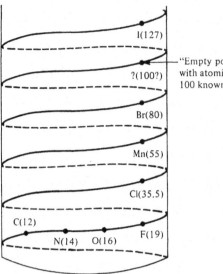

Figure 2 Part of Chancourtois' telluric helix. Dots indicate positions of elements. (For clarity, some dots and some elements are not shown.) Numbers in parentheses indicate atomic weights. Notice how the halogens fall in a vertical column.

Perhaps part of the reason was that Newlands had the poor judgement to name his table the *Law of Octaves* because he though he saw some analogy to the principles of music (each of his columns had eight elements and in music an octave has eight notes). One critic "kindly" suggested that even though the Law of Octaves wouldn't be useful in chemistry, maybe it could be useful in the field of music.

Mendeleev was one of those who thought that some sense could be made of charts like Figure 1. Notice that this graph has some slight similarity to the inter-course data of Section 3: there is a somewhat regular alternation of peaks and valleys. If Mendeleev had reacted like the modern sex researchers and believed the data to be incorrect because it was so erratic, then he would not have made the breakthrough he did. Luckily, he took the data at face value and, by a process of reasoning we need not describe, was led to the same idea others had come to: to make a two-dimensional list or table as a means of displaying both chemical similarity and atomic weight.

Table 1 shows one of his first published tables. As one reads across the rows one gets the elements in order of atomic weight. By strategic choice of when to start a new column and when to leave an occasional gap, Mendeleev managed to arrange things so that elements with similar properties were in the same column. For example, the halogens are together in column VII. The result is a somewhat ragtag table with columns of different sizes and many empty cells. It might have made a lesser man lose heart. It is not at all surprising that other chemists ignored it. However, Mendeleev turned weakness to strength by claiming that some of the

Table 1 Periodic classification of the elements (Mendeleev, 1872)[†,‡]

Series	Group → I R_2O	II RO	III R_2O_3	IV RO_2 H_4R	V R_2O_5 H_3R	VI RO_3 H_2R	VII R_2O_7 HR	VIII RO_4
Higher oxides and hydrides	—	—	—					—
1	H(1)							
2	Li(7)	Be(9.4)	B(11)	C(12)	N(14)	O(16)	F(19)	
3	Na(23)	Mg(24)	Al(27.3)	Si(28)	P(31)	S(32)	Cl(35.5)	
4	K(39)	Ca(40)	—(44)	Ti(48)	V(51)	Cr(52)	Mn(55)	Fe(56), Co(59), Ni(59), Cu(63)
5	[Cu(63)]	Zn(65)	—(68)	—(72)	As(75)	Se(78)	Br(80)	
6	Rb(85)	Sr(87)	?Yt(88)	Zr(90)	Nb(94)	Mo(96)	—(100)	Ru(104), Rh(104), Pd(106), Ag(108)
7	[Ag(108)]	Cd(112)	In(113)	Sn(118)	Sb(122)	Te(125)	I(127)	
8	Cs(133)	Ba(137)	?Di(138)	?Ce(140)	—	—	—	
9	—	—	—	—	—	—	—	
10	—	—	?Er(178)	?La(180)	Ta(182)	W(184)	—	Os(195), Ir(197), Pt(198), Au(199)
11	[Au(199)]	Hg(200)	Tl(204)	Pb(207)	Bi(203)	—	—	
12	—	—	—	Th(231)	—	U(240)	—	

† Atomic weights are shown in parentheses. Horizontal lines indicate "missing" elements which Mendeleev expected to be discovered in the future.

‡ Adapted from *Annalen der Chemie*, suppl. vol. 8, 1872.

135

empty cells in the table were not a shortcoming of the table but were inhabited by undiscovered elements.

For example, in row 5 and column IV there is an empty cell which Mendeleev claimed should be occupied by an element still undiscovered at the time. Furthermore, he predicted some of its properties, based upon the properties of the neighbors of this cell in the table: that its atomic weight would be 72, its density 5.5, and so on. As a matter of fact, in 1887 Winkler discovered a new element, now called germanium, which fit into that slot. It had atomic weight of 72.5, density 5.5, and many other properties predicted by Mendeleev. Together with earlier discoveries of gallium (Ga, atomic weight 69.9) and scandium (Sc, atomic weight 44.96), which filled in other gaps according to Mendeleev's predictions, this established the periodic table as a central tool of chemistry—useful for many things, including the discovery of new elements.

It may seem strange that one could make these predictions only on the basis of the position of an empty space in the table. Actually there is a little more to it than that. Mendeleev had in mind that the various chemical properties of an element were determined from its table position by certain formulas and rules of thumb. Thus the mathematical model consists really of three things:

1. The table
2. The assumption that the empty cells in the table corresponded to undiscovered elements
3. The formulas and rules about how table position determines chemical properties

For simplicity, we are leaving out the details of the third part of his model.

Comparing Mendeleev's work with that of his predecessors and contemporaries shows clearly the crucial role of his faith that there was mathematical regularity beneath the chaos that was chemistry at that time. Others, such as John Newlands, had arranged the elements in a two-dimensional table but had not been bold enough to leave gaps for undiscovered elements in order to preserve the mathematical patterns. Lothar Meyer had left gaps (and is therefore sometimes considered a codiscoverer of the table), but he stopped short of predicting the natures of the undiscovered elements based on their positions in the table. Only Mendeleev had the faith to go out on a limb. Table 2 shows how well his faith was rewarded.

This example of confusion being turned into clarity by the right way of looking at the facts (in this case, simply arranging the elements in a two-dimensional table) is not an isolated one in the physical sciences. Many other examples could be given. It is also possible to cite examples of areas in science today which are overwhelmed by confusing details and where the clarification has not yet arrived. Particle physics is an excellent example.

The fundamental idea of particle physics is that the material things in the universe are all built up out of fundamental building blocks. In Mendeleev's time these building blocks were called atoms and there were over 60 different varieties known (e.g., oxygen atoms, carbon atoms, chlorine atoms, etc.). As we have seen, every now and then additional types were discovered (for example,

Table 2

	Mendeleev's prediction	Actual properties of germanium
Atomic weight	72	72.5
Density[†]	5.5	5.469
Density of oxide[‡]	4.7	4.703
Density of chloride[‡]	1.9	1.887
Boiling point of chloride	Less than 100°	86°
Density of ethide[‡]	0.96	Slightly less than 1
Boiling point of ethide	160°	160°

[†] All densities measured in grams/cubic centimeter.
[‡] A type of compound formed from the element.

germanium in 1887). Although Mendeleev clarified the relationships of these different types of atoms, it was still a little confusing to have so many different kinds.

In 1911 Ernest Rutherford proposed that each of these different kinds of atoms was itself built of smaller particles and there were only three varieties of these smaller particles: electrons, protons, and neutrons. Each type of atom has its own recipe or formula for how many of each of the three fundamental particles is contained in the atom. It was a beautiful simplification. But the luxury of a universe built of only three kinds of things was too good to last. In the intervening years an ever increasing number of fundamental particles have been discovered. Today we have positrons, antiprotons, mesons, leptons, neutrinos, and other objects besides electrons, protons, and neutrons, and a good deal of confusion about how these things are related and how they fit together to form the physical world.

At various stages in the history of this subject the "Mendeleev trick" has been performed: someone has found a new theory or a new way to look at the facts and make it all seem neat and tidy. More and more these new theories have been heavily mathematical. Unfortunately, as new particles were discovered and new experiments were performed these theories have had to be revised. But new mathematical models have always been found. For this reason, physical scientists have a lot of faith in mathematical modeling as a way of trying to understand the world.

But what about the biological and social sciences? Although there have been successful mathematical models in these fields (and some of them will be found in this book), there are many areas which are confronted by mountains of apparently confusing facts.

Take, for example, the problem of crime. We have plenty of facts about it, but little good theory. In particular, the relation of crime to economic status is unclear. It is known that certain groups of poor people have high crime rates. But there are also groups of people who are very poor and have very low crime rates. Even among groups with high crime rates, there are impressively large numbers of individuals who are completely law abiding. Finally, there are plenty of people in the middle and upper classes who commit crimes. If we examine the relation of crime to other explanatory variables, such as law enforcement factors, the situation is also confusing. It is possible that what is lacking is not more data but a better theory for the information we have?

EXERCISES

1 In mathematics, the word "periodicity" (or "periodic") is sometimes applied to a function or its graph. If $f(x + T) = f(x)$ for all relevant x, $f(x)$ is periodic, with period T. For example, $\sin x$ is periodic, with the period 2π.

   (a) Give other examples of periodic functions.

   (b) The graph of Figure 1 appears nearly periodic. What is its approximate period?

2 Using the data given in the "Handbook of Chemistry and Physics," plot the melting points of the elements against the atomic weights in the style of Figure 1 for the elements from hydrogen (H) through manganese (Mn), inclusive. Are there periodicities? How are they related to the table?

3 Plot the ionization energy of the elements against the atomic weights in the style of Figure 1 for the elements from hydrogen (H) through manganese (Mn) inclusive. Consult the "Handbook of Chemistry and Physics" for data. Is your graph periodic?

4 Make a telluric helix (see Figure 2) out of paper or light cardboard as follows:

1. Use the left-most vertical line of a piece of graph paper to plot atomic weight.
2. Use the bottom line to index the columns. Divide this axis into eight equal segments. The points of division, starting at the left endpoint of the first segment (the origin), represent the columns of Mendeleev's table. The first eight points are labelled I, II, ..., VIII, and the last is unlabeled.
3. Plot the elements as points on the graph according to atomic weight and table position.
4. Roll the paper into a vertical cylinder with the unlabeled point and point I on the bottom line coinciding. How do family relationships in the table show up on the cylinder? (Perhaps you can visualize this without constructing the cylinder, or you might want to do only part of the table.)

5 Would it make sense to carry out some kind of adjustment of data or moving-average process, as described in Section 3, on the data of Figure 1? Justify your answer.

6 Some observers today claim that the social sciences can never be true sciences because they can't be mathematized. Assuming that it is true that they have not yet been effectively mathematized (a debatable point), what does the history of chemistry suggest about the prospects?

BIBLIOGRAPHY

Cassebaum, H., and G. B. Kaufman: The Periodic System of the Chemical Elements: The Search for its Discoverer, *Isis*, vol. 62, pp. 314–327, 1971.

Holton, Gerald: "Introduction to Concepts and Theories in Physical Science," 2d ed., Addison-Wesley, Reading, Mass., 1973.

Spronsen, J. W., von: "The Periodic System of Chemical Elements," American Elsevier, New York, 1969.

THREE

EVALUATION OF MATHEMATICAL MODELS

1 A BIRD'S-EYE VIEW OF EVALUATION—COLLEGE ENROLLMENT

Abstract In this section we put forward some ideas for determining whether a mathematical model is good or bad. We give an overview of these ideas, as applied to the problem of predicting college enrollments. The sections which follow this one deal with individual evaluation criteria in greater depth.

Prerequisites High school algebra.

In pure mathematics one doesn't often ask whether a theorem is good or bad. One theorem may be more or less beautiful than another, it may be more or less difficult to prove, or it may play a more or less central role in mathematics. But even these kinds of judgements are often left for cocktail party conversation. In many cases, evaluating mathematics is rightly considered a matter of personal taste.

This sort of casual attitude is less appropriate in the field of mathematical modeling. If one wants to use a mathematical model to shoot a rocket to the moon or reduce the cancer rate or cope with an inflationary economy, then it makes an important difference whether one has good models or bad ones. The model one has available may not be good enough to use. Or there may be a number of competing models available for use in a given situation. For these reasons it seems sensible for evaluation to be in the forefront of our thinking about mathematical modeling.

It would be nice to have some formula or theory by which to evaluate models, but unfortunately we know of no mechanical or systematic way of taking the measure of a model. But that doesn't mean that there are no principles to go by.

There are certain characteristics which models have to varying degrees and which bear on the question of how good they are. The characteristics we will discuss are:

Accuracy
Descriptive realism
Precision
Robustness
Generality
Fruitfulness

Sections 2 through 7 are designed to illustrate aspects of these characteristics individually and in detail. But, in advance of the details, it may be useful to have a quick overview of all these ideas at once. For this purpose, let's consider the problem of predicting how many students will be enrolled in college in the next year. To start with, let's compare two models devised to answer this question.

Model 1

We observe that this year there are 10 million people in the entire population who are in the age category from 18 through 22, which forms the bulk of the college student population. We also obtain the information that the number of college students is 5 million this year. From these figures we might conjecture that

$$E = 0.5P \tag{1}$$

where E = number of students enrolled and P = population aged 18 through 22 years. This equation is based upon the following two assumptions:

A1. Each college student is in the age category 18–22.
A2. In the 18–22 age group, one out of every two is enrolled in college.

Now if we determine from the Bureau of the Census that next year there will be 11 million people in the age category 18–22, i.e., $P = 11$ million, then we can use Equation (1) to compute

$$E = (0.5)(11,000,000)$$

$$= 5,500,000$$

If it turns out next year that there really are 5,500,000 students enrolled in college (or pretty near that number), we will be happy. In that case our model has the characteristic of accuracy.

Definition

A model is said to be *accurate* if the output of the model (the answers it gives) is correct or very near to correct.

Our model 1 has been tailor-made to be correct for this year, but we won't know till next year whether the 5,500,000 students it predicts for next year is correct or nearly correct. But who wants to wait till next year? The whole idea of this model is to predict next year's enrollment in advance so that plans can be made.

This is a common stumbling block of nearly all models. We might call it the Catch 22 of modeling: we build a model to find something we wouldn't ordinarily know without the model; but we can't tell whether the model is accurate without finding out this thing, that we couldn't ordinarily find out without the model, and then comparing with the prediction of the model. Is there any way out? Yes, there are a few ways.

First, some models have impressive track records. For example, there is a mathematical model of the motions of the heavenly bodies that will predict the time and place of the next eclipse of the sun. Of course, we can't be sure what the sun, moon, and earth are going to do in the future; so technically we can't evaluate the accuracy of the prediction. But this model describes accurately the times and places of all other eclipses that have ever been recorded in the past; so we are inclined to have faith in it. Unfortunately, few models have as good a track record as this.

With or without a good track record, the second approach to the Catch 22 is to supplement the consideration of accuracy with evaluations along the other dimensions (descriptive realism, precision, robustness, etc.). For example, model 1 has an immediately evident shortcoming in regard to what we call descriptive realism.

Definition

A model is said to be *descriptively realistic* if it is based on assumptions which are correct.

Assumption A1 is not correct. Although the bulk of college students comes from the age category 18–22, there are many older college students and some younger ones. If we examine available statistics for this year, we might find the following assumptions to be more reasonable (although probably not absolutely correct).

Model 2

A3. Students in college can be divided into three age categories.
  (*a*) Aged 18–22
  (*b*) Aged 23 and over
  (*c*) Aged under 18
A4. In each age category of the population, a certain percent are enrolled in college.
  (*a*) In the 18–22 age category, 30 percent
  (*b*) In the 23-and-over category, 3 percent
  (*c*) In the under-18 category, 1 percent

If we denote the sizes of these age categories by P_a, P_b, and P_c respectively, then this assumption produces:

$$E = 0.30P_a + 0.03P_b + 0.01P_c \tag{2}$$

We can't determine the accuracy of model 2, in regard to the prediction for next year, any better than we could judge the accuracy of model 1 (unless we are willing to wait till next year, which we already agreed was silly). But because model 2 is more descriptively realistic than model 1, we should be inclined to trust it more. In this case realism serves as a sort of a "stand-in" for accuracy. There are other stand-ins. For example,

Definition

A model is said to be *precise* if its predictions are definite numbers (or other definite kinds of mathematical entities: functions, geometric figures, etc.). By contrast, if a model's prediction is a range of numbers (or a set of functions, a set of figures, etc.), the model is imprecise.

Both models 1 and 2 are precise in the sense that both Equations (1) and (2) give definite values for E once the population figures are plugged into the right-hand side. For example, when we substituted $P = 11,000,000$ in Equation (1), we got the definite value $E = 5,500,000$. If our model had been differently constructed so as to a give a range of possibilities for E, say $5,000,000 \leq E \leq 6,000,000$, then we would have called the model *imprecise*.

Here is an example of a model which is imprecise. Suppose that, when we examine the statistics for past years, we discover that the ratio of college enrollment to persons aged 18 through 22 is not always 0.5 but varies somewhat, as in Table 1.

Then, to build a new model, we might replace assumption A2 in model 1 by:

A5. The fraction of the 18–22 age group which is enrolled in college in any particular year (not just years covered by Table 1) is always between 0.46 and 0.5.

Table 1

Year	Enrollment	Population (18–22)	Ratio
5 years ago	3,680,000	8,000,000	0.46
4 years ago	3,901,000	8,300,000	0.47
3 years ago	4,176,000	8,700,000	0.48
2 years ago	4,410,000	9,000,000	0.49
Last year	4,655,000	9,500,000	0.49
This year	5,000,000	10,000,000	0.5

Model 3

Using assumptions A1 and A5, we produce the following inequality [compare it with Equation (1) in model 1]:

$$(0.46)(11,000,000) \leq E \leq (0.5)(11,000,000)$$
$$5,060,000 \leq E \leq 5,500,000$$

Generally speaking, it is nice to have precision in a model. However, when we compare models 1 and 3 in the light of Table 1, it appears that, if we opt for precision by choosing model 1, we do so at the expense of descriptive realism.

Definition

A model is said to be *robust* if it is relatively immune to errors in the input data.

It stands to reason that, if our population estimates for next year are wrong by a few percent (which is certainly possible), then our predictions are also wrong by a few percent. But how many percent? If the population figures are off by 10 percent, then will our predictions also be off by 10 percent? Or more than 10 percent? Or less? More to the point, will one model be off by a smaller percentage than another?

It turns out that some models have the property of magnifying the percentage error. These models are said to be sensitive to error in the input data. On the other hand, there are robust models, in which the percentage error in the output is less than the percentage error in the input. As we shall now see, models 1 and 2 differ in their sensitivity to error in the input data (robustness). Before we investigate this, let's be clear what we mean by percentage error.

Definition

If the true value of something is v and the measured or predicted value is v', then we call $v' - v$ the *error*. Sometimes an error is expressed as a fraction of the true value by computing $(v' - v)/v$. We call this the *fractional error*. When we express this fraction as a percent (by computing $[(v' - v)/v] \times 100$ percent), we call this the *percentage error*.

For example, let's suppose that in the last census the number of people in the 18–22 age category was determined as 9,300,000, but, after a more careful and accurate recount, the number turned out to be 9,500,000. The error is $9,300,000 - 9,500,000 = -200,000$. The fractional error is (approximately) -0.021. The percentage error is -2.1 percent.

In discussions of error we frequently talk in terms of the absolute value of the percentage error rather than the percentage error itself (2.1 percent instead of -2.1 percent in the example just given). The reason for this is that, for many measurement processes, we have an idea of how close we might come to the correct value but can't tell if our estimate is likely to be an over- or underestimate. Think of target shooting, for example. An experienced shooter can tell you that he will come within a certain number of inches of the bull's-eye, but he can't tell you whether the shot will fall below, above, to the left, or to the right of the bull's-eye. In the remainder of our discussion of error, the term "percentage error" will usually be understood to be an absolute value.

Now suppose we decide, based on previous experience, that we must be prepared for an error of up to 5 percent in our prediction of next year's population figures. If the error is exactly 5 percent (and not −5 percent which has been converted to +5 percent by dropping the minus sign), the prediction of the 18–22 age category is 11,000,000, and the true value is P, then

$$\frac{11,000,000 - P}{P} = 0.05$$

$$1.05P = 11,000,000$$

$$P = \frac{11,000,000}{1.05}$$

$$= 10,476,190$$

When we apply Equation (1) to this true value, we predict

$$E = 0.5(10,476,190)$$

$$= 5,238,095$$

Recall that, in our earlier discussion of model 1, our predicted enrollment was 5,500,000. The percentage error in the output is

$$\frac{5,500,000 - 5,238,095}{5,238,095} = 0.05$$

or 5 percent, exactly the same percentage error we supposed in the input.

Naturally, this "let's suppose" type of calculation, where we pick a number out of the air for our input error (which is, of course, unknowable), is not totally satisfying. What if the input error is 4 or 3 or 3.24 percent? It turns out that, for *any* percentage error in the input, the output of model 1 shows the same percentage error. This can be proved algebraically, and we leave it for the reader as an exercise (see Exercise 11).

If our input error had been −5 percent, the same type of calculation would show an output error of −5 percent. (Try this as an exercise.)

This characteristic that "error in equals error out" is not true for all models. What's worse, it may not always be possible to be sure of what output error results from a given input error. Model 2 exhibits both of these features. It turns out that, if the input errors in model 2 [Equation (2)] are the same absolute percentage and also in the same direction, i.e., either all overestimates or all underestimates, then the percentage output error equals the percentage input error. But if some of the errors are in opposite directions (some overestimates and some underestimates), then their effects may cancel each other somewhat and leave a percentage output error that is smaller than the percentage input error.

We can investigate these statements algebraically. Suppose the unknown true values of the populations in the three categories are P_a, P_b, and P_c, while the predicted values are P'_a, P'_b, and P'_c. If we do a "let's suppose" calculation based on an assumption of an absolute percentage error of 10 percent, we get

$$P'_a = P_a \pm 0.1P_a$$
$$P'_b = P_b \pm 0.1P_b \qquad\qquad (3)$$
$$P'_c = P_c \pm 0.1P_c$$

The reason for the uncertainty about the signs is that we do not know whether any particular prediction will be an overestimate or an underestimate.

If it should turn out that all signs are + (all estimates are overestimates), then the following calculations show that "error in equals error out":

$$
\begin{aligned}
E' &= 0.30P'_a + 0.03P'_b + 0.01P'_c \\
&= 0.30(P_a + 0.1P_a) + 0.03(P_b + 0.1P_b) + 0.01(P_c + 0.1P_c) \\
&= 1.1(0.30P_a + 0.03P_b + 0.01P_c) \\
&= 1.1E
\end{aligned}
$$

Thus $(E' - E)/E = 0.1$. which signifies a 10 percent error in the output, the same magnitude of percentage error as in the input.

A case where the percentage output error is *less* than the percentage input error occurs when P_a is overestimated and P_b and P_c are underestimated. This means we have a + sign in the first equation of Equation (3) and − signs in the last two. Now, when we attempt to carry through the same type of algebraic calculation as before, there is a step where we get an inequality instead of an equality:

$$
\begin{aligned}
E' &= 0.30P'_a + 0.03P'_b + 0.01P'_c \\
&= 0.30(P_a + 0.1P_a) + 0.03(P_b - 0.1P_b) + 0.01(P_c - 0.1P_c) \\
&< 0.30(P_a + 0.1P_a) + 0.03(P_b + 0.1P_b) + 0.01(P_c + 0.1P_c) \\
&= 1.1E
\end{aligned}
$$

Thus $E' - E < 0.1E$ and $(E' - E)/E < 0.1$. Therefore the percentage error in the output is less than 10 percent.

Unfortunately, our algebraic calculations don't tell us how much less the percentage output error might be than the percentage input error. The amount less will depend on the particular values of P_a, P_b, and P_c. Table 2 gives some idea of the amount less we might obtain, based on the true values $P_a = 10$ million,

Table 2

I	II			III	IV	V	VI
Pattern of errors[†]	Predicted values[‡]			Model predictions based on wrong values, millions	Model predictions based on true values, millions	Error (Col III – Col IV), millions	Percentage error, percent
$P_a\ P_b\ P_c$	P'_a, millions	P'_b, millions	P'_c, millions				
+ + +	11	99	55	6.82	6.2	0.62	10
+ + −	11	99	45	6.72	6.2	0.52	8.4
+ − +	11	81	55	6.28	6.2	0.08	1.3
+ − −	11	81	45	6.18	6.2	−0.02	0.3
− + +	9	99	55	6.22	6.2	0.02	0.3
− + −	9	99	45	6.12	6.2	−0.08	1.3
− − +	9	81	55	5.68	6.2	−0.52	8.4
− − −	9	81	45	5.58	6.2	−0.62	10

[†] The + means the predicted value is 10 percent *higher* than the true value. The − means the predicted value is 10 percent *lower* than the true value.

[‡] These values are based on true values of $P_a = 10$ million, $P_b = 90$ million, and $P_c = 50$ million.

$P_b = 90$ million, and $P_c = 50$ million. The table shows that, if we are lucky, we might have a percentage error in the output of near zero. We may summarize the table and what it tells us about the robustness of models 1 and 2 as follows:

In two cases (lines 1 and 8) model 2 gives "error in equals error out." In these cases, model 2 has the same sensitivity to error as model 1. But in six cases the percentage error in output is less than the percentage error in input and model 2 is more robust than model 1. On the whole, we consider model 2 more robust than model 1.

Another criterion which can sometimes be used to compare models is generality.

Definition

A model is said to be *general* if it applies to a wide variety of situations.

To illustrate this concept, we introduce yet another model for the college-enrollment prediction problem.

Model 4

This model is based on the following two assumptions:

A1. All college students are in the age group 18 through 22.
A2. Each individual college will have its enrollment expand or decline by the

same ratio, that ratio being the ratio of next year's to this year's population in the 18–22 age category.

To implement this model, we first determine the size of the 18–22 age group this year; call it P. Then we obtain from the Bureau of the Census the prediction for the size of next year's 18–22 age group, say P'. We compute the ratio

$$r = \frac{P'}{P}$$

This is the ratio by which the "college-age population" will grow (or decline if $r < 1$).

Now we determine the enrollments this year of all of the individual colleges, say $E_1, E_2, E_3, \ldots$. In each case, we assume next year's enrollment E_i' is r times as large. That is

$$E_i' = rE_i \qquad \text{for all } i$$

If we denote the total enrollments this year and next year by E and E' respectively, we get

$$
\begin{aligned}
E' &= E_i' + \cdots + E_n' \\
&= rE_1 + \cdots + rE_n \\
&= r(E_1 + \cdots + E_n) \\
&= rE
\end{aligned}
$$

This model may be a little hard to apply because n, the number of colleges, is large and gathering the numbers $E_1, E_2, \ldots, E_n$ may be quite time-consuming. But model 4 has the advantage of greater generality than the other models. That is, this model could be applied to any individual college, whereas the others cannot. Or it could be applied to any selected group of colleges, for example, the subset of colleges that are private, have religious ties, or enroll between 4000 and 6000 students.

The last criterion for evaluation which we shall consider is fruitfulness.

Definition

A model is said to be *fruitful* if either:

1. Its conclusions are useful.
2. It inspires or points the way to other good models.

Part 1 of the definition of fruitfulness can often be evaluated even before we work out any of the details of the model. For example, our college-enrollment models could be useful planning tools (provided they are reasonably accurate) for a governmental agency administering a student loan program. Model 4 would also be useful to individual colleges in their planning.

Even if these enrollment models turned out to be too inaccurate for planning, they could still be fruitful, in the sense of part 2 of the definition of fruitfulness, if they inspire other good models. For example, reasoning by analogy, we could devise models to predict how many automobiles would be junked or abandoned in any given year. (Cars play the role of people, and, for each age or age group of car, there is a rate that can be applied.) Section 6 gives a more extensive and important example of this kind of fruitfulness.

The six criteria which we have discussed are not the only possible ones for deciding how good a model is. We might, for example, consider the time and expense in devising or using a model. Conversely, there are numerous excellent models that score poorly on one or more of these criteria. A number of our examples in the following sections show this.

Sidelight: The Advantage of Inaccuracy

Accuracy is often expensive, and this may make inaccuracy fairly attractive now and then. Consider the problem a beer company has in delivering beer to taverns and stores at fixed geographical locations, as in Figure 1. What is the shortest route that starts and ends at the brewery? If we use a random route, at each stage choosing the next stop at random, we will do much more driving than necessary. Applying a little common sense will give a better route, but when there are many stops it is not likely to give the very best route.

The optimum solution to this problem (called the traveling salesman problem, see Sections 1 and 4 of Chapter 4) for up to a few hundred stops can be found by a computer, but this will cost money. In the case of a beer company, it may not be worth it because the solution may not be usable every day. Customers run out of beer at irregular times; so the stops the salesman needs to make vary

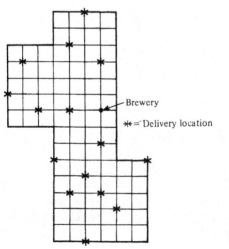

Brewery

✳= Delivery location

Figure 1

from day to day. Paying for a new computer solution every day might cost more than the amount saved in driving.

On the other hand, suppose you are delivering newspapers instead of beer. Now you are likely to have the same stores to visit every day. An expensive solution that gives a totally accurate answer to the shortest-route problem might pay for itself in the long run because it could be used to save money every day.

EXERCISES

Evaluation is an art, not a science. For this reason, the answers to many of the following exercises will be matters of opinion.

● **1** Is it possible for the assumptions of model 3 and those of model 2 to be simultaneously correct? Explain.

2 Are there any advantages or disadvantages to further subdividing the age categories in model 2, for example, by sex, race, more age divisions, or any other criterion you can think of?

● **3** Model 3 is based on Table 1, which shows an increasing ratio of enrollment to population. Can you revise model 3 to take this increase into account? Be specific in stating what assumptions your model is based on. What equations or inequalities result from your assumptions?

4 Compare models 2 and 4 with respect to descriptive realism.

5 Can you build a model which combines the best features of models 2 and 4? Be specific about your assumptions. State what equations or inequalities result from them. You may assume that, within reason, any data you need are available. You may even invent numerical values to use for the sake of illustration.

6 Answer question 5 but with models 2 and 3 in place of models 2 and 4.

7 Is it possible for a model to be accurate but not precise? Can it be precise without being accurate? Explain your answers. Give an example if you can. (It can be a made-up example.)

● **8** Is there any difference in generality between models 1 and 2?

9 Which models in this section are adaptable to the problem of predicting high school enrollments? Explain how you would adapt them. If any are not adaptable, explain why they aren't. Answer the same question about kindergarten enrollments.

10 Verify the calculation of each of the seven numerical entries of line 8 of Table 2.

11 (a) Prove algebraically that, in model 1, if the percentage error in input is 5 percent, then the percentage error in output is also 5 percent, regardless of the exact values of P and E.

(b) Now do the same proof for any arbitrary percentage input error.

12 (a) Using the equations in Equation (3), verify algebraically that, when all predicted populations are 10 percent under the true values P_a, P_b, and P_c, then the percentage error in output equals the percentage error in input.

(b) Now show that this is still true no matter what the exact percentage error is.

● **13** Suppose we are attempting to measure the area A of a square by measuring the side s and applying the well-known formula $A = s^2$. Suppose we measure s to be 100 but with an error that could be anywhere between 0 and 10 percent. Calculate the various percentage errors in A (the output) that result from the following errors in s (the input): 0, 1, 2, 3, ..., 10 percent. Can you formulate any rule of thumb about how percentage input error relates to percentage output error? For example, is it the case that "percentage error in equals percentage error out?"

14 It turns out that, when taking a census, underestimating the true population is more common than overestimating it. Can you think of reasons why this might be so?

15 Suppose you wish to apply model 4 to the task of predicting the total enrollment in all medium-sized colleges, where medium-sized means that the enrollment is between 4000 and 8000. Suppose you do this by identifying all colleges which are medium-sized this year and determining their enrollments this year, say $E_1, E_2, \ldots, E_R$. Then you find the growth or decline factor r, calculate $rE_1 + rE_2 + rE_3 + \cdots + rE_R$, and use this as your prediction of the total enrollment in medium-sized colleges *next* year. Explain why this calculation may be wrong even if it turned out that each assumption of model 4 was exactly correct.

Computer exercise

16 Write a computer program to carry out the percentage-error calculations in Table 2. Inputs to the program are the true values $P_a = 10$ million, $P_b = 80$ million, and $P_c = 50$ million and the absolute percentage error of 10 percent. Your program should generate the eight sign patterns of column I. Then for each such pattern compute all the table entries (columns II through VI) and print them out.

2 DESCRIPTIVE REALISM—SIMPLE LINEAR REGRESSION

Abstract It's always nice to know what makes things tick, and, if one can base a mathematical model on this knowledge, the model is said to be descriptively realistic. But descriptive realism is not always possible or necessary. We use the subject of regression analysis to illustrate this. In this section we stick to 1-variable (simple) linear regression. Statistical aspects of the theory are omitted.

Prerequisites The analytic geometry of straight lines in the plane. Minimization of a function of two variables occurs in the derivation of one formula. Access to a computer may be useful.

A mathematical model is said to be *descriptively relaistic* if it is deduced from a correct (or at least believable) description of the mechanism involved in whatever is being modeled. For example, observation indicates that a full moon occurs approximately every 29 days. Thus, if M_L is the date of the last full moon and M_N is the date of the next, then

$$M_N = M_L + 29 \tag{1}$$

This is a fairly accurate and useful model. But where is the mechanism that "makes it tick?" In fact, we did not derive the model from an underlying mechanism. Although useful and reasonably accurate, it is not a descriptively realistic model.

We could fairly easily build a realistic model to account for the phases of the moon. It turns out that the phases of the moon are caused by the motions of the moon and the earth in relation to one another and the sun. In different positions, different portions of the moon are "lit up" by the sun. The part of the moon we see is the lit-up part. If we describe the orbits of the earth and moon mathematically, we could derive Equation (1). Would such a realistic derivation give you greater confidence in Equation (1)?

Here is an example involving the mathematical concepts which we will study in this section.

Table 1

Month	Energy, millions of kilowatt hours	Cost, millions of dollars
1	1700	95
2	1750	98
3	1800	99
4	1930	105
5	2000	112

Example 1 Energy (Hypothetical Data)

An electric power company wishes to see how its operating costs are related to the amount of energy (in kilowatthours) generated. We will consider two ways to produce an equation relating the two quantities. One way, our first introduction to regression analysis, is not descriptively realistic: the equation expresses a pattern in the data, but there is no story to tell us why that pattern exists. The second model is descriptively realistic: it takes advantage of some details about how the power company operates and uses these details to show why there is a pattern in the data and find an equation describing that pattern.

Method 1 (regression) The data in Table 1 are produced from company records. A common and useful way of visualizing this information is to take the energy and cost figures from each line of the table and form an ordered pair, which we can plot on coordinate axes, as in Figure 1. The points which are obtained in this way form what is called a *scatter diagram*.

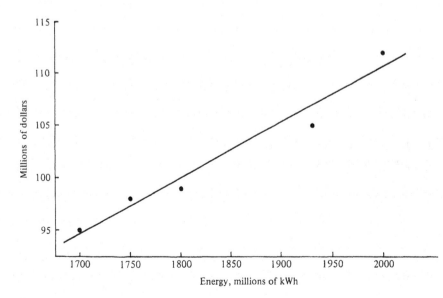

Figure 1

After plotting these points, we notice that they seem to fall nearly in a straight-line pattern. We will soon see how to calculate the equation of a best-fitting line. The line in Figure 1 is

$$C = 0.0529K + 4.75 \tag{2}$$

(C and K measured in units of 1 million).

But why a straight line? Is there any reason why operating cost should be related to the amount of power generated according to a linear equation like Equation (2)? The answer is yes. Read on.

Method 2 (a descriptively realistic model of fixed and variable cost) An important business principle is that operating costs can be divided into two categories:

1. *Fixed costs*, which don't change when the volume of energy generated increases or decreases moderately. These include office staff, real-estate taxes, insurance, interest on debt, some maintenance costs, and various other items.
2. *Variable costs*, which go up or down in proportion to the amount of energy generated. Fuel is the main variable cost.

Let us suppose that the fixed costs are \$4,700,000 per month. We will also suppose that, to generate energy, the company burns oil, which costs \$30 per barrel, and each barrel yields 570 kilowatthours of power. This makes the variable cost 30/570 dollars per kilowatthour. If we generate K million kilowatthours, the variable cost will total 30/570 K million dollars. Adding the fixed cost gives a total cost C (in millions of dollars):

$$C = 0.052632K + 4.7 \tag{3}$$

This is so close to Equation (2) that, in the practical spirit of applied mathematics, we will consider them the same.

Which derivation of Equation (3) do we have more confidence in: the descriptively realistic derivation based on the mechanism of energy generation or the one based upon fitting a line to data? Both have points in their favor. It could be accidental that the five points in Figure 1 fall nearly on a line. In this regard the descriptively realistic model is reassuring because it tells us that there is a good believable reason why we should expect the points to lie on a line. On the other hand the descriptively realistic model by itself is just theory, and it's always nice to have theory confirmed by data. The two models together make a strong combination.

Our next example shows how serious the lack of descriptive realism can be.

Example 2 Carcinogenesis (Real Data)

There are many factors involved in causing cancer, and one of them is exposure to certain toxic chemicals called carcinogens. Obviously the probability of getting cancer from exposure to a carcinogen depends on the amount of exposure, in other words the dose.

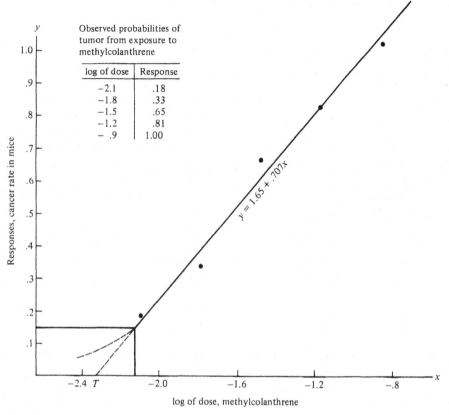

Figure 2

Observed probabilities of tumor from exposure to methylcolanthrene

log of dose	Response
−2.1	.18
−1.8	.33
−1.5	.65
−1.2	.81
− .9	1.00

$y = 1.65 + .707x$

log of dose, methylcolanthrene

Figure 2 shows some of the dose-response data obtained when mice were given injections of methylcolanthrene and then monitored for tumors throughout their lifetimes. Five separate groups of mice were used and each group got a different dose. In each group, the fraction of mice that developed tumors was determined. This fraction is the response listed in the table and plotted in Figure 2. If we plot this response against the log of the dose, we get five data points, which appear to be nearly along a straight line. We will shortly see how to compute the equation of this line. But for now the important thing to realize is that the method of finding this line is not based on any understanding of how methylcolanthrene produces cancer.

Can the line in Figure 2 be extrapolated down to the region of lower doses, the so-called low-dose rectangle? This is not an idle question. The doses we have data about (the five points in the figure) are very high, much higher than a normal mouse (or person) would encounter in daily life. Thus the low-dose rectangle might also be called the real-life rectangle. Unfortunately it is impractical to do the experiments needed to get data points in the low-dose rectangle. (One needs too many mice and too many dollars and too many years.)

At present, the answer to whether we can extrapolate the line into the low-dose rectangle depends on which cancer researcher you ask. Some would consider this reasonable. Notice that such an extrapolation would imply a positive threshold T, below which no cancer would occur (response = 0). Other researchers claim that the proper extrapolation is not to continue the straight line but to curve it, as shown in Figure 2. According to this point of view, the only way to reduce the incidence of cancer due to methylcolanthrene to 0 is to completely eliminate methylcolanthrene. At the time we are writing, neither side of this argument can prove its case with data. And, as we have already mentioned, it is highly unlikely that the data will ever become available. Therefore, the only way to get a reliable idea of what happens in the low-dose rectangle is to establish a descriptively realistic model of how methylcolanthrene acts in the body. If this model yields an equation for the dose-response curve, we might deduce what goes on in the low-dose rectangle.

The question posed here is a very urgent one for public policy. Federal regulatory agencies must set standards for how much methylcolanthrene (and other carcinogens) to allow in the environment, and they have no fully satisfactory way of doing it at present. Descriptively realistic models of carcinogenesis would help enormously.

With these examples in mind we now state the problem we wish to study in this section. We have a table of values which relates two variables x and y (for example, energy and cost or methylcolanthrene dose and cancer rate). We want to find an algebraic relation that holds, at least approximately, between x and y. As a first step, we think of the data as a set of ordered pairs: $(x_1, y_1), (x_2, y_2), \ldots,$ (x_n, y_n). Each line of the table gives one ordered pair. We plot these points in the xy plane. If the points seem to lie approximately in a straight line (this doesn't always happen), then we will fit a straight line to the points.

In order to find the best-fitting straight line, we first define the concept of the distance of a line to a set of points as follows:

If $P_1(x_1, y_1), P_2(x_2, y_2), \ldots, P_n(x_n, y_n)$ is a set of points, then the distance from this set of points to a line L is

$$D = \sqrt{d_1^2 + d_2^2 + \cdots + d_n^2}$$

where d_i is the vertical distance from P_i to the line L (see Figure 3). D is called the least-squares distance from the line to the points.

If we are given a line with equation

$$y = mx + b$$

we will now see how to compute D in terms of m, b, and the coordinates of the points P_i. It is convenient to rewrite the equation

$$y = m(x - \bar{x}) + b + m\bar{x}$$

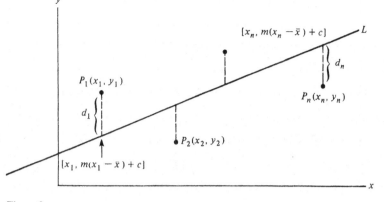

Figure 3

where $\bar{x}$ is the mean of the x_i's, i.e., $\bar{x} = (x_1 + x_2 + \cdots + x_n)/n$. Then, setting $b + m\bar{x} = c$, we have $d_i = |m(x_i - \bar{x}) + c - y_i|$, so that

$$D = \sqrt{[m(x_1 - \bar{x}) + c - y_1]^2 + \cdots + [m(x_n - \bar{x}) + c - y_n]^2}$$

Now to find the values of the slope m and intercept b which give the line closest to the points, we need to minimize D with respect to variables m and c. ($\bar{x}$, the x_i, and the y_i are known constants since the points are given.) It turns out that we get the same result, but in a simpler way, if we minimize D^2, a quantity we call S. To minimize S, we set

$$\frac{\partial S}{\partial m} = 2[m(x_1 - \bar{x}) + c - y_1](x_1 - \bar{x}) + \cdots + 2[m(x_n - \bar{x}) + c - y_n](x_n - \bar{x}) = 0$$

$$\frac{\partial S}{\partial c} = 2[m(x_1 - \bar{x}) + c - y_1] + \cdots + 2[m(x_n - \bar{x}) + c - y_n] = 0$$

These equations yield

$$m \sum_{1}^{n} (x_i - \bar{x})^2 + c \sum_{1}^{n} (x_i - \bar{x}) - \sum_{1}^{n} (x_i - \bar{x})y_i = 0$$

$$m \sum_{1}^{n} (x_i - \bar{x}) + nc - \sum_{1}^{n} y_i = 0$$

(4)

If we set $\bar{y} = (y_1 + y_2 + \cdots + y_n)/n$ and take into account the definition of $\bar{x}$, Equation (4) yields

$$m = \frac{\sum_{1}^{n} (x_i - \bar{x})y_i}{\sum_{1}^{n} (x_i - \bar{x})^2}$$

(5)

$$c = \bar{y}$$

Definition

The line $y = m(x - \bar{x}) + c$, where m and c are given by Equation (5), is called the *regression line* for the given data. Observe that it passes through the point $(\bar{x}, \bar{y})$. The process of finding and analyzing such lines is called *simple linear regression*.

The next example illustrates the arithmetic involved in computing m and c from real data points. As you study this example and the formulas in Equation (5), you will undoubtedly notice that, if n gets large, there will be quite a lot of arithmetic. For this reason regression lines are usually calculated by computer. Your computing center undoubtedly has a "canned program" for simple linear regression, which you can use. Some pocket calculators even have built-in routines to calculate the slope and intercept of a regression line: all you have to do is enter the co-ordinates of the points.

Example 3 Gas Efficiency of Cars (Real Data)

Current legislation requires automobiles to meet standards of gasoline efficiency e, measured in miles per gallon. Of course this figure varies with the speed s, measured in miles per hour, at which the car is driven. Thus, we expect there to be some function f, relating s and e, say, $e = f(s)$. What is the form of f? Is it linear $[f(s) = ms + b$ for suitable constants m and $b]$ or quadratic $[f(s) = as^2 + bs + c$ for suitable constants a, b, and $c]$ or something else entirely?

As an example, consider the data listed and plotted in Figure 4, which were obtained by the Federal Highway Administration in testing one particular car. (For the moment, disregard the asterisks.) Since the points appear to lie on a straight line, we will fit the points with a line, using the least-squares method, that is, plugging the data into Equation (5). In doing this, note that e plays the role of y in Equation (5) and s plays the role of x:

$$c = \bar{e} = \frac{18.25 + 20.00 + 16.32 + 15.77 + 13.61}{5}$$

$$= 16.79$$

$$m = \frac{\sum\limits_{1}^{n} (s_i - \bar{s})e_i}{\sum\limits_{1}^{n}(s_i - \bar{s})^2}$$

$$= \frac{(30 - 50)18.25 + (40 - 50)20 + (50 - 50)16.32 + (60 - 50)15.77 + (70 - 50)13.61}{(30 - 50)^2 + (40 - 50)^2 + (50 - 50)^2 + (60 - 50)^2 + (70 - 50)^2}$$

$$= \frac{-135.1}{1000} = -0.1351$$

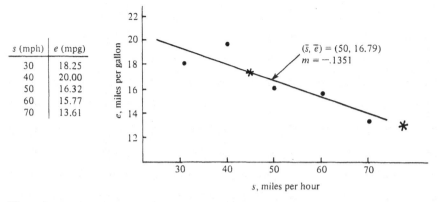

s (mph)	e (mpg)
30	18.25
40	20.00
50	16.32
60	15.77
70	13.61

Figure 4

Thus we plot a line of slope -0.1351 through the point $(\bar{s}, \bar{e})$, which is $(50, 16.79)$. To find the equation of the least-squares line, substitute m and c into $e = m(s - \bar{s}) + c$ to obtain

$$e = -0.1351(s - 50) + 16.79 \tag{6}$$

Notice that this model is not descriptively realistic because the factors which determine the car's performance, things like the number of cylinders, timing, carburetor adjustment, weight, and so on, do not appear in the model. In theory, it might be possible to deduce the speed-efficiency relationship by taking these factors into account and applying the laws of physics. This would be much more difficult, although possibly quite useful for auto companies.

Naturally, it is too much to expect that, any time we plot data points for a pair of variables, we will obtain a fairly neat linear relationship. Figure 5 shows data points for a statistical investigation of the 1836 presidential election made by Paul Lazarsfeld. The data were assembled to test a hypothesis put forth by Arthur Schlesinger, Jr., in "The Age of Jackson." Schlesinger contended that the tendency of voters to choose their party affiliation according to their economic class does not begin in the New Deal, but can be seen already in the election of Andrew Jackson.

Each data point represents a single election district. The horizontal coordinate is the average assessed valuation of property units in the district, and the vertical coordinate is the percentage of voters in that district who voted democratic. These data certainly show a tendency of persons lower on the economic scale to vote Democratic. But a cautious interpretation would avoid claiming a linear relationship: the least-squares line simply doesn't fit very well.

Sometimes a scatter diagram is not linear, but the points do appear to fall on some other type of curve. Figure 6 shows a graph of seasonal temperature change in Alaska. As shown, the points are fit well by a sine curve.

Figure 7a shows a scatter diagram in which the points appear to fall on a curve of the form $y = b + a \log x$. In this case, there are two approaches:

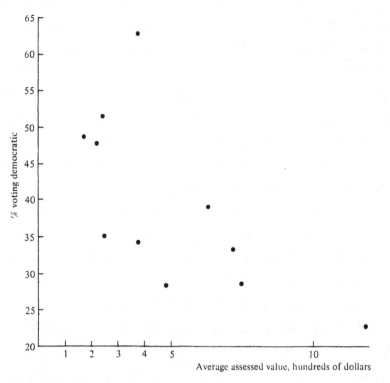

Figure 5 The 1836 presidential election.

1. One could develop formulas for the values of a and b that minimize the sum of the squares of the vertical distances of the points to the curve $y = b + a \log x$. This would involve minimizing

$$s = \sum_{1}^{n} (b + a \log x_i - y_i)^2 \tag{7}$$

by taking partial derivatives, setting them to zero, and solving for a and b.

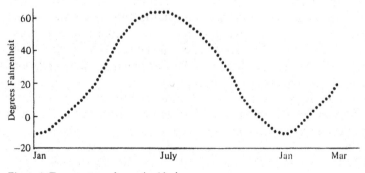

Figure 6 Temperature change in Alaska.

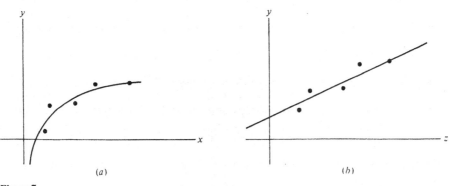

Figure 7

2. Here is a second approach which needs no new formulas. If a typical point (x_i, y_i) nearly satisfies $y = b + a \log x$, then (approximately)

$$y_i = b + a \log x_i \qquad i = 1, 2, \ldots, n$$

If we set $z_i = \log x_i$, then (approximately)

$$y_i = b + a z_i \qquad i = 1, 2, \ldots, n$$

Thus the points (z_i, y_i) lie approximately on a curve of the form $y = b + az$, a straight line. We can now seek the best-fitting line to the points (z_i, y_i), using linear regression by the least-squares method. The result will be a graph, like that of Figure 7b. In producing this graph, the horizontal axis is calibrated with

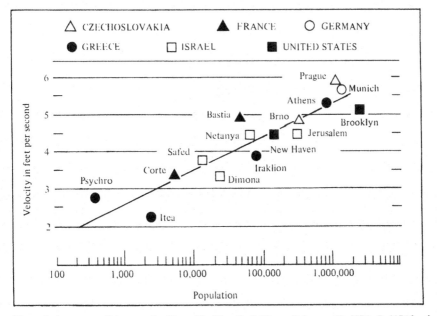

Figure 8 Average walking speeds. (*From The New York Times, February 29. 1976.* © *1976 by the New York Times Company. Reprinted by permission.*)

z units (1 square of graph paper = a fixed number of z units) and not x units. This is the procedure that was used in Example 2, where we compared the response to the log of the dose and not the dose itself.

In some cases where authors have converted from x to z units by means of a logarithm or some other function, the horizontal axis will be labeled in terms of x instead of z units. In this case the axis appears to have a nonuniform calibration (1 square of graph paper does not equal a fixed number of x units). This has been done in Figure 8, which attempts to show that people walk faster in bigger cities.

Example 4 Carbon Dating (Hypothetical Data)

Carbon 14 is a variety (isotope) of carbon which, over time, undergoes radioactive decay and becomes transformed into carbon 12. In the table in Figure 9 there are some data about the fraction f of an original amount of carbon 14 left after various numbers of years elapsed t.

Theory and experiment both suggest that there is a linear relationship between t and $\log f$. Therefore, we plot the ordered pairs $(t_i, \log f_i)$ in Figure 9. We set $y = \log f$; so these points can be thought of as (t_i, y_i). We now attempt to fit a regression line to these points, i.e., a line of the form $y = m(t - \bar{t}) + c$. We compute the mean of t and y: $\bar{t} = (5 + 6 + 7 + 8 + 9)/5 = 7$; $\bar{y} = (-0.62 - 0.76 - 0.87 - 0.99 - 1.1)/5 = -0.87$. According to formula (5) and with t in place of x,

$$
m = \frac{\begin{array}{c}(5 - 7)(-0.62) + (6 - 7)(-0.76) + (7 - 7)(-0.87) \\ + (8 - 7)(-0.99) + (9 - 7)(-1.1)\end{array}}{(5 - 7)^2 + (6 - 7)^2 + (7 - 7)^2 + (8 - 7)^2 + (9 - 7)^2}
$$

$$
= -0.12
$$

$$
c = \bar{y} = -0.87
$$

Therefore the line is

$$
y = -0.12(t - 7) - 0.87
$$
$$
= -0.12t - 0.03 \tag{8}
$$

This is the line shown in Figure 9. Notice that the vertical axis is marked with two different scales, one in y units and the other in f units. The f-units scale is not uniform: 1 square of graph paper doesn't equal a constant number of f units.

The relationship between f and t is obtained from Equation (8) as follows:

$$
y = \log f = -0.12t - 0.03
$$
$$
f = e^{-0.12t}e^{-0.03}
$$
$$
= 0.97e^{-0.12t} \tag{9}
$$

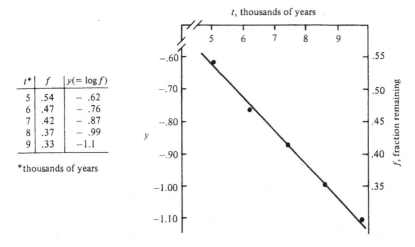

t^*	f	$y(= \log f)$
5	.54	$-.62$
6	.47	$-.76$
7	.42	$-.87$
8	.37	$-.99$
9	.33	-1.1

*thousands of years

Figure 9 Radioactive decay.

Equation (9) is not the scientifically accepted description of carbon-14 decay. The accepted relationship is

$$f = 1.00e^{-0.1210t} \tag{10}$$

This is fairly close to Equation (9), produced by regression. Our regression equation actually fits the data better than Equation (10). The reason Equation (10) is preferred has to do with the theory behind radioactive decay: it seems easier to believe that there were measurement errors in the numbers in the table than to give up the theory. For an outline of this theory see Example 8 of Section 3A and Exercise 11 of Section 3B in Chapter 5. The theory requires a formula of the form

$$f = 1.00e^{mt}$$

Alternatively

$$y = \log f = mt$$

Thus the points (t_i, y_i) should be fit by a line of the form $y = mt$ instead of the form $y = m(t - \bar{t}) + c$ used above. We leave the details of this approach as an exercise.

The radioactive decay of carbon 14 is the basis of the carbon-dating method, which is used to find out how old certain objects are. Suppose one has dug up a wooden totem pole in an archeological excavation and wants to know how old it is. Wood, like all other living matter, contains carbon 14. The decay of carbon 14 into carbon 12 begins when the living organism, in this case a tree, dies. Presumably the tree died when it was cut down to make the totem pole. The amount of carbon 14 present at that time can be determined by examining existing trees today. By measuring the amount of carbon 14 left in the totem pole, one can find f, the proportion left. Then one can use Equation (10) to solve for the age t.

What good is a regression line and its equation? Two common uses we will illustrate are:

1. To interpolate or extrapolate from known data points
2. To uncover associations between variables, which may sometimes be causative links.

Example 5

We return to our automobile testing in Example 3. Only five speeds were tested. Testing is, after all, time consuming and expensive, and there are many cars to test. If we want mileage efficiencies for speeds like 45 or 80 miles/hour, which were not tested, we get them from Equation (6).

When we do this for a value which falls between existing data values, such as 45 which falls between 40 and 50, we call this *interpolation*. For $s = 45$, we get

$$e = -0.1351(45 - 50) + 16.79$$
$$= 17.47$$

When we use Equation (6) to make a prediction of e for an s value, like 80, which lies beyond any s value we have data for, this is called *extrapolation*. When $s = 80$, we predict

$$e = -0.1351(80 - 50) + 16.79$$
$$= 12.74$$

The predicted data points (45, 17.47) and (80, 12.74) which we have just calculated, are shown by asterisks in Figure 4. It is important to realize that these are only predictions. If we actually tested these speeds, we would probably discover that the true values lie a little above or below the line, just as the five originally given data points do. (In the statistical aspects of the theory, we can calculate the probabilities of greater or lesser deviations from the regression line.) Extrapolation is usually considered a riskier endeavor than interpolation and should be done with great care. Figure 10 illustrates why.

The equation of the regression line for the fastest time for running the mile is (approximately)

$$\text{Minutes} = 16.562 - 0.00645(\text{calendar year})$$

This implies that, when calendar year $= 16.562/0.00645 = 2568$, it will take 0 minutes to run the mile—obviously a ridiculous prediction. A more conservative extrapolation to the year 2001 predicts a 3:40 mile for that year. Even this is pretty speculative.

The problems inherent in extrapolation are not always so amusing; in Example 2 (carcinogenesis) it is a life and death matter affecting all of us.

In the next example we illustrate the second common use of regression analysis, uncovering causative factors between variables.

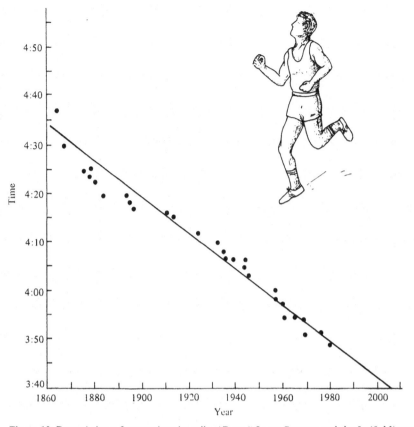

Figure 10 Record times for running the mile. (*From "Least Squares and the 3:40 Minute Mile," by C. O. Oakley and J. C. Baker, Mathematics Teacher, April 1977. Reprinted by permission of the National Council of Teachers of Mathematics.*)

Example 6 Juvenile Diabetes (Real Data)

Diabetes is presently (1983) a very serious health problem in the United States. It is the fifth leading cause of death and the second leading cause of blindness. Juvenile diabetes is an especially serious form of the disease. There are over 1 million juvenile diabetics, and almost all must take insulin each day. It would be a great boon to be able to prevent juvenile diabetes.

Lately there has been an increasing conviction on the part of medical researchers that the mumps virus plays a role in causing some cases of diabetes. Odd bits of evidence for this idea have been around since 1864, but only recently has the evidence become strong enough to lift the idea out of the "harebrained" category into the "very likely" category. This example presents some of the modern evidence.

When H. A. Sultz and his associates studied the incidence of mumps in Erie County, New York, and compared the national incidence of new cases of juvenile diabetes, they found that peak years of juvenile diabetes followed about 4 years

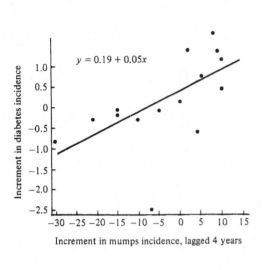

Year*	Yearly increments in incidence rates, per 100,000 people	
	Mumps	Diabetes
1948	−30	−0.9
1949	−15	−0.3
1950	−10	−0.4
1951	5	0.6
1952	10	1.0
1953	10	0.3
1954	0	0
1955	0	0
1956	9	1.2
1957	8	1.6
1958	2	1.2
1959	−5	−0.2
1960	−15	−0.2
1961	−7	−2.5
1962	−21	−0.4
1963	4	−0.7

*Years correspond to diabetes column. Mumps data are from 4 years earlier.

Figure 11 Increment in mumps incidence, lagged 4 years.

after peak years of mumps. It appeared that a jump in mumps incidence might cause a jump in diabetes incidence 4 years later. This suggests that, if D is the increase (or decrease if D is negative) of new diabetes cases over the previous year and M is the increase (or decrease if M is negative) of mumps cases over the year previous to 4 years ago, then there is some sort of relationship between M and D, say $D = f(M)$. In the table accompanying Figure 11 we list the M and D values for these diseases, with the diabetes data moved backwards 4 years so that the numbers we wish to compare appear on the same line of the table. From each line of the table we create the ordered pair (D, M) and plot the resulting points, as shown in Figure 11.

Notice that the points are not tightly clustered around the regression line. In a case like this, where the points are widely scattered and show only a weak tendency to follow the line, there are two possible interpretations.

1. The variables are not related. The small degree of clustering about the line is an accident or just wishful thinking on the observer's part.
2. The variables are related, but there are other factors that muddy the waters. In this example, it may be that there are other causes of juvenile diabetes that contribute to the variable D. Indeed, that is what medical researchers think: for example, they suspect that the measles virus is a causative agent for persons of certain genetic types.

In this case, we have other evidence that helps us decide in favor of hypothesis 2. Medical researchers have caught the virus in the act, so to speak. Prompted by data such as those in Figure 11, experimenters have cultured human pancreas cells in the laboratory and tried to infect them with the mumps virus. Intensive observation of the results provided evidence that the mumps virus can affect pancreas cells in a way that might trigger juvenile diabetes. This experiment does not give a complete picture of the development of juvenile diabetes: it doesn't explain why some people get juvenile diabetes without getting mumps or why some people don't get juvenile diabetes even though they have had mumps. Nevertheless, this experiment and others like it take us a definite step toward a descriptively realistic model of the causation of juvenile diabetes. Ironically, such experiments are often inspired by regression models which are not descriptively realistic.

Sidelight: What Is Real Anyhow?

Descriptive realism is a concept that gets queerer and queerer the closer one examines it. The problem is that people may disagree about what makes something tick and it may be impossible to prove who is right. Take a clock for example. We "know" that what makes it tick are gears, springs, and other mechanical parts and sometimes a battery or other electric power source. If we wished to describe the operation of all these parts mathematically, we would have a model that was descriptively realistic.

But suppose we imagined instead that our clock was inhabited by a little gremlin who turned the hands. We can suppose that the gremlin has some way of tuning in to some cosmic clock, so that the gremlin always knows the exact time, and has the physical capability of turning the hands at the required rate, in a perfectly smooth and even manner. You are probably thinking this is absurd. Of course it is, but why is it absurd? The gremlin model gives the "right answer," that is, it predicts that the clock will operate in exactly the way that is actually does operate. So we can't fault it on the grounds that its predictions contradict reality.

Perhaps you are thinking that we can reject the gremlin model because, when we open the clock, we see gears and machinery and not a gremlin. But we could add to the gremlin model the assumption that the gremlin turns into gears and machinery when viewed by a human being. For any objection you might make to the gremlin model, the gremlin could be given more magical properties that would overcome your objection.

This example may seem like something contrived just as a joke, but the fact is that, all through history, various sciences have been afflicted with problems similar to this.

1. Ancient people looked at the patterns of the stars in the heavens and thought they saw a bull (Taurus), a lion (Leo), an archer (Sagittarius), and so on. They invented stories about these figures that helped explain various natural phenomena. The early Greeks explained the rising and setting of the sun by supposing that Apollo pulled it across the sky. But after a while the gremlin (Apollo) was

taken out of the model, and people began to believe that the rising and setting of the sun was accomplished without anyone pulling anything.

2. In the nineteenth century, Charles Darwin proposed the theory of evolution, which held that higher life forms evolved from lower ones. The opposite view, called creationism, held that God created all living things in the form they have today. The argument still goes on. Extreme evolutionists find creation by God no more believable than the gremlin in our clock. Extreme creationists don't find it believable that our forefathers were little creatures swimming around in a pond of slimy muck. There are also shades of opinion between the extremes.

3. In physics, that most hardheaded of disciplines, there was a crisis of descriptive realism in the early twentieth century. The quantum theory held that the behavior of an electron is uncertain and unpredictable in principle (not just in practice). In effect, the electron flips a coin to see where it is going next. When first proposed, this uncertainty principle seemed unacceptable to many physicists because it seemed to fly in the face of one of science's most cherished assumptions—the assumption that if only we knew enough about the universe we could predict its future course. Gradually the uncertainty principle won adherents because it did a better job of fitting the experimental facts than any other model. But Einstein remained a significant holdout and never believed it. He simply didn't think the world really worked that way (although he did agree that it appeared to). He expressed his opposition in the well-known dictum: "God does not play dice with the Universe."

This problem in physics has never completely resolved itself. To a great extent physicists have reacted by abandoning the attempt to decide what is realistic and settling instead for models that do a decent job of prediction. Here is Niels Bohr, one of the defenders of the uncertainty principle, essentially admitting that he is not concerned about realism:

> It is wrong to think that the task of Physics is to find out how nature is. Physics concerns what we can say about nature.

By contrast, Einstein never abandoned the quest for descriptive realism, as expressed in the following remark:

> Physics is an attempt conceptually to grasp reality.

4. At present there is great debate in psychology about the nature of motivation. Why do we go to a soccer game? In ordinary language we might say that we do this because we *feel like it*, because we *like* soccer games and going is an exercise of our *free will*. But psychologists in the behaviorist school (for example, followers of B. F. Skinner) would say that using these internal mental states to predict behavior is not descriptively realistic. They do not deny the existence of feelings, but they say that it is unrealistic to think of them as the causes of our behavior. (Perhaps the following inexact analogy will clarify this: imagine a

clockmaker who believes in gremlins but doesn't think they are involved in making clocks work.)

It would be nice to think that science can eventually find out what is real and stop these arguments. But in the past it has not always worked out that way. Here is James Clerk Maxwell, the founder of the mathematical theory of electromagnetism, speaking about the "real" nature of light:

> There are two theories of the nature of light, the corpuscle theory and the wave theory; we used to believe in the corpuscle theory; we now believe in the wave theory because all those who believed in the corpuscle theory have died.

EXERCISES

● **1** The following data concern the growth of a plant after grafting:

Months after grafting	1	2	3	4	5	6
Height, inches	0.8	2.4	4.0	5.1	7.3	9.4

Fit a least-squares line and use it to predict the graft's height at $4\frac{1}{2}$ months. Now predict the height after 5 years. Which prediction is more reliable? Can you give specific reasons why one might be more reliable?

2 The table below shows winning speeds (to the nearest mile per hour) for the Indianapolis 500 Auto Race from 1961 to 1970. There is a slight increase in speed over the years, due to improvements in technology. Plot the ordered pairs (year, speed) and fit a regression line. You can save yourself some trouble by replacing the variable "year" by "year − 1960." Check last year's time from an almanac or newspaper. How does it compare with what you would extrapolate from the regression line?

Year	1961	1962	1963	1964	1965	1966	1967	1968	1969	1970
Winning speed	139	140	143	147	151	144	151	153	157	156

● **3** As one digs down into the earth, the temperature varies, a matter of great interest to geologists and mining engineers. The table below shows data taken from a well at Grenoble, France. The first point measured is 28 meters below the surface; this is partly because closer to the surface the temperature varies with the seasons. Temperatures are recorded as increases over the temperature at this basepoint. Find the regression line, expressing degrees as a function of depth.

Meters below surface	28	68	178	248	298
Degrees above basepoint temperature	0	1.2	4.7	9.3	10.5

4 During World War II the U.S. Navy compiled monthly reports on enemy submarines sunk. These reports were often not entirely accurate. (Can you think of reasons for that?) When compared with the numbers actually sunk, the following table emerges. Find the best-fitting regression line, expressing the number actually sunk as a function of the number reported sunk.

Month	1	2	3	4	5	6	7	8	9	10	11	12	13	14	15	16
Reported sunk	3	2	4	2	5	5	9	12	8	13	14	3	4	13	10	16
Actually sunk	3	2	6	3	4	3	11	9	10	16	13	5	6	19	15	15

5 In studying how birds fly, researchers find it useful to compare the length of the wingspan W and the length of the humerus H, which is one of the bones in the wing. The ordered pairs (W, H) were obtained for 139 birds and plotted. The curve whose equation is $H = 0.0419W^{1.138}$ gives a very good fit. What change of variables would you make in order to display this relationship as a linear equation? What linear equation do you obtain?

6 Aflatoxin, a substance which sometimes occurs in peanuts, beans, and other vegetables, is one of the most potent carcinogens known. The table below shows dose-response data collected at six different locations. Plot the six ordered pairs (log of dose, cancer rate) and find the regression line.

Location	Daily amount of aflatoxin eaten, nanograms per kilogram of body weight		Response, liver-cancer rate per 100,000 people
	Dose	Log of dose	
Kenya (high altitude)	3.5	0.54	0.7
Thailand 1	5.0	0.70	2.0
Kenya (middle altitude)	5.8	0.76	2.9
Kenya (low altitude)	10.0	1.0	4.2
Thailand 2	45.0	1.65	6.0
Mozambique	222.4	2.35	25.4

7 The stopping distance for an automobile depends on its speed at the time the brakes are applied. Here are some data for a particular car.

Speed, miles/hour	20	30	40	50	60
Stopping distance, feet	14.53	32.90	71.67	122.70	167.13

A straight-line fit is not very satisfactory. The laws of physics suggest that the square root of the distance ought to be linearly related to speed. Carry out a simple linear regression analysis along these lines (with speed on the horizontal axis).

8 Suppose the data points for the temperature variation example (Figure 6) are (d_i, t_i), d standing for date and t for temperature. Set up an equation, analogous to Equation (7), which expresses the distance of the points to a curve of the form $t = a \sin(b + cd) + e$. Find four equations which a, b, c, and e must satisfy in order to minimize the distance.

● **9** If we extrapolate the dose-response regression line for methylcolanthrene (see Example 2) into the low-dose region, what is the point where it crosses the horizontal axis (the threshold T in Figure 2)? What dose does this correspond to?

10 Suppose it turns out that every positive dose of methylcolanthrene (see Example 2), no matter how small, has a nonzero probability of causing cancer in a person (or mouse). What does this mean about where the dose-response curve crosses the horizontal axis? (*Be careful:* The scale on the horizontal axis represents the log of the dose, not the dose itself.)

● **11** Suppose you collected data from various workers in a factory on how long it takes them to drive to work and how many miles away they live. You plot the ordered pairs of the form (distance, time) and find the regression line. Would you expect it to go through the origin? Explain your answer.

12 Suppose you were looking for a regression line to fit a set of points $P_i(x_i, y_i)$, $i = 1, \ldots, n$, and you were only interested in considering lines of the form $y = mx$ (lines through the origin). The search for the best-fitting line of this form becomes a maximization problem in just one variable, m. Carry out this analysis and show that the formula for the optimal m is

$$m = \frac{\sum\limits_{1}^{n} x_i y_i}{\sum\limits_{1}^{n} x_i^2}$$

13 There are many causes of leukemia, among them exposure to radiation. The leukemia rate from all causes *except* radiation is something like 60 cases per year per million people. Suppose one collects data on leukemia rates from 20 communities with different radiation exposures (on account of different altitudes or proximity to sources of radiation, such as uranium mines, weapons testing areas, nuclear power plants, etc.). You intend to plot the data and fit a curve by regression. How do you cope with the fact that not all leukemia is due to radiation?

14 In finding the best-fitting line for a set of points, we assume that this line has equation $y = mx + b$, where m and b are to be determined. But there is an infinite family of lines which are being left out because they don't have this form. The ones left out are the vertical lines, those with equations of the form $x =$ constant. Is it reasonable to leave these out of the analysis? If so, why?

● **15** The way in which we defined the distances d_i in our definition of the least-squares distance D seems a little arbitrary. For example, we might instead define d_i to be the horizontal distance from P_i to the line.

(*a*) Under what conditions on the slope of the line would this alternative definition of d_i give the same numerical value of D as the definition we actually use?

(*b*) Under what conditions will the alternative definition always give larger values of D?

(*c*) Can you see any reason why the alternative definition should not be used instead of the one we do use?

16 Suppose we take the data on gas efficiency and speed from Example 3 and form a scatter diagram in a slightly different way. Instead of taking speed as the first coordinate and plotting it on the horizontal axis, suppose we take speed as the second coordinate and plot it on the vertical axis. If we now calculate the regression line in the standard way, will we get the same equation relating efficiency and speed as we obtained in Example 3? If not, then how would you choose which procedure to use?

17 For which example in this section would you say it is most important to have a descriptively realistic basis for the regression equation? For which example is it least important?

Computer exercise

18 Write a computer program to calculate the equation of a regression line. Input to the program should be a table of values for two variables (e.g., two arrays) and a description of which variable is to be expressed as a function of the other.

BIBLIOGRAPHY

Freedman, David, Robert Pisani, and Roger Purves: "Statistics," W. W. Norton, New York, 1978. Part III introduces the statistical side of regression (which we have left out), but with next to no mathematical technicalities.

Hoel, Paul G.: "Introduction to Mathematical Statistics," John Wiley & Sons, Inc., New York, 1971.

Lando, B. M.. and C. A. Lando: Is the Graph of Temperature Variation a Sine Curve? *Math. Teacher*, vol. 20, no. 6, September, 1977.

Oakley, Cletus O.. and Justine C. Baker: Least Squares and the 3:40-Minute Mile, *Math. Teacher*, vol. 70, no. 4, April, 1977.

3 DESCRIPTIVE REALISM—CORRELATION IS NOT CAUSATION

Abstract A correlation between two variables doesn't prove that one is a cause of the other. There are other possible explanations for the correlation. This section gives examples of the other possibilities.

Prerequisites Some knowledge of scatter diagrams (as in the previous section).

One of the main purposes of statistics is to find cause and effect links between variables. But this can be treacherous. For example, reliable evidence has been collected in northern Europe to show that, in years with many births, the stork population was higher than in years with few births. Do storks bring babies after all? There is, of course, a different explanation. Read on!

In this section we discuss the various conclusions we may draw when two variables x and y show an *association* or *correlation*. This means that either of the following situations applies:

1. High values of x tend to be associated with high values of y. This is called *positive correlation* (see Figure 1a).
2. High values of x tend to be associated with low values of y. This is called *negative correlation* (see Figure 1b).

Sometimes an association between two variables is displayed just by a table of values, as in Example 1 below. In other cases, as in Example 2, the association may be displayed by a scatter diagram of the points (x_i, y_i). In addition, the regression equation may be computed. Finally, it is also possible to compute a measure of the strength of the association, called the *correlation coefficient* (see Section 4), denoted R, or its square R^2. We report R^2 for some of the examples below. R^2 is always a number between 0 and 1. A value near 0 indicates a weak association, while a value near 1 shows a strong association.

No matter how the data is displayed or what else is computed, the problem remains: what conclusion can we draw about causation?

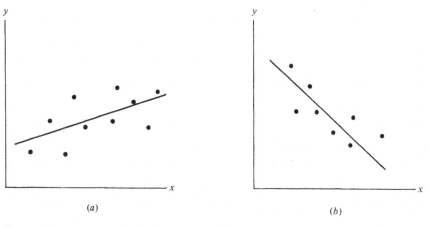

(a)   (b)

Figure 1

Example 1 Heart Disease and Cholesterol (Real Data)

Cholesterol is a substance naturally produced by the human body, which has been much studied in connection with coronary heart disease (CHD). There are various kinds of cholesterol. The type called low-density lipoprotein cholesterol (LDL) seems to play a role in clogging up the arteries in persons with CHD. By contrast, high-density lipoprotein cholesterol (HDL) does not appear harmful and may even be helpful. The data below concern HDL. There seems to be a negative correlation between HDL cholesterol level and CHD rate.

Men were divided into groups according to HDL cholesterol levels in their bloodstream. In each group the percent with CHD was determined. Table 1 shows a tendency for high HDL levels to be associated with low CHD levels. Does this mean HDL cholesterol prevents heart disease?

Table 1 Prevalence of CHD at different HDL cholesterol levels in men aged 50 through 69

HDL cholesterol level, milligrams/deciliter	CHD rate, per 1000
Less than 25	180
25–34	123.6
35–44	94.6
45–54	77.9
55–64	77.9
65–74	79.1
75 plus	85.5

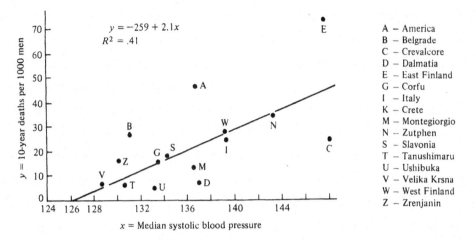

$y = -259 + 2.1x$
$R^2 = .41$

A — America
B — Belgrade
C — Crevalcore
D — Dalmatia
E — East Finland
G — Corfu
I — Italy
K — Crete
M — Montegiorgio
N — Zutphen
S — Slavonia
T — Tanushimaru
U — Ushibuka
V — Velika Krsna
W — West Finland
Z — Zrenjanin

$y = $ 10-year deaths per 1000 men

$x = $ Median systolic blood pressure

Figure 2 Mortality and blood pressure, an international comparison. (*From "Seven Countries," by Ancel Keys, Harvard University Press, 1980. Reprinted by permission.*)

Example 2 Heart Disease and Blood Pressure (Real Data)

Figure 2 is a scatter diagram, displaying an association between high blood pressure and coronary heart disease (CHD). Each point represents a different geographical area. Within each of the 16 areas a large group of men was tested to determine the average systolic blood pressure for the group x and the (age-standardized) rate y at which these men developed CHD within 10 years. The points represent the ordered pairs (x, y) for the 16 areas. There seems to be a positive correlation.

Does this scatter diagram demonstrate that high blood pressure causes CHD? Or may it be that CHD causes high blood pressure? Does the calculation of the regression line and R^2 help you answer these questions?

Here are the various possible reasons two variables may be associated with a positive or negative correlation.

I. Simple Causality

The level of the x variable determines the level of the y variable; i.e., x causes y. We represent this by an arrow leading from x to y as in Figure 3. As a slight variant of this, there may be a chain of causation starting at x and involving intermediate variables before getting to y as in Figure 4.

In Example 2 it seems likely that we have a chain of causation from high blood pressure to CHD. The evidence includes the fact that, in experimental animals, when we raise their blood pressure the animals develop atherosclerosis. Atherosclerosis involves clogging of the arteries with cholesterol and other substances. When the coronary artery, which supplies the heart with oxygen, gets too clogged up, CHD results from the lack of oxygen. See Figure 5. In addition to this experi-

x ———————→ y **Figure 3**

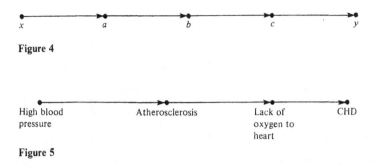

Figure 4

Figure 5

mental evidence, there is a believable mechanism which has been proposed to show how high blood pressure may cause atherosclerosis. This mechanism is the filtration hypothesis, which holds that higher blood pressure forces cholesterol (or some kinds of it) into the wall of the artery, where it becomes embedded and serves as the base for more deposits.

II. Reverse Causality

The level of the y variable determines the level of the x variable; i.e., y causes x. See Figure 6. In Example 2 reverse causality would mean that CHD causes high blood pressure, instead of high blood pressure causing CHD.

x y **Figure 6**

In a way, there is no difference between reverse causality and simple causality: in both cases one variable determines the level of the other. But sometimes there is a psychological difference. Often you have a preconceived notion that one thing causes another. To try to prove this, you might assemble statistics about the alleged cause x and the alleged effect y. If you find a correlation, you may think it helps prove your suspicion. Jumping to such a conclusion is psychologically tempting but not always wise. It may be that what you think is the cause is really the effect and vice versa. Here is an example.

Example 3

Primitive people living in the New Hebrides Islands of the South Pacific were aware of the fact that healthy people generally had body lice, while those in poor health didn't. Although numerical data were not collected, if they had been, one might have obtained a graph like that of Figure 7.

The islanders concluded that body lice were a cause of good health. More sophisticated scientific observers came to the opposite conclusion: good health caused the presence of body lice. Body lice were so common that most people had some: those in poor health had fewer because lice are fussy and prefer healthy hosts.

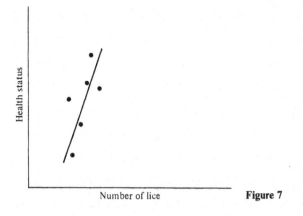

Number of lice   **Figure 7**

It seems likely that the correlation between the human birth rate and the stork population is a case of reverse causality: babies bring storks. More precisely, in years of high birth rates, more houses were built to accommodate the growing population. Storks nest under the eaves of houses; so the more houses there are, the more storks that can be accommodated. The availability of shelter, therefore, causes an increase in the stork population.

III. Mutual Causality

This is the case where changes in the x variable will produce changes in the y variable and vica versa. This is also called *feedback*. See Figure 8.

x y **Figure 8**

Example 4 Sales and Advertising

If you collect figures on total sales S and advertising budget A from 100 automobile dealers, you will undoubtedly discover a positive correlation: higher values of S go together with higher values of A. But why? One possibility is that advertising raises sales. There must be some truth to this, or else firms wouldn't advertise. But it is also true that firms which sell lots of cars are more easily able to afford more advertising. There is no need to choose between these explanations. It seems likely that both are true.

IV. Hidden Variable Causes Both

In this case there is a third variable z which causes changes in both x and y as in Figure 9. Changes in x which leave z unaffected will also leave y unaffected. Changes in y which leave z unaffected will also leave x unaffected.

Figure 9

Example 5 Liquor and Religion

Suppose you collect and plot two pieces of data from each of 20 cities: number of churches C and number of liquor stores L. The result will look like Figure 10.

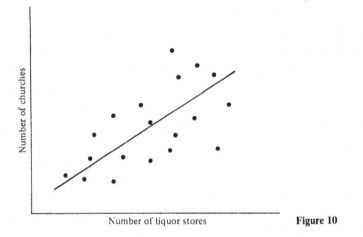

Number of liquor stores   **Figure 10**

It doesn't seem likely that religion drives people to drink or that demon rum drives people to church. The solution is that both L and C are determined by a third variable, the population size of the city. See Figure 11. Larger cities have more liquor stores and more churches.

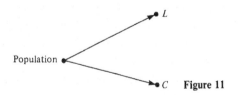

Population   **Figure 11**

Example 6 Hay Fever and the Price of Corn

It has been discovered that the price of corn in the United States in the fall of the year is negatively correlated with the severity of hay-fever cases. It turns out that neither of these factors directly influences the other. The explanation is that both are caused by weather conditions. Weather that is good for growing things will produce large corn crops, which makes corn cheaper. But these weather conditions

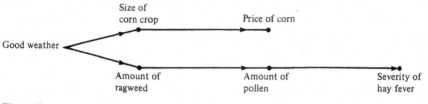

Figure 12

that are good for corn are also good for ragweed and other plants responsible for producing the pollen that causes hay fever. See Figure 12.

V. Complete Accident

It is possible that none of the above explanations applies and it is best to think of variables x and y as being totally unrelated as in Figure 13.

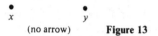

(no arrow)   **Figure 13**

Example 7 Rolling Dice

Suppose you roll a pair of dice five times and record the outcomes as in the tables in Figures 14 and 15. Figure 14 and the accompanying table show the results the first time I tried this. A quick glance at Figure 14 suggests that the correlation is low (calculation shows $R^2 = 0.035$), which is what we expect since there is no causal relationship involving the two dice. But we carried out 99 more trials of this nature

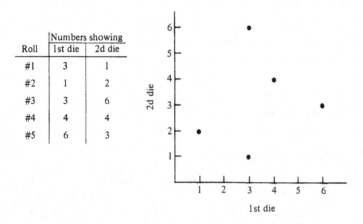

| Roll | Numbers showing | |
	1st die	2d die
#1	3	1
#2	1	2
#3	3	6
#4	4	4
#5	6	3

Figure 14

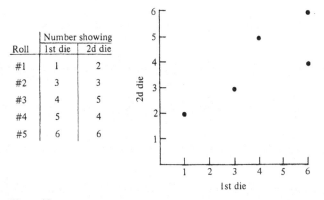

Roll	Number showing 1st die	2d die
#1	1	2
#2	3	3
#3	4	5
#4	5	4
#5	6	6

Figure 15

(simulating the process on a computer), and among the 100 trials there were some that showed a strong correlation (value of R^2 near 1). Figure 15 and the accompanying table show the best trial, one in which $R^2 = 0.672$.

A value this close to 1 normally makes us take notice and speculate about whether there may be some cause and effect relationship between the outcomes of the two dice. But this obviously doesn't fit into any of the four categories previously considered (simple, reverse, mutual, and hidden-variable causality). The correlation is just accidental.

Accidents like this can clearly happen even when there are no dice involved. For example, if we select five students and record their heights and social security numbers, we might discover a correlation that wouldn't hold up if we selected a different group of students. The advanced study of correlation and regression includes techniques to measure the likelihood of such chance correlations. As you might expect, the likelihood is less when many observations are taken. For example, if you use 100 students, you are much less likely to get accidental correlation.

Deciding which of explanations I through V fits a particular case is often very difficult. There are no cut and dried ways to make that decision: what is needed is good judgement based on wide knowledge of the variables which are correlated.

Regardless of which of these explanations applies, it is important to keep in mind that there may be other factors that you are completely unaware of that play a role, as in the following example.

Example 8 HDL and CHD

In Example 1 we explained that it seems possible (though not certain) that high levels of HDL in a person's bloodstream lower the risk of CHD. Suppose for a minute this is true. It surely is not the only factor. Exercise, genetic factors, stress, and other things we are not yet aware of may be involved as well. Thus Figure 16*b* is possibly a better summary of the situation than Figure 16*a*.

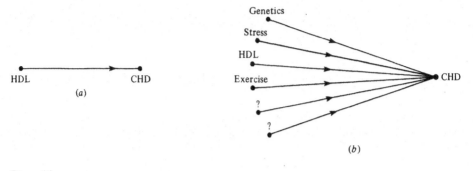

Figure 16

Of course it may not always be practical to find every contributory cause and all the chains of causation involving these causes . Consider the following analysis by Lecomte Pierre du Nouy:

> Let us take a cannon shot, for instance. Shall we say that the firing of a shell is "caused" by the small percussion cap, or by the movement of the soldier's hand which has pulled the string? Shall we say that the cause is the charge of powder?...
>
> Neither can we say that the workers who manufactured the powder, or the chemical engineers who invented it, or the builders of the factory, or the capitalists who gave the funds to build it, or their parents or their grandparents, etc., are responsible. And yet each one of them . . . shoulders part of the responsibility, which gradually crumbles away, without ever disappearing totally, and reaches back to the origin of the world.

This is fine from a philosophic point of view, but not so fine if you have the practical task of preventing the cannon from being fired. In that case you should be concerned with finding out which causes are the most important and which you can exert the most influence on. For a more interesting modern example, consider the problem of reducing crime rates. The cause of crime is criminals. Tracing back further, the things which cause people to become criminals undoubtedly include certain kinds of family backgrounds, poor social conditions, etc. Making a complete list of all the chains of causation is much less useful than finding a few on which one could have some leverage for change.

EXERCISES

In many of the following exercises it is hard to be sure what the correct answer is. Therefore your answer should include the reasoning behind it. Answers should be judged on the basis of this reasoning.

● **1** In a study of traffic flow through the Lincoln Tunnel in New York it was discovered that there was a negative correlation between the speed and density of the traffic stream: the faster the traffic, the fewer vehicles per mile. The table below has some of the data collected.

5-minute interval, p. m.	Average density, vehicles/mile	Average speed, miles/hour
4:30–4:35	69	16.8
4:35–4:40	88	14.8
4:40–4:45	65	19.1
4:45–4:50	84	14.9
5:10–5:15	63	21.6
5:30–5:35	53	25.7
6:05–6:10	82	16.8
6:10–6:15	53	25.1
6:15–6:20	70	19.7
6:20–6:25	34	31.8

(a) Which of explanations I through V discussed in this section seems most reasonable?

(b) Calculate the regression line.

2 If you interview a group of fathers and determine, in each case, the father's height and the height of his (full-grown) son, you will probably discover a positive correlation. In one such study the R^2 value was 96.7 and the regression equation turned out to be

$$\text{Son} = 0.53 \text{ father} + 32.8$$

(a) Which of explanations I through V discussed in this section seems best to account for this?

(b) What other variables are contributory causes?

● **3** The statistician of a professional basketball team makes a table which records, separately for each player, the number of fouls committed last season and the number of points scored last season. Suppose the statistician discovers a positive correlation. What would be the most reasonable explanation?

4 In trying to interpret the mumps and diabetes data of Example 6 on page 163, it seems reasonable to rule out diabetes as a cause of mumps. Reread the example and explain why.

● **5** In the study of social strife, data have been collected from various countries to estimate, for each country, the levels of two variables: x = the number of person-days of strife and y = the number of deaths due to strife. There is a positive correlation. An obvious explanation is simple causality: strife causes violence which causes death. Can you make an argument that reverse causality also operates here?

6 X is said to be a *necessary cause* of Y if Y never occurs without X occurring, e.g., the ignition of gasoline is a necessary cause of my car starting. X is said to be a *sufficient cause* of Y if the occurrence of X will, by itself, cause Y; e.g., teasing a rattlesnake is a sufficient cause of being attacked (but it is not a necessary cause). Examine Examples 1 through 8 and Exercises 1 through 5 of this section and see if there are any cases where a necessary or sufficient cause is operating.

Computer exercise

7 (a) Write a computer program to simulate rolling a pair of dice five times (using a random-number generator). Have the program print out a table like the one in Figure 14.

(b) Modify the program so it computes the coefficients a and b in the regression line $s = af + b$, where s = the outcome of the second die and f = the outcome of the first die. (See Section 2 for the procedure.)

(c) Modify the program so it computes R^2 by the formula

$$R^2 = \frac{\sum_{1}^{5} (\hat{s}_i - \bar{s})^2}{\sum_{1}^{5} (s_i - \bar{s})^2}$$

where s_i = the second die outcome on the ith roll

$\bar{s}$ = mean of the s_i's

$\hat{s}_i = af_i + b$, where f_i is the first die outcome on the ith roll and a and b are as found in part (b)

(d) Modify your program further so that it loops through steps (a), (b), and (c) 100 times (without printing results) and then prints the table which gave the highest value of R^2.

BIBLIOGRAPHY

Campbell, Stephen K.: "Flaws and Fallacies in Statistical Thinking," Prentice-Hall, Englewood Cliffs, N. J., 1974.
Huff, Darrell: "How to Lie with Statistics," W. W. Norton, New York, 1954.
Nouy, Lecomte Pierre, du: "Human Denstiny," Longmans, Green, and Co., New York, 1947.

4 ACCURACY—MULTIPLE LINEAR REGRESSION

Abstract A mathematical model that predicts numerical answers which are correct (or nearly so) is said to be *accurate*. In regression analysis it is often possible to improve on the accuracy of a simple linear regression model by considering additional variables. The potential virtues of this are obvious. But there are pitfalls too, and these are also emphasized.

Prerequisites Three-dimensional coordinate geometry. Multivariable calculus is used to derive the formulas. A computer is essential for the numerical exercises.

It may be true that "thou canst not stir a flower without troubling of a star" but the computer program for guiding a space capsule does not, in fact, have to take my gardening into account.

R. Lewontin

Can you weigh a person with a tape measure? We all know that taller people often are heavier; so, if you knew a person's height, maybe you could come close to predicting that person's weight. Table 1 shows heights and weights for 10 students. The best-fitting regression line (Figure 1) turns out to be

$$\text{Weight} = 7.07 \text{ height} - 333 \tag{1}$$

But back up for a minute, forget the mathematics, and think about whether this makes sense. It is surely possible to have two people with the same height and different weights (see students 4 and 5 in the table). This demonstrates that there is no way to predict weight from height. In particular, Equation (1) predicts a weight of 147.76 when we substitute 68 for the height, and this is a mediocre fit to the true weights of students 4 and 5.

Table 1

Student	Shoe size	Waist size	Collar size	Height	Weight
1	9	34	15.5	68	160
2	10	32	15.5	70	160
3	10.5	31	16	71	150
4	7.5	29	14.5	68	120
5	8	34	16	68	175
6	10.5	34	15.5	76	190
7	12	38	16.5	73.5	205
8	12	34	17.5	75.5	215
9	11	36	16.5	73	185
10	9.5	32	15.5	72	170

The mathematical modeler's response to this is that the model can be improved to give more accuracy. As you might suppose, students 4 and 5 can be distinguished by their waist measurements: one is fat and the other skinny. This leads to the notion that, if we measure the waist sizes of all students, we might use this, in combination with height, to predict weight. The outcome might be an equation like

$$\text{Weight} = a\,\text{height} + b\,\text{waist} + c \tag{2}$$

for suitable constants a, b, and c. Or, for even greater accuracy, we might also add collar size to our model and look for an equation like

$$\text{Weight} = a_1\,\text{height} + a_2\,\text{waist} + a_3\,\text{collar} + a_4$$

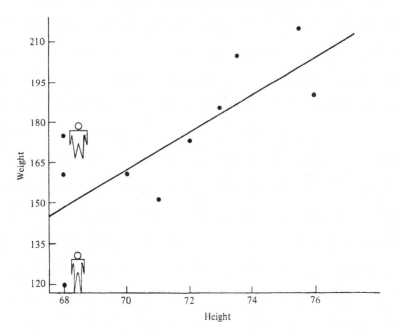

Figure 1

In principle we can take any number of variables into account. The technique of doing this is called *multiple linear regression*.

The variables on the right side of a multiple linear regression equation, such as Equation (2), are called *explanatory variables* because their function is to explain or account for the value of the variable on the left side. In this section we will restrict ourselves to two explanatory variables in describing the theory, because then the situation can be visualized as a geometry problem in 3-space.

Here is how the mathematics of it looks. Suppose we are given a table of values for three quantities, such as waist measurement, height, and weight for various individuals or, instead, air-pollution indices, per-capita cigarette consumption, and lung-disease rates for various cities. Let the three variables be called x, y, and z so that each line of the table can be thought of as a triple (x_i, y_i, z_i), which can, in turn, be thought of as a point in 3-space. Based on our understanding of the real-world situation, we have in mind "explaining" one of these variables, say z, on the basis of the other two, which are then called explanatory variables. We shall consider the case where we are looking for a linear explanation, that is, an equation of the form

$$z = ax + by + c \tag{3}$$

which is, to a good approximation, satisfied by all the triples (x_i, y_i, z_i). Since Equation (3) is the equation of a plane in 3-space, we can interpret the problem geometrically: we are looking for a plane which comes as close as possible to passing through the points (Figure 2).

If d_i represents the vertical distance from (x_i, y_i, z_i) to some plane, we define the distance from the set of points to the plane as

$$D = \sqrt{d_1^2 + d_2^2 + \cdots + d_n^2}$$

For algebraic convenience we will deal instead with the square of this distance,

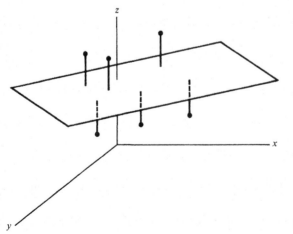

Figure 2

which we call S. If the plane has equation $z = ax + by + c$, then

$$S = \sum_1^n [z_i - (ax_i + by_i + c)]^2 \tag{4}$$

Our task is to choose values for a, b, and c so as to make S as small as possible; in this way we will also be minimizing D. Therefore these values of a, b, and c will correspond to the closest-fitting plane.

To minimize S, we set the partial derivatives with respect to a, b, and c equal to 0:

$$\frac{\partial S}{\partial a} = -2 \sum_1^n x_i(z_i - ax_i - by_i - c) = 0$$

$$\frac{\partial S}{\partial b} = -2 \sum_1^n y_i(z_i - ax_i - by_i - c) = 0 \tag{5}$$

$$\frac{\partial S}{\partial c} = -2 \sum_1^n (z_i - ax_i - by_i - c) = 0$$

A little algebra transforms these to the following equations in variables a, b, and c:

$$a \sum x_i^2 + b \sum x_i y_i + c \sum x_i = \sum x_i z_i$$
$$a \sum x_i y_i + b \sum y_i^2 + c \sum y_i = \sum y_i z_i \tag{6}$$
$$a \sum x_i + b \sum y_i + cn = \sum z_i$$

These equations are called the normal equations for a, b, and c. They comprise three linear equations in a, b, and c and can be solved by various standard means, such as gaussian elimination, provided the three equations are independent. This is usually the case. The next example illustrates how to find these equations. You will notice that, even for the highly simplified data set we use, there is lots of arithmetic. Consequently, computer programs have been prepared to compute a, b, and c from the points (x_i, y_i, z_i). If you want to work out the numerical exercises in this section, it is essential to use some computer assistance.

Example 1

Let us consider a shortened version of Table 1, consisting of just the first five students. (This cuts down the arithmetic.) We need to compute 12 coefficients for substitution into Equation (6). As before, $x = $ waist measurement, $y = $ height, and $z = $ weight. For the first equation we need:

$$\sum x_i^2 = (34)^2 + (32)^2 + (31)^2 + (29)^2 + (34)^2 = 5{,}138$$
$$\sum x_i y_i = (34)(68) + (32)(70) + (31)(71) + (29)(68) + (34)(68) = 11{,}037$$
$$\sum x_i = 34 + 32 + 31 + 29 + 34 = 160$$
$$\sum x_i z_i = (34)(160) + (32)(160) + (31)(150) + (29)(120) + (34)(175) = 24{,}640$$

We now have the coefficients of the first equation, which is

$$5{,}138a + 11{,}037b + 160c = 24{,}640$$

We skip the details of the calculations for the next two equations. The way system (6) comes out is

$$
\begin{aligned}
5{,}138a + 11{,}037b + 160c &= 24{,}640 \\
11{,}037a + 23{,}813b + 345c &= 52{,}790 \\
160a + 345b + 5c &= 765
\end{aligned}
\tag{7}
$$

The solution of this system (correct to two decimal places) is

$$a = \frac{259}{27} = 9.59$$

$$b = \frac{38}{9} = 4.22$$

$$c = \frac{-12{,}023}{27} = -445.30$$

Therefore the best-fitting plane is

$$z = 9.59x + 4.22y - 455.30 \tag{8}$$

If the procedure in the last example is carried out for the whole of Table 1 instead of the first five entries, we get a different system of equations in place of Equation (7) and the following best-fitting plane:

$$z = 6.35x + 4.59y - 368 \tag{9}$$

Equations (1) and (9) give two separate models that enable us to make a prediction of weight with a tape measure. Which is better? Common sense indicates that Equation (9) is better because it has two explanatory variables. But we need something better than common sense to settle the question. We need a mathematical way to measure the goodness of fit.

A number of measures have been described and are in common use. We describe one of them, called the *multiple correlation coefficient*, denoted R.* Before defining it, we make the following notational definition.

Definition

In a regression equation where z is expressed in terms of one or more explanatory variables, $\hat{z}_i$ is the value of z predicted by the equation when we substitute the ith set of values for the explanatory variables (e.g., x_i and y_i).

---

* There is a related statistical quantity called the *sample correlation coefficient*, denoted r. Sometimes $R = r$, but not always. We deal only with R.

Example 2

Suppose we want to find $\hat{z}_1$ for regression Equation (8). The subscript 1 refers to the first data point (the first student in Table 1). So, in effect, we are asking, "What weight does the equation predict when we substitute the waist and height of the first student." These waist and height values are denoted x_1 and y_1 and have values 34 and 68. So

$$\begin{aligned}
\hat{z}_1 &= 6.35x_1 + 4.59y_1 - 368 \\
&= (6.35)(34) + (4.59)(68) - 368 \\
&= 160.02
\end{aligned}$$

Likewise

$$\begin{aligned}
\hat{z}_2 &= 6.35x_2 + 4.59y_2 - 368 \\
&= (6.35)(32) + (4.59)(70) - 368 \\
&= 156.5
\end{aligned}$$

But, if we use regression Equation (1), then

$$\begin{aligned}
\hat{z}_1 &= (7.07)y_1 - 333 \\
&= (7.07)(68) - 333 \\
&= 147.76
\end{aligned}$$

This last value of $\hat{z}_1$ compares poorly to the actual z_1 value of 160.

Definition

The multiple correlation coefficient is defined as

$$R = \sqrt{1 - \frac{\sum (z_i - \hat{z}_i)^2}{\sum (z_i - \bar{z})^2}} \tag{10}$$

The square root sign in the formula normally means positive square root. However, in the case of one explanatory variable, where the scatter diagram is in two dimensions, some authors take the sign to agree with the sign of the slope of the regression line. None of this matters very much because, in almost all cases, the computation of the square root is omitted and the (positive) value of R^2 is reported. We follow this custom.

Notice that the denominator $\sum (z_i - \bar{z})^2$ doesn't depend on the regression equation since the z_i and $\bar{z}$ (the mean of the z_i) are measurements known at the outset. (In Table 1 they are the heights of the students.) The numerator $\sum (z_i - \hat{z}_i)^2$ is what we called S in Equation (4). It has been minimized by choosing a, b, and c to be the solutions of the normal equations. In the event this numerator is 0, then each $z_i = \hat{z}_i$, which means that our regression equation predicts the actual z values perfectly, and so one has the maximum in accuracy. But when that numerator is 0, it means $R^2 = 1$. Therefore, $R^2 = 1$ corresponds to perfect accuracy, and the closer R^2 gets to 1, the closer the regression equation fits the data. In other words, the closer R^2 gets to 1, the better fit we have.

Some additional information, helpful in calculating and interpreting R^2, is contained in the following theorem.

Theorem 1

1. An alternate formula for R^2 is

$$R^2 = \frac{\sum (\hat{z}_i - \bar{z})^2}{\sum (z_i - \bar{z})^2} \tag{11}$$

2. $0 \leq R^2 \leq 1$.

Proof Formula (11) follows immediately from the definition of R^2 provided we can show

$$\sum_1^n (z_i - \bar{z})^2 = \sum_1^n (z_i - \hat{z}_i)^2 + \sum_1^n (\hat{z}_i - \bar{z})^2 \tag{12}$$

By squaring and simplifying, we see that this is equivalent to

$$-2\bar{z} \sum_1^n z_i = -2 \sum_1^n z_i \hat{z}_i + 2 \sum_1^n \hat{z}_i^2 - 2\bar{z} \sum_1^n \hat{z}_i \tag{13}$$

To prove Equation (13), we make use of the normal equations, Equation (6). Since $\hat{z}_i = ax_i + by_i + c$, when $a, b,$ and c are chosen to satisfy the normal equations, these equations can be rewritten

$$\sum_1^n x_i(z_i - \hat{z}_i) = 0$$

$$\sum_1^n y_i(z_i - \hat{z}_i) = 0$$

$$\sum_1^n (z_i - \hat{z}_i) = 0$$

From the last of these we see that $\sum \hat{z}_i = \sum z_i$. Furthermore, $\sum z_i = n\bar{z}$; so Equation (13) is equivalent to

$$\sum_1^n \hat{z}_i(\hat{z}_i - z_i) = 0 \tag{14}$$

If we take a times the first normal equation (as rewritten above) plus b times the second plus c times the last, we obtain

$$\sum_1^n (ax_i + by_i + c)(z_i - \hat{z}_i) = 0$$

which is the same as Equation (14). This proves part 1 of the theorem.

For part 2 notice first that formula (11) shows that R^2 is nonnegative. To show that $R^2 \leq 1$, we observe that Equation (12) implies

$$\sum_1^n (\hat{z}_i - \bar{z})^2 \leq \sum_1^n (z_i - \bar{z})^2$$

Therefore $R^2 \leq 1$.

Example 3

We calculate R^2 for regression Equation (9):

$$\hat{z}_i = 6.35x_i + 4.59y_i - 368$$

$$\bar{z} = \frac{160 + 160 + 150 + \cdots + 170}{10} = 153$$

Using these formulas, we compute

$(z_1 - \hat{z}_1)^2 = (160 - 160.02)^2 = 0.0004$ $(z_1 - \bar{z})^2 = (160 - 153)^2 = 49$

$(z_2 - \hat{z}_2)^2 = (160 - 156.5)^2 = 12.25$ $(z_2 - \bar{z})^2 = (160 - 153)^2 = 49$

$$\vdots \qquad\qquad\qquad\qquad \vdots$$

$(z_{10} - \hat{z}_{10})^2 = (170 - 165.68)^2 = 18.6624$ $(z_{10} - \bar{z})^2 = (170 - 153)^2 = 289$

Therefore

$$R^2 = 1 - \frac{0.004 + 12.25 + \cdots + 18.6624}{49 + 49 + \cdots + 289}$$

$$= 0.862$$

For comparison, we carry out the calculation for regression Equation (1), which attempts to explain the same data with only y (height) as the explanatory variable.

$$\hat{z}_i = 7.07y_i - 333$$

$\bar{z}$ is the same as before, and so is $\sum (z_i - \bar{z})^2$. Therefore we only need to calculate

$$(z_1 - \hat{z}_1)^2 = (160 - 147.76)^2 = 149.8176$$

$$\vdots$$

$$(z_{10} - \hat{z}_{10})^2 = (170 - 176.04)^2 = 36.4816$$

Consequently

$$R^2 = 1 - \frac{149.8176 + \cdots + 36.4816}{49 + \cdots + 289} = 0.594$$

Judging by the two values of R^2 (0.594 versus 0.862), we see there has been a considerable gain in accuracy by taking the additional variable of waist measurement into account. Height and waist together, via Equation (9), do a more accurate job of predicting weight than height alone, via Equation (1).

In the next example we will see that taking an extra variable into account can change the conclusions we draw from data.

Example 4 Cancer and Fluoridation (Real Data)

It is well known that adding fluoride to water supplies prevents dental cavities. But is there a price to be paid for this blessing? Some critics, mostly outside the scientific community, have maintained that fluoride causes cancer. Is there any evidence for this view? There seems to be no evidence that scientists consider persuasive, but this does not mean that one can't find any evidence at all. (If this seems confusing, keep in mind that, while statistics is a mathematical science, the use of statistics is not a science and requires judgement.) To illustrate this, we now consider two models for the fluoride-cancer link. Both models are based on the data in Table 2.

The simplest approach to this data is to look for a relation between cancer mortality rates and years of fluoridation, without taking account of the age profile of each city. In doing this, we will use the log of the number of years of fluoridation, since experience shows that in dose-response cancer studies the log of the dose gives a better linear fit. Thus we plot the ordered pairs (L, C) and look for a re-regression line. In the language of this section, we use L as the only explanatory variable to predict C, the cancer mortality rate. Simple linear regression yields

$$C = 27.1L + 181$$

with $R^2 = 0.047$. Since the sign of the coefficient of L is positive, this seems to say that the higher L is (the more years of fluoridation), the higher C will be. If we are satisfied with this model—and we really shouldn't be—it seems to suggest the possibility that fluoride causes cancer.

There are at least two reasons to be suspicious of this model: first, no demonstrated mechanism has been put forward to explain how fluoride could cause cancer; second, the value of R^2 is very low—low enough to make a sensible statistician nervous about this model.

A better approach to this question is to take other explanatory variables into account. It is well known that older folks are more prone to getting cancer, and this leads to the idea that the variation in cancer mortality from one city to another may be due more to the fact that these cities vary in the proportion of older citizens than to the variation in fluoridation. The way to test this is to do a multiple linear regression in which L and A (age profile) are used as explanatory variables for C. When we do this, we obtain

$$C = 0.566L + 10.6A + 85.8$$
$$R^2 = 0.493$$

The R^2 value is a great improvement over the one for the previous model. The coefficient of L is still positive, suggesting a positive link from fluoride to cancer.

Table 2

City	Cancer mortality rate	Number of years fluoridated	L	Percent age 65 or more
New York	215	5	0.70	12.1
Chicago	204	14	1.15	10.6
Philadelphia	217	16	1.20	11.7
Detroit	213	3	0.48	11.5
Baltimore	223	17	1.23	10.6
Dallas	191	4	0.60	7.9
Washington	200	18	1.26	9.4
Cleveland	219	14	1.15	10.6
Milwaukee	189	16	1.20	11.0
San Francisco	249	17	1.23	14.0
St. Louis	207	14	1.15	14.7
Pittsburgh	243	17	1.23	13.5
Denver	157	16	1.20	11.5
Buffalo	248	15	1.18	13.3
Minneapolis	228	12	1.08	15.0
Fort Worth	169	5	0.70	9.6
Oklahoma City	170	15	1.18	9.8
Louisville	230	18	1.26	12.4
Miami	266	18	1.25	14.5
Tulsa	159	16	1.20	9.1

But this coefficient of 0.566 is much smaller than the coefficient of 27.1 found in the previous model; so the alleged link between fluoride and cancer is about 50 times weaker in the more sensible model.

But this is not the end of the story. Other variables could also be taken into account, which would further weaken the evidence. The story is too long to be told here, but the upshot of the many statistical investigations which have been done is that the great majority of scientists feel there is no convincing case that fluoride causes cancer.

If the previous example convinced you that adding explanatory variables to increase R^2 is a good strategy, then the next piece of news may seem disturbing: adding explanatory variables is sometimes a bad idea. The problem is that R^2 has an important shortcoming as a measure of fit or accuracy: it is always possible to increase R^2 a little bit by adding another variable to the model, even a variable which has nothing at all to do with what you are trying to explain, as in the following example.

Example 5 Time and Distance (Real Data)

Table 3 contains data collected to attempt to predict how long it takes students to drive to Adelphi University from their homes, when the number of miles

Table 3

Miles	Time, minutes	Soc. sec. No.
2.7	10	95
4	15	09
25.8	50	10
9	24	43
5.8	20	35
2.8	11	62
15.2	50	15

involved is given. The last column, containing the last two digits of the students' social security numbers, appears utterly irrelevant to the problem.

When we carry out a linear regression on the first two columns, using distance as the explanatory variable, we obtain:

$$\text{Time} = 1.89 \text{ miles} + 8.05$$
$$R^2 = 0.867$$

Although it makes no realistic sense, we can go through the motions of performing a regression in which social security number is used as a second explanatory variable. This yields

$$\text{Time} = 1.7 \text{ miles} - 0.0872 \text{ social security number} + 13.2$$
$$R^2 = 0.883$$

Notice that in this silly model the R^2 value is slightly higher than before.

It is possible to demonstrate theoretically that, when we add another explanatory variable, the value of R^2 never decreases and generally increases. In order to visualize the argument, we carry it out only for the increase from one explanatory variable x to two, x and y. The variable to be explained is z.

First observe that no matter what regression equation we use, the denominator $\sum (z_i - \bar{z})^2$ in formula (10) doesn't change because it depends only on the z_i, not on any explanatory variables. Therefore, to show R^2 doesn't decrease as we add a variable, it suffices to show that the numerator $\sum (z_i - \hat{z}_i)^2$ doesn't increase.

Let us suppose that we have a least-squares line $z = mx + b$ that has been fit to the set of points (x_i, z_i), as in Figure 3a. Next imagine that for each i, we add the additional variable y. Thus we have a set of ordered triples (x_i, y_i, z_i), and these can be plotted in a three-dimensional space. We call these points Q_i. It will make comparisons convenient if we create this 3-space by adding a dimension to our existing 2-space. In our new 3-space the coordinate plane $y = 0$ is the original 2-space. Thus a point $P_i(x_i, z_i)$ of our original 2-space becomes $P_i(x_i, 0, z_i)$ in 3-space.

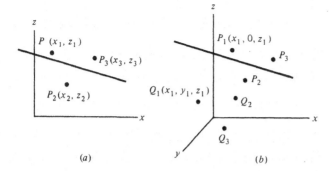

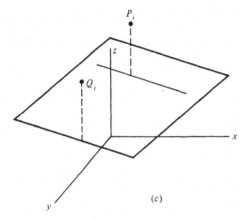

Figure 3

Now the equation of the least-squares line $z = mx + b$ may be interpreted as $z = mx + 0y + b$, which is the equation of a plane in 3-space. The vertical distance of a point $P_i(x_i, z_i)$ to the line $z = mx + b$ in the two-dimensional space is the same as the vertical distance of the corresponding point $Q_i(x_i, y_i, z_i)$ to the plane $z = mx + 0y + b$, namely $z_i - (mx_i + 0y_i + b)$ in both cases.

The upshot is that the least-squares distance from the P_i to the best-fitting line can be equaled by the least-squares distance from the Q_i to at least one plane, a plane which is probably not the best-fitting plane. Therefore the least-squares plane must do at least as well. In other words the least-squares distance from the points in 3-space to the best-fitting plane is at least as good as the least-squares distance from the points in 2-space to the best-fitting line.

The practical importance of this is that the value of adding another variable to a regression model cannot be demonstrated by merely observing that R^2 increases. Of course the size of the increase can be persuasive. In addition there are other statistical quantities that can be computed to measure the value of adding (or retaining) some particular explanatory variable. Finally, never to be forgotten, there is common sense. In the previous example, adding the social security number doesn't make a descriptively realistic model, and we don't need complicated computations to tell us that it's a bad idea. However, our common-sense judgement about descriptive realism isn't always reliable either. This is the subject of the next example.

Example 6

In the middle of the 1960s the U.S. Office of Education tried to find out what makes students succeed in school. In particular, it wanted to know whether the quality of the school system, as measured by the money spent on education, quality of the teachers, and similar factors could explain why some students, especially minority students, showed lower scores on standardized tests.

To study these questions, the government financed a major statistical investigation, headed by James Coleman, which involved hundreds of thousands of grade school pupils and tens of thousands of teachers. The results, summarized in the "Coleman Report" contain some surprises.

Table 4 shows a small part of the data collected. Each number in columns 2 through 7 is an average value for the sixth grade in the elementary school involved. Here is what the columns mean:

SLRY—average teacher salary per pupil
WHTC—percent of students with white-collar fathers
SES—a measure of students socioeconomic level
TCHR—verbal test scores of teachers
MOM—educational level of pupils' mothers (1 unit = 2 years)
SCOR—verbal test scores of pupils

Why were these particular variables chosen in preference to others, such as religion, ethnic identity, average class size, race, etc.? Actually, the Coleman Report was very extensive and other parts of it did investigate variables beside the ones in Table 4. As far as the variables in Table 4 are concerned, the story behind them is in two parts:

1. WHTC, SES, and MOM measure the family background of the students. Many previous studies had shown that they account for a lion's share of the variation in pupil performance.
2. SLRY and TCHR are included because these factors can be changed if it turned out that it would improve pupils' performances to do so. For example, it was hoped that, by paying higher salaries to teachers, one might get better teachers or improve the morale of the existing teachers and that this would help improve pupils' scores. This is common-sense realism; but we will see that the numbers raise questions about this common sense.

Table 4

School number	SLRY	WHTC	SES	TCHR	MOM	SCOR
1	3.83	28.87	7.20	26.60	6.19	37.01
2	2.89	20.10	−11.71	24.40	5.17	26.51
3	2.86	69.05	12.32	25.70	7.04	36.51
4	2.92	65.40	14.28	25.70	7.10	40.70
5	3.06	29.59	6.31	25.40	6.15	37.10
6	2.07	44.82	6.16	21.60	6.41	33.90
7	2.52	77.37	12.70	24.90	6.86	41.80
8	2.45	24.67	−0.17	25.01	5.78	33.40
9	3.13	65.01	9.85	26.60	6.51	41.01
10	2.44	9.99	−0.05	28.01	5.57	37.20
11	2.09	12.20	−12.86	23.51	5.62	23.30
12	2.52	22.55	0.92	23.60	5.34	35.20
13	2.22	14.30	4.77	24.51	5.80	34.90
14	2.67	31.79	−0.96	25.80	6.19	33.10
15	2.71	11.60	−16.04	25.20	5.62	22.70
16	3.14	68.47	10.62	25.01	6.94	39.70
17	3.54	42.64	2.66	25.01	6.33	31.80
18	2.52	16.70	−10.99	24.80	6.01	31.70
19	2.68	86.27	15.03	25.51	7.51	43.10
20	2.37	76.73	12.77	24.51	6.96	41.01

Our first approach to accounting for SCOR might be to use all the other variables as explanatory variables. This yields

$$SCOR = 19.95 - 1.79SLRY + 0.04WHTC + 0.65SES$$
$$+ 1.11TCHR - 1.81MOM \quad (15)$$
$$R^2 = 0.9063$$

The R^2 value is high, which is satisfying. But do we really need all five explanatory variables? For example, since SES, WHTC, and MOM all measure the pupils' home backgrounds, maybe only one of them would be needed. Here is how the regression equation comes out if we delete WHTC and MOM:

$$SCOR = 12.12 - 1.74SLRY + 0.55SES + 1.04TCHR$$
$$R^2 = 0.9007 \quad (16)$$

The R^2 value is only a trifle less than in the previous model; so it seems reasonable to do without the variables MOM and WHTC. Putting them back would add only $0.0056 (= 0.9063 - 0.9007)$ to R^2, and, since we can get a small increase in R^2 even by adding irrelevant variables (like the social security number in Example 5), it seems reasonable to consider MOM and WHTC unnecessary.

We could try simplifying further by eliminating SLRY. This gives

$$SCOR = 0.54SES + 0.75TCHR + 14.58$$
$$R^2 = 0.8873 \tag{17}$$

Once again, we have had to pay only a small price in R^2 for this simplification; so maybe it's a good idea. Observe that this would also sidestep a peculiar feature of Equation (16), the negative coefficient for SLRY. This negative sign seems to say that, if we increase teachers' salaries, keeping other explanatory variables the same, we will decrease pupils' scores. If this offends your sense of realism, then this would be a reason to prefer Equation (17). [Likewise, if you are a hired gun, working for a teacher's union, Equation (17) will seem more attractive to you.]

Any way you look at the data in Table 4 shows that increasing teachers' salaries is unlikely to raise SCOR *by itself*. But what if raising salaries attracts new teachers with higher verbal abilities so that TCHR increases? Our data give no insight into this extremely important question.

Regression analysis, as presented so far, is entirely a matter of making numerical predictions from numerical data. But not all data are numerical. Think of how many important characteristics of you as a person cannot be measured numerically but are a matter of belonging to one category or another: race, sex, religion, and so on. It is possible to create models, a little like regression models, to predict one category from given data about other categories. For example, a sociologist can make a pretty fair prediction of a person's occupational category (white or blue collar) from that person's race, educational level, and father's occupational category.

In some problems the given data may include both numerical and non-numerical information: for example, the sociologist can do an even better job of predicting occupational category if the person's IQ score (numerical data) is known, in addition to race, educational level, and father's occupation.

The fancy methods for dealing with categorical data are beyond the scope of this book. But sometimes simple approaches suggest themselves, as in the following example.

Example 7 Aging and Fitness (Real Data)

Researchers in physical education have been interested in finding out whether leading a physically active life slows down aging. One specific version of this general question is this: do runners and joggers who have a high level of cardiovascular fitness show less increase of reaction time with advancing age? We have two numerical variables, age and reaction time, and one categorical variable, fitness level (high and low). We would like to account for reaction time, using age and fitness as explanatory variables.

Researchers measured these variables and determined the fitness category for 64 subjects. These subjects were grouped into 10-year age categories, and the average reaction time was computed for each combined age-fitness category. In Figure 4 the hollow dots represent the average reaction times in the low-fitness

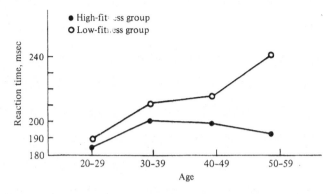

Figure 4

category and the solid dots represent reaction times in the high-fitness category. Now each series of dots could be fit with its own regression line, which would show the relation of age and reaction time.

Instead, the investigators settled for the simpler method of simply connecting the dots to show the two separate trends. Whichever method is used, the key idea is to treat the two fitness categories separately and compare the two trends. What emerges from Figure 4 when we do this is that, in the low-fitness group, there is a steady deterioration in reaction time as age increases but, in the high-fitness group, this is not so. (Before dashing out to buy some running shoes, be sure to read Section 3 on correlation and causation.)

EXERCISES

● **1** Show that, if a least-squares plane (or line) passes through each data point, then $R^2 = 1$. Also show the converse, that, if $R^2 = 1$, then the least-squares plane (or line) passes through each data point.

2 (a) Suppose you want to explain variable z and have two other variables x and y available to use as explanatory variables. Suppose you have only three data points (x_1, y_1, z_1), (x_2, y_2, z_2), and (x_3, y_3, z_3). Show that, by using both x and y as explanatory variables, you get $R^2 = 1$.

(b) Suppose x, y, and z stand for totally unrelated variables: e.g., x = a student's height, y = the student's social security number, and z = the student's IQ. In this case, what would your reaction be to getting $R^2 = 1$? Does this suggest any principle about how many data points to use?

● **3** How do you think Exercise 2 generalizes to the case where there are n explanatory variables?

4 Suppose we define r_i, the ith residual, to be d_i, the distance from the ith data point to the regression plane (Figure 2) when P_i is above the regression, and to be $-d_i$ when the ith data point is below the regression plane. Show that $\sum r_i = 0$. (*Hint*: Find a formula for r_i; then examine the normal equations.)

● **5** If you use three explanatory variables to predict a fourth variable, how many equations and how many unknowns will there be in the normal equations?

6 (a) Suppose we have a set of data points (x_i, y_i, z_i), $i = 1, \ldots, k$, and, for each value of i, $y_i = ax_i + b$. Suppose you attempt to use both x and y as explanatory variables to account for z. Show that one of the normal equations is then a linear combination of the other two.

(b) Can you give a geometric interpretation of this?

7 Fill in the missing algebraic steps in the derivation of formula (11) for R^2.

● **8** Calculate R^2 for the example in the previous section, involving speed and gas efficiency of cars.

9 If you want to restrict yourself to one tape measurement in order to predict a person's weight, should you use height, waist, collar size, or shoe size? (In other words, which 1-variable regression model, based on Table 1, gives the best R^2?)

10 Here is some fake data for 10 students concerning scores on a certain exam. Find a linear regression equation in which x and y are used as explanatory variables.

Hours studied	IQ	Exam grade
x	y	z
8	98	56
5	99	44
11	118	79
13	94	72
10	109	70
5	116	54
18	97	94
15	100	85
2	99	33
8	114	65

11 Here is a list of diamond prices (real data). Can you use regression analysis to show some relationship between carats and price? Notice that there are different styles of cut that may also affect the price: marquise, round cut, etc. How would you take this into account?

Diamond prices (1976)

Carats	Shape	Price, $
0.29	Marquise	155
0.33	Round cut	240
0.44	Round cut	320
0.49	Pear	320
0.68	Pear	450
0.69	Round cut	470
0.90	Marquise	540
0.91	Pear	640
0.98	Round cut	1025
1.00	Marquise	1400
1.38	Marquise	2250
1.80	Pear	2500

12 Kippy and Alex are going into the lawn-mowing business this summer and need a quick convenient way of measuring the area of a lawn so the customer can be charged properly. If lawns were exactly rectangular, one could measure the length and width and multiply. Most lawns are approximately but not exactly rectangular; so $A = l \times w$ will work only approximately. (Here l is the maximum length and w is the maximum width. See Figure 5.) They decide to try the formula $A = al^b w^c$ to get a better fit. To study the matter, they carefully measure l and w and the exact area, A, of six lawns and

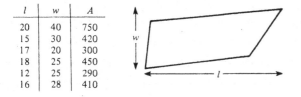

l	w	A
20	40	750
15	30	420
17	20	300
18	25	450
12	25	290
16	28	410

Figure 5

obtain the data in the table. Can you help them find the best values for a, b, and c? (*Hint*: This is not a linear regression problem, but you can convert it to one.)

13 Here is a shortcut approximation for R^2 which is useful for scatter diagrams in two dimensions:

1. Enclose the points of the scatter diagram as closely as possible in an ellipse. (A freehand drawing will do.)
2. Measure the length L and width W of the ellipse (Figure 6).
3. $R^2 \approx (1 - W/L)^2$.

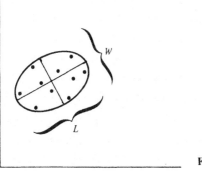

Figure 6

(*a*) Use this method to calculate R^2 for Exercises 1, 2, 3, 6, and 7 of Section 2.

(*b*) Compare these approximate values with the true values. Use the results to devise a rule of thumb for what the percentage error may be when using the shortcut.

Computer exercise

14 Write a computer program that computes the best-fitting linear regression equation and its R^2 value. Your program should accept as input the data points (x_i, y_i, z_i) and use z as the variable to be accounted for and x and y as explanatory variables.

BIBLIOGRAPHY

Cohen, Jacob, and Patricia Cohen: "Applied Multiple Regression/Correlation Analysis for the Behavioral Sciences," Lawrence Erlbaum Associates, Hillsdale, N.J., 1975.

Draper, Norman, and Harry Smith: "Applied Regression Analysis," John Wiley & Sons, Inc., New York, 1966.

5 PRECISION—MALTHUS AND THE DISMAL THEOREM

Abstract Other things being equal, we like models to give precise answers. For example, a model that predicts that the next rain will occur 2 days from now is preferable to one which predicts that it will occur between 1 and 5 days from now. (Worst of all would be a model that predicts it will rain eventually.) Nevertheless, imprecise models can be valuable. In this section we give an imprecise model which has turned out to be one of the most influential mathematical models in the social sciences.

Prerequisites High school algebra (logarithms, laws of exponents).

To get land's fruit in quantity
Takes jolts of labor evermore.
Hence food will grow like one, two, three . . .
While numbers grow like one, two, four . . . .

> *Song of Malthus: A Ballad on Diminishing Returns Anonymous*

It is ironic that one of the most influential mathematical models ever devised outside the physical sciences is also highly imprecise. This model is Thomas Malthus' (1766–1834) model of the relation of population to its food supply. In one form or another, Malthus' ideas have been a source of controversy for almost 200 years, and still today his name is often a fighting word. In this section we examine the ideas of Malthus and some modern revisions, with particular focus on the lack of precision these models contain. Some specific questions we have in mind are:

1. Why is there so much imprecision? Is it inevitable?
2. Are the models valuable despite the imprecision?
3. Why did Malthus choose to be as imprecise as he was?

Malthus' key idea was that population has a tendency toward what he called *geometric* growth—what we nowadays call *exponential* growth.

Definition

Suppose t measures time and $t = 0, 1, 2, \ldots$ represents a series of equally spaced instants. The time between successive t values may be 1 second, 1 generation, or 1 century—indeed any unit that is convenient for the problem. Let some quantity, like population, be measured at each of these instants and let its value at t be $P(t)$. Then the series $P(0), P(1), P(2), \ldots$ is said to display geometric or exponential growth at rate r if successive values are always in the same fixed ratio r. That is,

$$P(1) = rP(0)$$
$$P(2) = rP(1)$$
$$\vdots$$
$$P(t + 1) = rP(t) \tag{1}$$
$$\vdots$$

Example

The series 1, 2, 4, 8, 16, ... displays exponential growth with $r = 2$. So does the series 3, 6, 12, 24, ... The series 1000, 100, 10, 1, 0.1, 0.01, ... displays exponential growth with $r = 0.1$. In this last series "growth" is a misnomer; "decline" is a better word since $r < 1$. However, when we have a sequence whose r value is unspecified, so it could show either growth or decline, we use the term "growth" to cover both possibilities.

Theorem 1

If $P(t)$ displays exponential growth with rate r, then $P(t) = P(0)r^t$ for all $t \geq 0$.

Proof Substituting the first equation of Equation (1) into the second gives

$$P(2) = rP(1) = r[rP(0)] = r^2P(0)$$

Likewise

$$P(3) = rP(2) = r[r^2P(0)] = r^3P(0)$$
$$\vdots$$
$$P(t) = rP(t - 1) = r[r^{t-1}P(0)] = r^tP(0)$$

Figure 1 shows some exponential growth curves. All start with the same value of $P(0)$, but the r values differ. The figures are meant to be a geometric reminder of the following theorem.

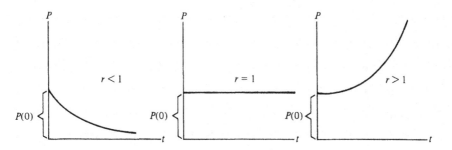

Figure 1 Three types of exponential curve.

Theorem 2

1. If $r > 1$, $r^t \to \infty$ as $t \to \infty$.
2. If $r = 1$, $r^t = 1$ for all t.
3. If $0 < r < 1$, $r^t \to 0$ as $t \to \infty$.

Proof Part 2 is obvious. For parts 1 and 3 see Exercises 2 and 4.

Malthus based his theory of population on the interplay of three population curves, of which two are purely theoretical and won't ever occur in actual fact. Here they are, briefly introduced in advance of their detailed descriptions:

1. The unchecked population curve $U(t)$. This is the curve that population would follow if food were unlimited.
2. The maximum supportable population curve $S(t)$. This is the curve that population would follow if food supply is limited, but growing in time as fast as Malthus thought possible, and shared out equally in amounts just enough for a person to barely subsist on.
3. The real population as it actually will grow in time $P(t)$.

The unchecked population $U(t)$ Since food supply is always limited, there can be no such thing as an unchecked population in the real world. But Malthus felt that during his lifetime the population of the American colonies was a good approximation because food was so plentiful there. By examining census figures for the years 1790, 1800, 1810, and 1820, he concluded that the American white population appeared nearly to be growing exponentially with rate $r = 1.35$ per 10-year period. Thus, if t measures decades, we have, approximately, $P(t) = P(0)(1.35)^t$. (See Table 1.)

Does this table prove that an unchecked population grows exponentially and with rate 1.35? Malthus conceded that the rate of 1.35 might not be an accurate estimate of the rate with which an unchecked population would grow. The unchecked rate would, he argued, be even higher. It is not at all necessary to be concerned about its precise value because all Malthus needed for his argument was that it is greater than 1; and this he felt to be demonstrated—with plenty to spare— by the American population. Do you agree?

Table 1

Decade t	Year	American white population	Ratios
0	1790	3,164,148	
1	1800	4,312,841	$P(1)/P(0) = 1.363$
2	1810	5,862,092	$P(2)/P(1) = 1.359$
3	1820	7,861,710	$P(3)/P(2) = 1.341$

But Malthus did not rest his case for exponential growth in an unchecked population entirely on the evidence from America. He also had theoretical grounds as indicated by his words: "All animals according to the known laws by which they are produced, must have a capacity of increasing in geometrical progression." What he undoubtedly meant is something like this.

Suppose we have, at $t = 0$, an initial group of $G(0)$ women born. Suppose that one generation later, at $t = 1$, they all simultaneously bring forth an average of r children apiece. Then $G(1)$, the size of the first generation is $G(1) = G(0)r$. At $t = 2$, each of these children is mature and gives birth to r children, who will be grandchildren of the original women, numbering $G(0)r^2$ in all (see Figure 2). Imagine that this synchronized reproduction of new generations goes on indefinitely and $G(t)$ denotes the size of the tth generation. Since the ratio of each generation to the previous one is r, we have exponential growth and, by Theorem 1, $G(t) = G(0)r^t$.

Notice that this model describes the sizes of various generations and does not produce a formula for the total population at any time. For example, if women always survive past the births of their grandchildren, but not to the births of any great grandchildren, then at $t = 2$ there will be $3 + 6 + 12 = 21$ people in the population shown in Figure 2. In general, the population $U(t)$ will be $G(0)r^t + G(0)r^{t-1} + G(0)r^{t-2}$. Alternatively, if women die after the births of their children but before their grandchildren, then at $t = 2$ there will be $6 + 12 = 18$ people in the population of Figure 2. [Can you find the formula for $U(t)$?] Thus our model of generation size cannot be converted into a model of total population without some assumption about mortality. However, since the total population is greater than any one-generational subset, it follows that, in this highly idealized model, population grows *at least* exponentially.

Let us pause to examine the consequences if a population grows exponentially with rate 1.35 per decade. By numerical calculation we see that after 50 years the population has multiplied itself by $(1.35)^5$, which is about 4.5. Taking the more conservative estimate of 4 in place of 4.5, let us ask ourslves what would happen to a population which quadruples itself every 50 years. Suppose $P(0) = 1$, i.e., we are starting with 1 female (Eve?), and we let growth proceed at this rate for 25 of

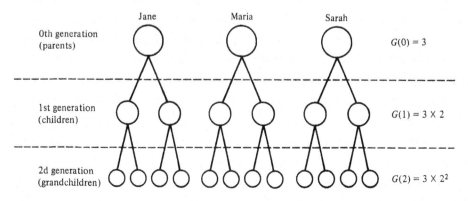

Figure 2 A simple females-only model for reproduction.

these 50-year periods (a total of 1250 years). Then the population is

$$4^{25}$$

How big is this? Perhaps you will be surprised that

$$4^{25} = 1{,}127{,}899{,}906{,}866{,}624$$

over 1 quadrillion. That's about one-quarter million times as many people as are actually alive today. This disparity is dramatic evidence that populations rarely grow unchecked for very long. To investigate the reason why, we turn now to the maximum supportable population.

The maximum supportable population $S(t)$ $S(t)$ is the number of people who could be supported with the food resources actually available at time t if these resources were shared out equally in amounts just barely sufficient to support each recipient. Thus, if the food supply stays constant year in and year out, then $S(t)$ stays constant also. On the other hand, if the food supply increases by a constant amount per unit of time, then $S(t)$ will increase by a corresponding constant amount per unit of time. In the event food supply is multiplied by the same constant r each year, then $S(t)$ will do the same and grow exponentially. A crucial part of Malthus' model is the assumption that food supply will *not* grow exponentially; the most that he felt we could expect was a constant addition to food supply in each unit of time. This would mean $S(t) = S(0) + At$, where $S(0)$ is the maximum supportable population at $t = 0$ and A is the additional number which can be supported because of the extra productivity that can be managed in 1 time unit. Malthus did not bother to determine what the particular values of A and $S(0)$ might be for any particular society because he did not need this level of precision.* However he did insist that $S(t)$ is a linear function of t.

The actual population $P(t)$ The actual population at time t, denoted $P(t)$, can never be greater than the maximum supportable population; i.e., $P(t) < S(t) = S(0) + At$. As far as the validity of Malthus' argument is concerned, it is not necessary to have any more precise details than this inequality (Figure 3).

We come now to the conclusion. First, Malthus claimed that under the assumptions he made, no population can grow unchecked indefinitely. This is the content of Theorem 3 below. Second, he claimed that the way in which checks are applied to a population are very unpleasant.

Theorem 3

If t is sufficiently large,

$$U(t) = U(0)r^t > S(0) + At = S(t)$$

provided $r > 1$. (See Figure 4.)

---

* He did suggest that, if t measured a 25-year period, $A \leq S(0)$.

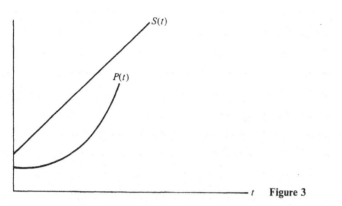

Figure 3

Proof If two lines have different slopes, the line with larger slope eventually crosses and stays above the one with smaller slope. The graph of $S(t)$ is a straight line with slope A. If the graph of $U(t)$ were also a straight line (which it is not), we would only have to show that the slope of $U(t)$ is larger than the slope of $S(t)$.

We can use this idea, however, provided we can find a line $y = L(t)$ which is, for every t, below $y = U(t)$. Then if $L(t)$ crosses $S(t)$, it forces $U(t)$ to do so also. See Figure 4. We shall look for such a line $L(t)$ among the tangents to $U(t)$.

Set $r = e^s$ so that $U(t) = U(0)e^{st}$. Since $r > 1$, $s > 0$. The derivative $U'(t) = sU(0)e^{st}$ is an increasing function of t. This means that $U(t)$ is *concave up* and so always lies above any tangent to itself. If we choose t_0 sufficiently large that $U'(t_0) = sU(0)e^{st_0} > A$ {we can do this by picking $t_0 > (1/s) \ln A/[sU(0)]$}, then we will have a tangent whose slope is greater than the slope of $S(t)$. This will do as the line $L(t)$ we are looking for.

In words this theorem says that any exponential curve with $r > 1$ rises above any straight line. Since an unchecked population is an exponential with $r > 1$, we see that an unchecked population "bumps into" the maximum supportable population curve. Notice that the theorem is a very general one. No conditions are imposed on the slope A or intercept $S(0)$ of the line, and the only restriction on

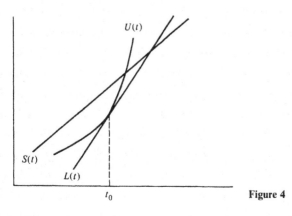

Figure 4

r is $r > 1$. This is what allows us to be so unconcerned about the exact values of r, A, and $S(0)$. We are spared paying any price for our imprecision.

Now what does this theorem mean for the actual population curve $P(t)$? Here's what Malthus would predict for the United States, beginning with the situation in about 1800. At that time, which we may take to be $t = 0$, existing food resources in North America were more than enough to feed the population, and so the maximum supportable population was greater than the actual population, i.e., $S(0) > P(0)$ (see Figure 5). Furthermore, the continent was very unpopulated; so the opportunity for further increases in food supply was great, i.e., A was large. Consequently, there was a period of time in which the actual population could grow unchecked, i.e., $P(t) = U(t)$ for the early t values. But according to Theorem 3, $U(t)$ will eventually cross $S(t)$, say at t^*. When that happens (and probably before), $P(t)$ can no longer follow the $U(t)$ curve. Consequently the actual population will be checked or restrained beginning at t^* at the latest. Figure 5 shows what might occur. Summarizing this without mathematics: any real population, although it may grow in an unchecked way for a time, in the long run becomes and remains checked.

At this point we lay aside the mathematics and get down to the bad news. According to Malthus' early writings, the checks to population were entirely unpleasant: miseries such as war, starvation, and disease and practices such as infanticide, abortion, and sodomy, which Malthus called "vices." In later writings, Malthus allowed that "moral restraint" (abstinence from intercourse) might play some role in checking population growth and thereby lessen the amount of misery and vice required. Contraceptive devices, which were neither widespread nor effective in his time, were not considered in his argument.

Thus Malthus was led to what has been called the *Dismal Theorem* (of course it is not a theorem in the usual mathematical sense of the word).

Dismal Theorem

In the long run, populations reach a stage where misery and vice are constantly present.

How much of this do you believe? In particular, do these ideas apply to the United States population depicted in Figure 6? We observe that the exponential curve which fits the early years does not fit later; there seems to be a slowing of the population growth. Do we conclude then that misery and vice are operating to restrain the American population?

It is worth noting that the Dismal Theorem doesn't specify when the misery and vice will arrive (imprecision again). This is relevant to some modern arguments about whether Malthus was right or wrong. Critics say he was wrong because in the last 200 years an ever-growing population has enjoyed an ever-better quality of life and misery and vice have been kept at bay. (For example, see the persuasive book by Julian Simon, listed in the Bibliography.) Others say that Malthus was right, but the "long run" just hasn't arrived yet.

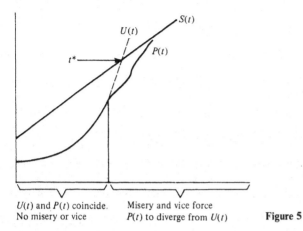

$U(t)$ and $P(t)$ coincide. No misery or vice

Misery and vice force $P(t)$ to diverge from $U(t)$

Figure 5

Our personal view of Malthus is composed of good news, bad news, and a dose of nervous hope. The good news is that food supply over the last 2 centuries has not been restricted to a linear growth curve. The chief reason for this is a factor Malthus could not have foreseen, namely the fact that scientific farming plus capital investment (farm machinery, fertilizers, etc.) have made it possible for food production to grow exponentially.

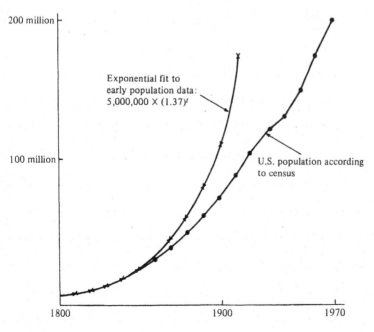

Exponential fit to early population data: $5,000,000 \times (1.37)^t$

U.S. population according to census

Figure 6

The bad news is that there may be other unpleasant checks to population that Malthus did not foresee, such as:

1. Pollution-related illnesses
2. Shortages of natural resources (oil, metals, etc.)
3. Effects of crowding—disease, social strife, etc.

The nervous hope is that intelligent use of family-planning methods might make it possible to control population growth without relying on the painful methods that Malthus referred to as misery and vice. Unfortunately family planning has not caught on in the underdeveloped nations to the extent that it has in the United States (as of 1983); so it's anybody's guess whether it will play a great role.

In the early 1970s attempts were made to build *world models* to project future population in a more detailed and precise way than Malthus did. These models took into account the fact that food production depended on capital investment, population (as population grows there is less land for agriculture), and pollution (which reduces the productivity of the land). They also took into account the dependence of population on natural resources, pollution, and crowding.

The world models presented by Jay Forrester and his colleagues contain five main variables: population, capital investment in industry, capital investment in agriculture, natural resources, and pollution. In addition there is a host of subsidiary variables. (Food supply becomes a subsidiary variable in Forrester's system!) Figure 7 shows some of the cause and effect relationships in one such model. This model had over 100 variables and constants, linked up with over 50 equations. Needless to say, in order to study this model, it is necessary to translate it first into a computer program.

Each cause and effect relationship in Figure 7 has a specific equation describing it, although these aren't shown in the figure. Each of these equations is quite precise; none of Malthus' vagueness about what the growth rate and other numbers actually are. The reason for including all this detail and precision in the building of the model is to produce projections of the future which would be more precise and accurate than anything we could obtain from Malthus' model. Figure 8 is typical of the sorts of the results which were obtained. Precise curves are plotted for the future for each of the five main variables. These curves appear to give answers to questions like: Will population continue to increase? If not, in what year will it start to decline? How steep will the decline be?

If you find Figure 8 depressing, you might be happy to hear that a substantial number of scientists consider the world dynamics models to be worthless because their predictions, despite their precision, are likely to be highly inaccurate. Many of these critics would be happy to have more precise models of the population-resource problem that would overcome the oversimplification and lack of precision in the basic Malthusian model. However, they feel that we do not yet know enough to formulate the correct equations. In particular, they feel that there is only a small probability that the Forrester models are based on correct equations, and therefore

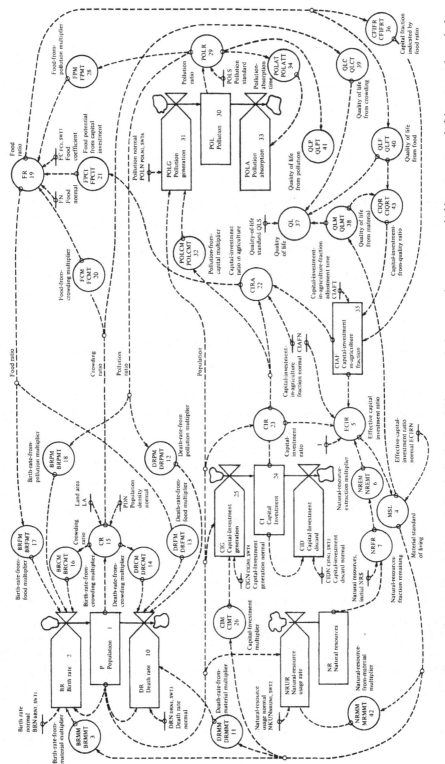

Figure 7 Complete diagram of the world model, interrelating the five level variables—population, natural resources, capital investment, capital-investment-in-agriculture fraction, and pollution. (*From "World Dynamics," by Jay W. Forrester, Wright-Allen Press, 1971. © MIT Press. Reprinted by permission.*)

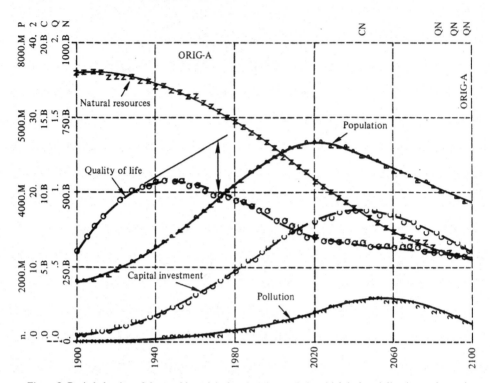

Figure 8 Basic behavior of the world model, showing the mode in which industrialization and population are suppressed by falling natural resources. (*From "World Dynamics," by Jay W. Forrester, Wright-Allen Press, 1971. © MIT Press. Reprinted by permission.*)

only a small chance that the predictions are accurate. Basically these critics are in the position of preferring vague and imprecise models (like that of Malthus) until such time as models can be built which are accurate as well as precise.

Sidelight: Life on Mars

In 1976 two Viking spacecraft made landings on Mars to search for life. In the hope of finding microscopic organisms, soil was scooped into a test chamber and observed with various instruments, including a geiger counter. One particular experiment involved squirting the soil with a liquid containing chemicals which martian microbes might use for food. The chemicals in this microbial "chicken soup" involved the element carbon which had been made deliberately radioactive.

On earth, living things eat carbon compounds. Some of the carbon in the food gets combined with oxygen to make carbon dioxide, which is exhaled as a gas. If the carbon has been radioactively tagged, this exhaled gas can be detected by a geiger counter.

Therefore, if the geiger counter on board a Viking lander detected radioactive carbon dioxide coming from the soil—which it did!—this might be a sign of life. But there is another interpretation. There are ordinary chemical reactions, which don't involve living things, but which can move carbon atoms from the "chicken soup" into gases in the atmosphere above the soil in the test chamber. So there's a puzzle: is the geiger counter detecting life or ordinary chemistry?

Here's one way out of the puzzle: even on Mars, reproduction of organisms should give rise to exponential growth in their numbers. This should also give rise to exponential growth in the amount of radioactive gas produced. As it happened, no clear exponential trend appeared in the data. For this and other reasons, the consensus of scientific opinion is that the Viking mission did not find life on Mars.

EXERCISES

● 1 (a) Use the substitution $r = e^s$ to find a formula for the derivative of r^t with respect to t.

(b) Show that r^t is a monotonically increasing function of t provided $r > 1$.

2 Show that, if $r > 1$, $r^t \to \infty$ as $t \to \infty$. (*Hint*: Modify the proof of Theorem 3, expressing r^t as e^{st} with $s > 0$.)

● 3 Show that, if $0 < r < 1$, $r^t \to 0$ as $t \to \infty$. (*Hint*: $r^t \to 0$ if and only if $1/r^t \to \infty$.)

4 Suppose food supply could grow according to the second power of time, so that the supportable population were $S(t) = S_2 t^2$. Show that, even in this case, there is a t_0 where $U(t) \geq S(t)$ for $t \geq t_0$. [*Hint*: Use Exercise 1 to write $U(t) = U(0)e^{ct}$ and express this as a Taylor series.]

● 5 Show that, if $r > 1$, there is a value T, which depends on r, with the property that, for any t, $r^{t+T} = 2r^t$. T is called the *doubling time* for the exponential growth curve r^t. Express T as a function of r.

6 Show that an approximate formula for the doubling time (see Exercise 5) is $T = 0.69/(r - 1)$. (*Hint*: Replace the denominator of the formula found in Exercise 5 by the constant and linear term of its Taylor-series expansion about the point $r = 1$.)

● 7 If population numbers A_1 one year and A_2 the next, then the percentage growth rate is defined as $[(A_2 - A_1)/A_1] \times 100$. Suppose a population follows the exponential law $P(t) = P_0 r^t$, where t is measured in years. Calculate the percentage growth rate (as a function of r) from:

(a) Year 0 to year 1

(b) Year 1 to year 2

(c) Year t_0 to year $t_0 + 1$

8 Suppose $P(t) = S_0 + S_1 t$. Show that it will not have the same percentage growth rate from one year to the next.

9 Suppose a population following the growth curve $P(t)$ has the property that its percentage growth is always the same from year to year, say g. Show that this implies that $P(t)$ displays exponential growth.

10 If g is the percentage growth rate for an exponentially growing population, then show that the doubling time (see Exercises 5 and 6) is given by approximately $69/g$.

11 Table 2 shows annual percentage growth rates for some countries as of 1973. Assuming that the population is growing exponentially in each case, calculate the doubling time for each country.

12 Theorem 3 gives no information about the value of t^* and how it depends on r, A, $S(0)$, and $U(0)$. Suppose we take $A = S(0) = 2U(0) = 100,000$. For each value r in the table below, determine t^*, the first integer value of t for which $U(t) \geq S(t)$, by computing powers of r.

r	1.01	1.02	1.05	1.10	1.20	1.30
t^*						

Table 2

Country	Percent growth	GNP,[†] $	Doubling time
Algeria	3.3	300	
Port. Guinea	1.1	250	
Kuwait	9.8	3760	
Japan	1.2	1920	
United States	0.8	4760	
Argentina	1.5	1160	
Ecuador	3.4	290	
Bahamas	4.6	2300	
Sweden	0.3	4040	
W. Germany	0.0	2930	
Albania	2.8	600	

[†] GNP = per capita gross national product.

● **13** (a) What would happen in our theoretical model of population size presented in the discussion surrounding Figure 2 if parents survived past the birth of their children but not to the birth of their grandchildren? Work out the total population at time $t = 0, 1, 2, 3,$ and 4 if we start with four women at $t = 0$.

(b) Can you find the formula for the total population at time t, based on the assumptions in part (a)?

14 Revise Exercise 13 by assuming that half of all women die before the birth of their children and the other half die after their children are born but before their grandchildren are born.

● **15** Suppose we modify the model accompanying Figure 2 so that Maria and her descendants all have three female children who survive to have children of their own.

(a) Express the formula for the population descended from Maria, Jane, and Sarah in the tth generation (those just born at time t).

(b) What is the ratio of Maria's descendants to all others in the tth generation. Compute this for $t = 0, 1, 2, \ldots, 20$. What do you think happens to the ratio as $t \to \infty$? Can you prove it?

16 The following problem was proposed by the thirteenth century mathematician Fibonacci. Suppose a pair of rabbits, jusy born at $t = 0$, will themselves produce a pair of baby rabbits when they are 2 months old ($t = 2$) and another pair each month thereafter.

(a) If rabbits never die, how many pairs are there at time t for $t = 0, 1, \ldots, 10$. (These numbers form what is called the Fibonacci sequence.)

(b) If f_t denotes the tth term of the Fibonacci sequence, show that $f_t = f_{t-1} + f_{t-2}$.

(c) Use a computer to compute the values of $(1 + \sqrt{5}/2)^t + (1 - \sqrt{5}/2)^t$ for $t = 0, 1, \ldots, 20$ and show that this formula yields the terms of the Fibonacci sequence for $t = 0, 1, \ldots, 20$.

17 Answer the questions posed in this section after the statement of the Dismal Theorem.

18 Explain why an initially unchecked population may begin to be restrained before t^* (see Figure 5).

19 What do you think Malthus would expect the relation to be between a country's rate of population growth and its wealth? Explain. Do the figures in Table 2 bear this out?

20 What flaws do you see in the arguments leading up to the Dismal Theorem?

21 Is Theorem 3 necessary to arrive at the Dismal Theorem or would it suffice to use part 1 of Theorem 2?

Computer exercises

22 Use a computer program to work out the t^* values for the table in Exercise 12. Do this by computing $U(t)$ and $S(t)$ for $t = 0, 1, \ldots$, until you arrive at a t value where $U(t) \geq S(t)$.

23 (*a*) How many people have ever lived up to and including the *i*th generation in the model accompanying Figure 2? Call this number $T(t)$. Does $T(t)$ show exponential growth? How about in the long run? In particular, does $T(t + 1)/T(t) \to r$ as $t \to \infty$? Use a computer program to get some evidence about this. Use $G(0) = 3$ and a variety of positive r values.

(*b*) Notice that $T(t)$ is the sum of finitely many terms of a geometric series. Use this to find a formula for $T(t)$. Can you use this to get information about $\lim T(t + 1)/T(t)$ as $t \to \infty$?

BIBLIOGRAPHY

Cole, H. S. D., et al.: "Models of Gloom," Universe Books, New York, 1973. A criticism of the world dynamics models of Forrester and others.

Forrester, Jay W.: "World Dynamics," Wright Allen Press, 1971. An ambitious computer simulation of the whole world.

Malthus, Thomas: "An Essay on the Principle of Population," Anthony Flew, ed., Penguin Books, Baltimore, 1970. The long introduction by Flew is an excellent summary of Malthus' theory and its historical context.

Roberts, Fred S.: "Discrete Mathematical Models," Prentice-Hall, Englewood Cliffs, N. J., 1976. Chapter 4 describes an interesting method for describing systems as complex as in the world dynamics models but without the need for hard-to-obtain precision.

Simon, Julian: "The Ultimate Resource," Princeton University Press, Princeton, N.J., 1981. A vigorous attack on Malthusian thinking. Simon argues that population growth is highly desirable at the present time.

6 GENERALITY AND FRUITFULNESS—FROM ASTROLOGY TO TECHNOLOGY WITH KEPLER AND NEWTON

Abstract A model which applies to many things is said to be very general. If model 1 explains everything that is explained by model 2 and more besides, then we call model 1 a *generalization* of model 2. Although we prefer general models, it is not easy to find them. In this section we present one of the most striking examples in the history of science where one model generalizes another: Newton's laws as a generalization of Kepler's laws. We also show how one of Kepler's laws helped verify one of Newton's laws and may have even inspired it.

Prerequisites Plane geometry.

I remember a sense almost of intoxication when I first read Newton's deduction of Kepler's second law from the law of gravitation. Few joys are so pure or so useful as this.

Bertrand Russell

Johannes Kepler (1571–1630) is sometimes considered the first of the modern scientists. Perhaps it is a measure of how far science has come that his famous three laws seem more like curiosities than cornerstones of science today. Compared with the work of Isaac Newton, less than a century later, Kepler's discoveries seem to be completely eclipsed because of the much greater generality of Newton's laws. As we shall see, Kepler's laws describe the motions of the planets and nothing else. But Newton's laws explain the motions of the planets and an enormous lot more besides.

Nevertheless, Kepler's laws are of fundamental importance for the history of science. A major reason is their fruitfulness. Because they provided the first account, in the history of the world, of the planetary motions which was both simple and precise, they gave scientists a very digestible kind of food for thought. In particular, Kepler's laws seem to have had some role in inspiring Newton to formulate his far more general laws. Unfortunately, Newton was closemouthed about exactly how he got his ideas; so we can't be too sure how much credit is due Kepler for setting Newton on the right course. However, one thing is clear. In his great work, the "Principia," Newton shows in detail how his laws imply Kepler's and how Kepler's can be used to deduce some of his. Thus Kepler's laws may rank among the most fruitful mathematical models in the history of science.

Maybe you are wondering why anybody cared about describing the motions of the planets. In the modern world, only space scientists concern themselves with how the planets behave for the very good reason that the planets have little impact on our daily lives. But in the medieval world nearly everyone, including Kepler and other scientists, believed in astrology and thought that, if we could understand the heavens, we would thereby understand human affairs as well.

Before listing Kepler's laws, a word about terminology is in order. Kepler's and Newton's laws are, in modern terminology, just modeling assumptions. Calling them "laws" is a historical accident and does not mean that these modeling assumptions express some absolute truth. Here are Kepler's laws:

K1. The orbit of each planet is an ellipse with the sun at one focus (Figure 1).
K2. For any given planet, the velocity with which it travels its elliptical course is not constant. Instead, the velocity varies in such a way that the *radius vector* (line segment) from the sun to the planet sweeps out a constant area in each unit of time. For example, in Figure 2, since the two shaded areas are equal, the planet will traverse the shorter arc $P_1 P_2$ in the same time as the longer arc $P_3 P_4$.
K3. The ratio of the cube of half the major axis of a planet's orbit (CT in Figure 1) to the square of the period (the time required for one revolution around the sun) has the same value for each planet in the solar system.

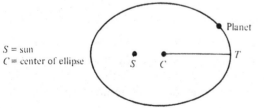

S = sun
C = center of ellipse

Figure 1 Kepler's first law.

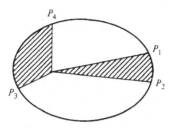

Figure 2 Kepler's second law.

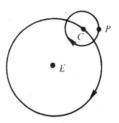

Figure 3 Two circular motions creating an epicycle.

To the modern mind, Kepler's first law seems as reasonable a way for the planets to behave as any other. But in Kepler's time scientific thought was burdened by certain incorrect assumptions and mystical ways of thinking that had an almost sacred status. One of these mistaken notions was the assumption that the movements of the planets were either uniform (i.e., constant velocity) circular motions or composed of such motions.

Thus it was permissible to believe that a heavenly body revolved in a circle about the earth, or, after Copernicus, the sun. Or, if this didn't fit the facts very well, one might add an *epicycle*. Epicyclic motion with one epicycle (see Figure 3) can be visualized as being composed of two separate circular motions. First imagine that the body P revolves around point C at a constant rate v_1. But now suppose that C moves around E in a circle and at a constant rate v_2 and drags P with it. As a physical manifestation of this, imagine a spinning top whose point follows a circular path around E. The planet would be represented by a point on the edge of the top of the top. Depending on the relative sizes of the velocities v_1 and v_2, the radii of the circles and the directions of rotation of the circles, one could get a composite motion of the planet P around E like that of Figure 4a, b, or c.

In some cases the ancient astronomers found it necessary to arrange a whole series of epicycles, one upon the other, to produce a motion that accorded with the observations made on the positions of a given planet in the sky on different days of the year. On the whole, the ancients did a reasonable job of fitting the data with this system, called the Ptolemaic system, based on uniform circular motion and epicycles. However, theoretically it was a dead end because it did not lead to any further insights. From the practical point of view it was a large pain in the neck. Epicycles are hard to understand, visualize, or calculate with, even for the best

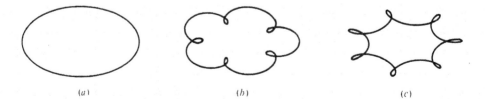

(a)   (b)   (c)

Figure 4 Three epicycles.

horoscope makers. They were a stiff price to pay for being fond of circles. (If you are having trouble understanding epicycles, take heart: The whole point of Kepler's work was to do away with them, and we won't need them further.)

As you can see, Kepler's laws simply describe what the state of affairs is. They do not say *why* the planets move as they do. Furthermore the laws are not derived from other principles with elegant or complicated mathematics: They were obtained by Kepler largely by exhaustive examination of the existing data and by searching for a concise, mathematically elegant way of describing what the planets were doing.

Newton's great achievement was to deduce the planetary motions from a set of principles of motion which applied equally well to apples falling from trees or to the revolutions of the planets. To accomplish this, he relied in part on the concept of force and a set of rules for how forces influence motion.

One of Newton's key laws, sometimes called the law of inertia, is:

N1. (*a*) A body at rest (velocity = 0) tends to remain at rest unless acted upon by a force.
(*b*) A body in motion (velocity ≠ 0) tends to remain in motion at the same speed and in the same direction, unless acted upon by a force.

As this law shows, Newton thought about a force as something capable of influencing a velocity. For example, when you kick a stationary soccer ball you impart a force to it, cause it to move, and change its velocity from 0 to something else. (But forces don't always produce changes in velocity. When a car drives onto a bridge, it provides a downward force on the bridge, but the bridge, because of its supports, provides an equal force in the upward direction; so there is no change in the zero velocity of the bridge.)

Although this law may seem uncontroversial, it had a revolutionary aspect because Newton asserted that it held for heavenly bodies as well as for those on earth. Scientists had always believed that straight-line motion was the "natural" motion of objects on earth, but they maintained that the natural motion of a heavenly body was a circle or, after Kepler, an ellipse. In a word, physics in the heavens was expected to be different from physics on earth. Newton upset the applecart by claiming that the laws of physics were the same up there as down here.

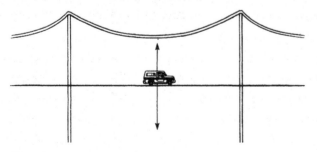

Figure 5

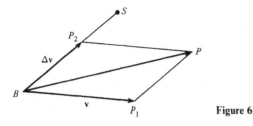

Figure 6

Mathematically, velocities and forces are represented by vectors. The direction of the vector represents the direction of the force or velocity and the length of the vector represents the magnitude of the force or velocity. Figure 5 shows vectors representing the downward force of a car and the upward supporting force of a bridge canceling each other out. We designate the vector from point P_1 to P_2 by $\overrightarrow{P_1P_2}$ or by a single boldface letter such as **v** (see Figure 6, where $\overrightarrow{BP_1}$ is also designated as **v**).

Here is another of Newton's key laws:

N2. When a force changes the velocity of a body, the amount of the change is proportional to the size of the force and is in the direction of the force.

This needs a bit more explanation. Suppose (see Figure 6) a body B has a velocity $\mathbf{v} = \overrightarrow{BP_1}$ and a force is applied in direction $\overrightarrow{BS}$. Then the *change* in velocity is represented by some suitable vector $\Delta\mathbf{v}$ in the direction of $\overrightarrow{BS}$, say $\Delta\mathbf{v} = \overrightarrow{BP_2}$. But $\Delta\mathbf{v}$ is not the new velocity. To obtain the new velocity, we need to combine the two velocities **v** and $\Delta\mathbf{v}$ in some manner. Newton asserted that the method of combination was the parallelogram law, which is:

N3. To combine a velocity **v** and a change in it (due to a force) $\Delta\mathbf{v}$, as in Figure 6, complete a parallelogram with **v** and $\Delta\mathbf{v}$ as two sides. The final velocity is represented by the vector $\overrightarrow{BP}$ from B to the fourth point of the parallelogram.

Example 1

Imagine a billiard ball rolling on a billiard table in a straight line and at constant velocity. (In reality the velocity won't be constant because of the force of friction with the table and the air, but we're going to ignore that.) At B it is hit by another ball with a force that causes a change in velocity $\Delta\mathbf{v}$. Figure 7 shows two cases of such a situation. In each case we can use the parallelogram law to find the resulting velocity $\overrightarrow{BC}$. Notice that in case a the ball winds up with a smaller magnitude velocity than it had before the other ball struck it, while in case b the new velocity has a greater magnitude. This is owing to the differing angles at B. Can you determine which angle leaves the magnitude of the velocity the same after as before (assuming **v** and $\Delta\mathbf{v}$ have the same magnitude)?

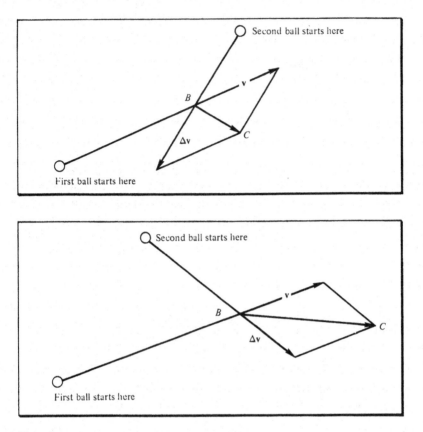

Figure 7

In the previous example we had a case of a momentary force, which acts just for an instant. But many forces, such as gravity, are continually acting ones, and these demand different treatment. To understand the difference, suppose we roll a steel billiard ball on a table that has a powerful electromagnet along one cushion, which we can activate by pushing a button. The harder we push, the stronger the force. If we push the button hard just for an instant, the path of the ball will be deflected just as if another ball had crashed into it. The path will be a broken line,

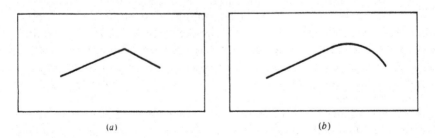

(a)   (b)

Figure 8

as determined by the parallelogram law and illustrated in Figure 8*a*. But if we hold the button down gently for a few seconds, we get a curved path, as in Figure 8*b*.

Finding the equation of such a curved path from the parallelogram law doesn't seem very easy. Because the planets move in curved orbits, the force making them curve must be a continually acting one, and we are faced with the same difficulty. Newton helped invent the calculus largely to help out with these problems involving continually acting forces. But he also showed how to think about them with a more elementary technique, involving the parallelogram law and the idea of approximation. We will follow his tracks in this elementary argument.

We can illustrate the idea of approximation with our magnetic billiard table (see Figure 9). Suppose that, when the ball reaches *B*, we press the button, then let it up, then press it, etc. This will produce a broken-line path that approximates the curved path that would result if we held the button down all the time. The nice feature is that each bend in the broken line can be computed by the parallelogram law. The price we pay for this convenience is that we only have an approximation to the curved path. But if we jiggle the button up and down more frequently, we get a better approximation. In the limit our approximations become identical to the curved path.

We turn now to Newton's thoughts about planets orbiting the sun. In view of the curved orbit and law N1 there must be a continually acting force. We can think of this force as being approximated by a series of momentary forces that flick on at $t = 0, 1, 2, 3, \ldots$. But this doesn't tell us the nature of the force. Newton's bold step was to assert that

N4. The force that keeps the planets in orbit is a force emanating from the sun.

The idea of a force emanating from the sun was hard for people to accept in Newton's day because there was no mechanical connection from the sun to the planets. Had there been a great cosmic chain connecting the sun to a planet, there would have been no difficulty, but "action at a distance" seemed quite improbable. Newton himself said that it seemed weird to him but it was the best idea he could think of.

Perhaps a modern analogy can be drawn to the concept of extrasensory perception. Over and over, honest researchers report that in a well-designed experiment subject *A* can "read the mind" of subject *B*, even though they are separated by a great distance; over and over, others will be skeptical because there was no known communication channel between the subjects. We do not mean to suggest that extrasensory perception is as well-established a phenomenon as gravitation.

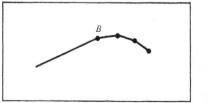

Figure 9

(Far from it!) We just mean that perhaps it is no more bizarre to a mind not previously inclined to think that way.

Newton went further and claimed that this attraction the sun exerted was the same kind of attraction the earth exerted on objects to make them fall, namely *gravity*. He also asserted that all objects in the universe exert a force of gravity upon one another, and he gave a formula to calculate it.

As far as the sun was concerned, Newton was able to use Kepler's second law to corroborate assumption N4. What he shows is:

Theorem 1

N1, N2, N3, and K2 imply that the force on a planet points toward the sun (and therefore probably emanates from the sun).

Proof We approximate the force by an intermittent one acting momentarily at $t = 0, 1, 2, 3, 4$, etc., at which times the planet is at points P_0, P_1, P_2, P_3, P_4, etc. (See Figure 10.)

Because the time step is one unit, the vector $\overrightarrow{P_1 P_2}$ represents the velocity the planet has at P_1. Since there are no forces acting between the "switch on" at $t = 1$ and the "switch on" at $t = 2$, the velocity of the planet at P_2 before the switch on is a vector of the same direction and length, namely $\overrightarrow{P_2 P_3'}$, where

$$\text{Length } \overrightarrow{P_1 P_2} = \text{length } \overrightarrow{P_2 P_3'}$$

However, when the force goes on at $t = 2$, according to law $N2$, we get a change in velocity $\Delta v = \overrightarrow{P_2 Q}$, whose direction and magnitude are, at the moment, unknown to us (see Figure 11). Our objective is to show that this vector points to S, and we can accomplish this by showing Q lies on SP_2 (in other words, Figure 11b is correct and Figure 11a is not).

After applying the parallelogram law (N3) to **v** and Δv we produce a vector $\overrightarrow{P_2 P_3}$ where P_3 is the planet's location at $t = 3$. Because $P_2 P_3' P_3 Q$ is a parallelogram, $P_3 P_3'$ is parallel to $P_2 Q$. Therefore, to show that Q lies on SP_2, it suffices to show that $P_3 P_3'$ is parallel to SP_2. Here's how we do this (see Figure 12):

Since the bases $P_1 P_2$ and $P_2 P_3'$ of triangles $SP_1 P_2$ and $SP_2 P_3'$ are equal and both triangles have the same altitude (SR):

$$\text{Area } \Delta SP_1 P_2 = \text{area } \Delta SP_2 P_3'$$

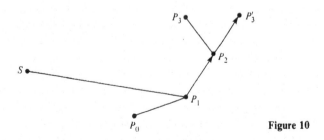

Figure 10

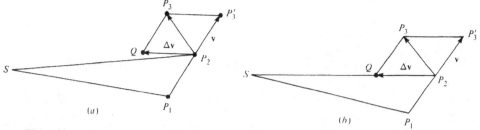

Figure 11

By Kepler's second law (K2),

$$\text{Area } \Delta SP_1P_2 = \text{area } \Delta SP_2P_3$$

Therefore

$$\text{Area } \Delta SP_2P_3 = \text{Area } \Delta SP_2P_3'$$

But since these last two triangles have the same base SP_2, the altitudes from P_3 and P_3' to this base must be equal in order for the areas to be equal. Now if the perpendicular distances from P_3 and P_3' to SP_2 are the same, $P_3'P_3$ must be parallel to P_2S. Consequently, Q lies on P_2S, which means that the force acting on the planet points to the sun.

The argument in the theorem can be reversed to show that, if the force points toward the sun and we assume N1, N2, and N3, then we can deduce Kepler's second law (K2).

The theoretical connection between Kepler's second law and Newton's laws, which we have been discussing, is also discussed in Newton's own writings. Unfortunately, we cannot tell from these writings what role Kepler's laws played in guiding Newton's thought processes. It may be that Newton started from Kepler's laws and arrived at the sun's gravitation (and other newtonian concepts) by a process of deduction, rather like our proof of Theorem 1. Or it may

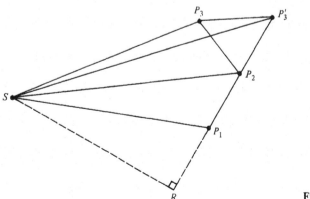

Figure 12

be that Newton arrived at the sun's gravitation some other way and then noticed that it could be derived as in Theorem 1. Either way, Kepler's laws played a fruitful role in the development of physics. It is only the exact details of this fruitfulness that can be disputed.

But the case for the greater generality of Newton's laws rests on a more massive and completely undisputable basis. Every textbook in physics and engineering written since his time is based on his ideas concerning the interactions of forces and velocities. These ideas are used to calculate the best designs for bridges, artificial hearts, automobiles, and almost every other element of our modern technology. Newton's idea that gravitational attraction is a characteristic not only of the earth but of every other body in the universe is the starting point for everything we know about astronomy. But universal gravitation is relevant to earth-bound phenomena as well. For example, Newton was able to explain the existence of ocean tides as a result of the moon's gravitational pull on the oceans. Nothing in Kepler's laws bears on that question.

Indeed, it is safe to say that, when Newton's work became widely known and accepted, Kepler's laws had much less practical or theoretical value for science. However, the value of his mathematical model for planetary motion lies precisely in that it was a stepping stone toward Newton's creation of a more general model.

Sidelight: Making the Most of a Mistake

Science is not always a simple march of progress. Sometimes attractive ideas turn sour. Sometimes weird-sounding ideas bear good fruit. Here is a situation where an attractive idea turned out to be wrong but bore fruit anyhow.

In 1772 the astronomers Johann Titius and Johann Bode discovered that the distances from the sun to the planets known at the time showed an interesting numerical pattern. If we take the distance from the sun to Saturn (the farthest planet known at the time) to be 100, then the distances are approximately:

Sun to Mercury	4
Sun to Venus	$4 + 2^0(3)$
Sun to Earth	$4 + 2^1(3)$
Sun to Mars	$4 + 2^2(3)$
?	$4 + 2^3(3)$
Sun to Jupiter	$4 + 2^4(3)$
Sun to Saturn	$4 + 2^5(3)$

Mercury doesn't fit the pattern exactly. [It should be $4 + 2^{-1}(3)$ to fit the pattern.] But the real problem is that there is no planet with distance $4 + 2^3(3)$. Bode insisted that there really was a planet at that distance but we just hadn't observed it yet. Astronomers of the time believed his argument, and, in Germany, a group of them called the *celestial police* were formed to look for the missing planet.

In 1801 Guissepe Piazzi found, not a planet, but a smaller body called an asteroid at the distance of $4 + 2^3(3)$. Before long the celestial police and others found many more asteroids at about that distance. It appeared that these were fragments of a planet that somehow got smashed into pieces.

The discovery of the asteroid belt was a great success for Bode's law and encouraged enthusiasts to search for planets beyond Saturn at distances which were calculated by extending the progression in the table. The next planet should be at a distance of $4 + 2^6(3)$. Sure enough, that's where Uranus turned up when it was discovered. But when Neptune was found, its predicted distance was about 30 percent greater than the true distance. Later, when Pluto was discovered with a predicted distance almost 100 percent greater than the actual distance, Bode's law was regarded as thoroughly demolished.

May it rest in peace! Right or wrong, it was fruitful anyhow because it spurred the discovery of the asteroid belt.

EXERCISES

● **1** In Figure 4a and c, one figure has both circles which make up the epicycle (Figure 3) going clockwise and the other has one going clockwise and the second going counterclockwise. Can you tell which is which?

2 (a) What would an epicycle look like if
   (i) P started out collinear with E and C and
   (ii) Each circle spun with the same angular velocity (degrees per unit time)
   (b) Which of the pictures in Figure 4 most resembles the epicycle that results when the circle centered at P doesn't spin at all?

● **3** Which of Figure 4a or b corresponds to a larger ratio of v_1/v_2?

4 In an epicyclic motion in which the circles at E and C have the radii r_2 and r_1 respectively, what will be the largest and smallest distances of the planet P to the earth E?

● **5** Suppose a body at A has velocity represented by vector $\mathbf{v} = \overrightarrow{AB}$ and suppose a force imposes a change in velocity $\Delta \mathbf{v}$ represented by vector $\overrightarrow{AC}$. Suppose the lengths of $\mathbf{v}$ and $\Delta \mathbf{v}$ are equal. Under what condition on angle BAC will the new velocity vector have greater magnitude than $\mathbf{v}$?

6 In Figure 12, does Kepler's second law determine the exact position of P_3? If so, prove it. If not, what is the locus of possible positions for P_3 which are consistent with Kepler's second law?

● **7** Does the proof that the force acting on a planet points to the sun require Kepler's first law, or would it still be valid for some different type of orbit?

8 Prove that, if the force acting on a planet points toward the sun and N1, N2, and N3 hold, then this implies Kepler's second law (K2).

Computer exercise

9 Program a computer to create an epicyclic orbit. (For best results use a graphics terminal because you will need a lot of points and hand plotting coordinate pairs generated by a computer will be difficult.) Use radii of 4 and 1 for the circles. Express the velocities of C and P as follows: let a time counter I (the index of a loop) take on values $I = 1, 2, 3, \ldots, 100$; specify the velocity $\mathbf{v}_1$ of C by selecting a number of degrees which C turns through in each unit increment of I; specify the velocity of P similarly, say $\mathbf{v}_2$. By choosing different values for $\mathbf{v}_1$ and $\mathbf{v}_2$, try to produce pictures like Figure 4.

BIBLIOGRAPHY

Cohen, I. Bernard: Newton's Discovery of Gravity, *Sci. Amer.*, vol. 244, no. 3 p. 166, March, 1981. A historical study.

Fink, A. M.: "Kepler's Laws and the Inverse Square Law," UMAP Module 473, available from COMAP, 271 Lincoln St., Lexington, MA 02173.

Simmons, George F.: "Differential Equations with Applications and Historical Notes," McGraw-Hill Book Company, New York, 1972. Section 21 of chapter 3 has a calculus-based discussion of Newton's and Kepler's laws.

7 ROBUSTNESS—THE UPS AND DOWNS OF ANCIENT ASTRONOMY

Abstract A robust model is one which is not too sensitive to errors in the data the model is based on. Robustness is not a characteristic of the data but rather of the model. In this section we study two models in which the errors in the data are of the same order of magnitude but the errors in the answers are very different because one model is robust and the other is not. The fate of one of these models may help explain Christopher Columbus' discovery of America.

Prerequisites Plane geometry, trigonometry, some differential calculus.

Most mathematical models are at least partly based on measurements made in the real world. Naturally, the accuracy of any conclusions obtained from the model will depend on the accuracy of the observations. But some models are more sensitive than others to errors in the data upon which they are based. In this section we illustrate this with two examples from one of the oldest sciences, astronomy.

Aristarchus of Samos (310–230 B.C.), an early Greek astronomer and geometer, may well deserve a prize for the most outrageously wrong answer produced by an otherwise excellent mathematical model. His model was designed to find the ratio of the distance from the earth to the sun to the distance from the earth to the moon. Of course, we know that, since the earth goes around the sun in an elliptical orbit, it doesn't have a constant distance to the sun. Likewise there is no one distance from the earth to the moon. But Aristarchus didn't know this. He assumed that the earth's orbit was circular, with the sun at the center, and that the moon's orbit was also circular, with the earth as center. Although these assumptions are wrong, they aren't far wrong; so it's not a bad idea to suppose that the distances are constant and therefore have some constant ratio.

To carry out his calculation, Aristarchus waited for the earth, moon, and sun to arrange themselves into a right triangle. (See Figure 1.) This is easier said than done. Think for a minute. How would you determine when angle EMS was $90°$? Before we plunge into this question, observe that, if Aristarchus could then measure the angle θ, he could determine the ratio $r = ES/EM$ by

$$\frac{ES}{EM} = \sec \theta \tag{1}$$

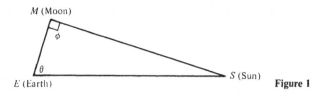

Figure 1

To find θ, you will need to have the sun and moon visible to you simultaneously. Then look in the direction of the sun and see how many degrees you must turn to see the moon. Naturally this is best done with some kind of instrument. (Do you think you could design such an instrument?) Aristarchus made θ out to be 87°. His instrument was not very good because the true value is 89°50'. Under some circumstances a discrepancy of less than 3° would not be serious. However

$$\sec 87° = 19.11 \qquad \text{Aristarchus' value}$$
$$\sec 89°50' = 343.77 \qquad \text{True value}$$

Thus the true ratio of distances is about 344, approximately 18 times the value Aristarchus got. The reason that a few degrees makes such a big difference is best seen by observing how steep the graph of the secant function becomes in the neighborhood of 90°. (See Figure 2.)

How did Aristarchus determine when the three bodies formed a right triangle? The answer is related to the question of why there are phases of the moon. Since the moon has no light of its own, the difference between a full moon, half-moon, or no visible moon at all lies in two factors:

1. How much of the moon is illuminated by the sun.
2. How much of the illuminated portion of the moon we can see from the earth.

The portion of the moon illuminated by the sun depends on only two factors: first the relative sizes of the sun and moon (compare Figure 3a and b, where the sizes differ but the distance apart is the same); second, their distance apart (compare Figure 3b and c). In examining these figures note that none of them gives an accurate representation of the true sizes and distances. Unfortunately, it is important to draw a figure with an accurate scale which shows these sizes and distances. If we were to take 4 inches on the page to represent the distance from sun to moon,

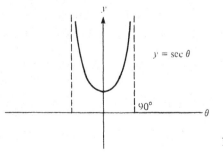

Figure 2

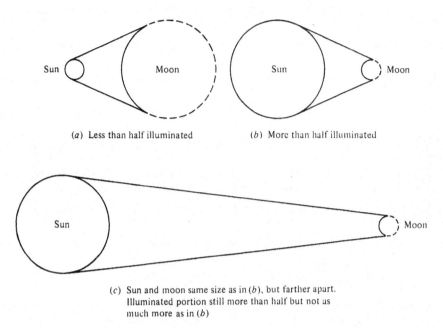

(a) Less than half illuminated   (b) More than half illuminated

(c) Sun and moon same size as in (b), but farther apart.
Illuminated portion still more than half but not as
much more as in (b)

Figure 3 How the portion of the moon that is illuminated depends on the relative sizes of the bodies and their distance apart.

then the circle representing the sun would have to have a diameter of 1 millimeter, which is less than the size of a small letter "o." The moon's circle, being about one four-hundredth of that, would be totally invisible on the page.

In any case, since the sizes of the sun and moon are constant and Aristarchus assumed that the distance from the moon to the sun was nearly constant, he concluded that the portion of the moon that was illuminated was nearly always the same. Therefore, to find this portion, it was merely a matter of finding a favorable time to make the measurement. Exercise 15 shows how this was done. The upshot was that Aristarchus convinced himself that it would cause very little error to suppose that the moon is always half illuminated. (As Exercise 15 shows, he had a very beautiful way to estimate the maximum discrepancy from this estimate of one-half.) Another way to think of this conclusion is this: the sun illuminates the moon with a band of rays, which are, for all practical purposes, parallel (as shown in Figure 4).

Although the portion of the moon illuminated is always one-half, at least in Aristarchus' model, the amount an observer on earth can see depends on the relative positions of earth, sun, and moon (see Figure 4, where three positions of the earth are shown). From this figure it is clear that, when we see a half-moon, the earth will be lined up with the diameter AB of the moon and the angle of the sun-moon-earth triangle at the moon will be 90°.

This, then, is how Aristarchus determined when the three bodies made a triangle with a right angle at the moon. He simply waited for a half-moon.

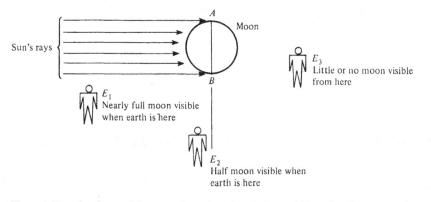

Figure 4 How the phases of the moon depend on the relative positions of earth, moon, and sun (not to scale).

In studying Aristarchus' model, we can find a number of ways in which inaccuracies can creep in. Here is a list of some of them. Do you think it is a complete list?

1. The measurement of θ
2. The assumption of a constant distance from sun to moon
3. The estimation that exactly one-half of the moon is illuminated by the sun
4. The determination of which day the moon appears to be an exact half-moon

We have seen earlier that the model is very sensitive to errors in measuring θ. A modern mathematician would be well aware of this since a modern mathematician is well aware of the shape of the secant curve and can easily calculate what difference a degree makes when calculating the secant of an angle near 90°. Consequently, a modern mathematician would insist on measuring θ with an instrument of high accuracy. But Aristarchus was not a modern mathematician, and he worked under formidable obstacles. In the first place he did not have the luxury of being able to demand a precise measuring instrument. In the second place, analytic geometry had not been invented nor had trigonometric functions been tabulated; so he didn't have a picture like Figure 2 to guide his thinking.

Having discovered that the model is very sensitive to errors in θ, the first item on our list of sources of inaccuracy, we could now investigate its sensitivity to the other three items on the list (see Exercises 17 and 18). However, we leave this as a project for the interested student.

An interesting contrast to Aristarchus' model is an attempt by Eratosthenes (280–195 B.C.) to measure the earth's circumference. Naturally, in his time, he could not do this by actually traveling around it and counting his steps, any more than Aristarchus could get out into space with a long ruler.

The circumstances under which Eratosthenes made his calculation are shown in Figure 5. A and B represent the cities of Alexandria and Syene, respectively, and O is the center of the earth. The sun does not appear in our figure because it is too

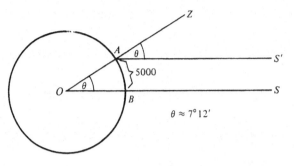

Figure 5

far away. However, the lines AS' and BS are meant to be lines drawn to the sun. These, of course, are the lines along which the sun's rays travel to the earth, and they are approximately parallel. In the model we consider them to be exactly parallel.

The figure illustrates the moment at which the sun is "directly overhead" at Syene. Eratosthenes ascertained this moment by noticing that the sun shone directly down a well at Syene. Assuming the well to be vertical, this means the line SB passes through the center of the earth at O.

Since AS' and BS are parallel, the angle θ at O, which cannot be directly measured, equals angle ZAS'. Eratosthenes was able to measure $\measuredangle\ ZAS'$ and obtained $7°12'$ ($= 7.2°$). (How do you think he did it?) He was also able to determine that the distance from A to B along a longitude line was 5000 stadia; allegedly he did this by counting the steps a camel took in going from A to B. Therefore

$$\frac{\text{Circumference}}{5000} = \frac{360°}{7.2°} = 50$$

so the circumference of the earth is 250,000 stadia.

We can't be sure what a stadium amounts to in miles; so there is uncertainty about how this stacks up against the modern estimate of 24,860 miles. Estimates of what 250,000 stadia means in miles have ranged from 25,000 to 28,903 miles. In the worst of these cases, Eratosthenes' estimate is off by 4,043 miles or 16 percent of the true value. This is not impressive accuracy by modern standards, but it is enormously better than Aristarchus' error.

The equation for circumference C which we obtain from Eratosthenes' model is

$$C = 5000\,\frac{360°}{\theta} \qquad \text{when } \theta \text{ is in degrees} \qquad (2a)$$

$$C = \frac{10{,}000\pi}{\theta} \qquad \text{when } \theta \text{ is in radians} \qquad (2b)$$

Now suppose we had a 3-degree error in measuring θ; e.g.,

$$\text{True } \theta = 10.2°$$
$$\text{Measured } \theta = 7.2°$$

(This was probably not what really happened, but it makes for a convenient comparison with Aristarchus' model since there was an error of about 3° in that model.) Based on these assumptions,

$$\text{True } C = 176{,}471 \text{ stadia}$$
$$\text{Predicted } C = 250{,}000 \text{ stadia}$$

The difference between these numbers is considerable: 250,000 is about 1.4 times as large as 176,471. But even this discrepancy shines by comparison with Aristarchus' model where the true value was about 18 times the predicted value.

How do we account for the fact that two mathematicians, working in the same historical period, having essentially the same instruments and mathematical theory available to them, and both among the best minds of antiquity, nevertheless had such different degrees of success? At least part of the answer lies in the fact that Eratosthenes' model is less sensitive to errors in the measurement of θ. Here is a formal analysis of this.

Suppose the value of θ we measure is θ_0 and the value of C which corresponds to this is C_0. Now the true value of θ will probably be different, say $\theta_0 + \Delta\theta$. Let the C value which the model predicts for this θ be $C_0 + \Delta C$. This is the true value of C. Then ΔC is the error in our prediction of C, and it is caused by the error in θ, which is $\Delta\theta$. Since we are generally interested in percentage errors, we compute these by

$$\frac{\Delta\theta}{\theta_0 + \Delta\theta} \times 100 \quad \text{and} \quad \frac{\Delta C}{C_0 + \Delta C} \times 100$$

It is important to realize that the percent error in θ is something we might be able to determine by testing and studying the instrument we use to measure θ. But the percent error in C is beyond our ability to measure directly; so we need some theory that connects it with the percent error in θ. Actually, the percent error in C is not of direct interest to us. For example, if we had a large error in C that was produced by a large error in θ, this would not necessarily mean that the model was sensitive to errors in θ. But if the same large error in C is due to a small error in θ, this would indicate sensitivity. Therefore, what is of interest to us is the ratio

$$\frac{[\Delta C/(C_0 + \Delta C)] \times 100}{[\Delta\theta/(\theta_0 + \Delta\theta)] \times 100}$$

which simplifies to

$$\frac{\Delta C}{\Delta\theta} \frac{\theta_0 + \Delta\theta}{C_0 + \Delta C} \tag{3}$$

The numerical absolute value of this expression tells us whether input errors are magnified or diminished by the model. Large values indicate large magnifications of percentage error, while a value close to zero indicates a shrinking of the percentage error. An absolute value of 1 is the dividing line: in this case the size of the percentage output error equals the size of the percentage input error.

If we are lucky and have a small enough $\Delta\theta$, then it may be acceptable to replace $\Delta C/\Delta\theta$ by $dC/d\theta$ (evaluated at $\theta = \theta_0$). Likewise ΔC and $\Delta\theta$ may be ignored in the remaining portion of formula (3). This leads us to define the sensitivity of C to small errors in θ, denoted $S(C, \theta)$ by

$$S(C, \theta) = \frac{dC}{d\theta}\frac{\theta_0}{C_0}$$

Example 1

In Eratosthenes' model, using radians, $C = 10{,}000\pi\theta^{-1}$, and so

$$S(C, \theta) = \frac{-10{,}000\pi\theta_0^{-2}\theta_0}{10{,}000\pi\theta_0^{-1}} = -1 \qquad (4)$$

This means that a small overestimate of θ, say 1 percent, gives an underestimate of C of about the same percentage.

The reasoning we carried out for Eratosthenes' model is applicable to any model which yields an equation of the form $y = f(x)$. Thus we make the following definition.

Definition

If $y = f(x)$, then the sensitivity of y to small errors in measuring x, denoted $S(y, x)$, is

$$S(y, x) = \frac{dy}{dx}\frac{x}{y}$$

The sensitivity is usually determined for a particular measurement of x, say x_0. In this case dy/dx is evaluated at x_0 and $x_0/f(x_0)$ is substituted for x/y.

Observe that the sensitivity of C to θ in Eratosthenes' model is the same for all θ_0. However, in our next example, involving Aristarchus' model, the sensitivity does depend on the independent variable.

Example 2

In Aristarchus' model, the ratio of distances r is given by

$$r = \sec \theta$$

Therefore

$$S(r, \theta) = \frac{dr}{d\theta}\frac{\theta}{r} = \frac{\theta \sec \theta \tan \theta}{\sec \theta}$$

$$= \theta \tan \theta$$

(provided we measure θ in radians so the usual differentiation formulas apply).

At 1.5184 radians, which is about equal to the 87° which Aristarchus measured, we obtain the value 28.953. This is an enormously greater sensitivity to small errors than we found for Eratosthenes' model. It means that the percentage error in output is about 29 times the percentage error in input.

Is an error of about 3° small enough to justify the use of the formula in Definition 1? The next example studies this for Eratosthenes' model.

Example 3

We return to formula (3) and calculate this quantity directly for $\theta_0 = 7.2°, \Delta\theta = 3$, and formula (2a). Substituting 7.2 and then 10.2 into Equation (2a) gives

$$C_0 = 250,000$$
$$C_0 + \Delta C = 176,470.59$$

By subtraction, $\Delta C = -73,529.41$. Substituting into Equation (3) gives

$$\frac{\Delta C}{\Delta\theta} \frac{\theta_0 + \Delta\theta}{C_0 + \Delta C} = \frac{-73,529.41}{3} \frac{10.2}{176,470.59}$$

$$= -1.416$$

This is different from the -1 obtained from the small error sensitivity formula $S(C, \theta)$; however, these sensitivities are not drastically different. For further insight into these matters see Exercises 23 and 24.

Often one has a model which involves many variables, for example,

$$y = f(x_1, x_2, \ldots, x_n)$$

We can try to determine the sensitivity of y to any single one of these, say x_i, in more or less the same way as before. As long as we imagine all other variables to have no errors, the appropriate formula is

$$S(y, x_i) = \frac{\partial y}{\partial x_i} \frac{x_i}{y}$$

For example, if $y = 2x_1 - 3x_2^2$, then $S(y, x_1) = 2x_1/(2x_1 - 3x_2^2)$. Now if we substitute the numerical values of x_1 and x_2, this expression gives the sensitivity of y to errors in x_1, assuming no errors in x_2.

Example 4

In Aristarchus' model, in addition to errors in measuring θ, another source of inaccuracy in the final result for r would be error in determining when ϕ is a right angle (see Figure 1). But Equation (1) doesn't even involve ϕ. So how could we calculate $S(r, \phi)$? The reason Equation (1) doesn't contain ϕ is because we assumed the definite value of 90° for it. Since we now wish to consider ϕ values other than

90°, we will have to find the equation which relates r to ϕ as well as θ. The law of sines implies

$$r = \frac{ES}{EM} = \frac{\sin \phi}{\sin [180 - (\phi + \theta)]} = \frac{\sin \phi}{\sin (\phi + \theta)}$$

Therefore

$$S(r, \phi) = \frac{\partial r}{\partial \phi} \frac{\phi}{r}$$

$$= \frac{\cos \phi \sin (\phi + \theta) - \sin \phi \cos (\phi + \theta)}{\sin^2 (\phi + \theta)} \frac{\phi \sin (\phi + \theta)}{\sin \phi}$$

which simplifies to

$$S(r, \phi) = \phi[\cot \phi - \cot (\phi + \theta)]$$

Substituting values $\theta = 87°$ and $\phi = 90°$ (in radian measure) gives

$$S(r, \phi) = 1.5708[0 - (-19.0811)]$$
$$= 29.9726$$

Recall that we earlier calculated that $S(r, \theta) = 28.953$, which is about the same. We thus discover that r appears to be equally sensitive to errors in θ as errors in ϕ.

We end this section with a historical note. Eratosthenes' model met the sad and undeserving fate of being ignored by most influential scholars. One of the first was Poseidonius, who made his own determination of the circumference of the earth. He observed that on a certain day the star Canopus appeared to just graze the horizon when viewed from Rhodes, whereas at Alexandria it appeared $7\frac{1}{2}°$ above the horizon. Taking the distance between Rhodes and Alexandria as 3,750 stadia (see Figure 6) and assuming lines AC and RC to be essentially parallel, Poseidonius obtained 180,000 stadia for the circumference (recall that Eratosthenes got the larger figure of 250,000 stadia).

Ptolemy (about 130 B.C.), who wrote the most influential work of the time on geography, chose to believe Poseidonius rather than Eratosthenes. Consequently,

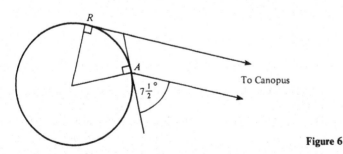

Figure 6

the circumference of the earth was generally underestimated, at least until the time of Columbus.

Christopher Columbus had started to think about the possibility of sailing westward to reach Asia at an early age. Reading Ptolemy's geography and other works, but ignoring Eratosthenes, he was able to convince himself that the earth was much smaller than it actually is and that Asia stretched farther to the east than it actually does. Consequently, Columbus figured that the east coast of Asia was located just about where the east coast of Mexico actually is. Consequently, when, in his famous voyage of 1492, he reached land in the Caribbean, this seemed all according to his calculations: he felt that he had reached Asia. This of course accounts for his use of the term "Indian" to describe the native Americans he encountered. Had Columbus believed Eratosthenes, he might have chosen a better name. But then again, had he believed Eratosthenes he would perhaps not have set sail at all. He would have known that Asia was a good deal farther to the west of Europe than Ptolemy's geography suggested.

Sidelight: Catastrophe Theory

Not every model yields a finite sensitivity because not every model involves an equation of the form $y = f(x_1, \ldots, x_n)$ for a smooth function f.

Here is a gruesome example. You are driving toward a cliff in an automobile, starting at x_0 feet from the edge. You plan to stop in exactly x feet. Naturally it is advisable to make $x \le x_0$. Suppose you measure x_0, possibly with some error, and choose your x equal to this measured value or just a bit less. In this case, the state of your health h will be infinitely sensitive to even a small overestimate of x_0.

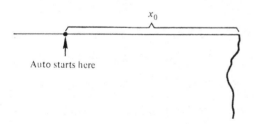

What makes this example work is that there is a catastrophic (discontinuous) change in h that results from a rather small change in the variable x. This particular example is a little fanciful, but there are realistic and important situations which seem to have this catastrophic nature. It is said that World War I was set off by a single assassin's bullet which killed Archduke Ferdinand at Sarajevo. Some claim that the Great Depression was triggered by a single bad day on the New York Stock Exchange. At a more personal level, we have probably all had the experience of having one extra frustration send us "over the edge" from tension to rage or tears or some other emotional state which was sharply different from the one which preceded it.

The existence of these situations is recognized by the proverb about the straw that broke the camel's back. These examples do not especially concern measurement, but they become problems of measurement if we have the task of measuring how many straws the camel can tolerate.

Recently some mathematicians have put forward a theory of discontinuous changes picturesquely called *catastrophe theory*. Although it surely captures a germ of truth, the theory has been strongly criticized for failing, so far, to make many useful predictions. Even if it does become more useful in the future, there is one great difficulty that it will always have to confront—its enormous sensitivity to errors in our measurement of how far the system is from a catastrophe (e.g., a cliff).

EXERCISES

● **1** In Eratosthenes' model the circumference of the earth C also depends upon the distance from Alexandria to Syene, which we may call d, according to the equation:

$$C = 360 \frac{d}{\theta}$$

Find $S(C, d)$, assuming d and θ to be about 5000 and $7°12'$.

2 How do you think Eratosthenes satisfied himself that the well was vertical? How would you do it?

3 How would you actually determine the direction AZ and measure angles (like ZAS') from it? See Figure 5.

● **4** Suppose we flew a plane around the earth at the equator at a constant altitude of h miles and discovered that the whole trip takes m miles. Can you deduce the circumference C from m and h? What is the formula?

● **5** (a) Find $S(C, h)$ for the model of Exercise 4. Assume m and h are measured to be 24,866 and 1 respectively. Compare to $S(C, m)$.

(b) Suppose that the maximum percentage error in the instrument which measures m is the same as the maximum percentage error in Eratosthenes' instrument for measuring θ. Suppose also that all other quantities can be measured with no error at all. Can you decide, in advance of actually calculating C by the two methods, which method is more likely to give the best result?

6 Work out the details of Poseidonius' model (see Figure 6). Is it correct to assume that, when an observer on earth looks in the direction of the horizon, the line of sight is tangent to the earth at the point where the observer is standing?

● **7** Two ancient cities, Lysimachia and Syene, have two different stars in their zeniths (directly overhead), and the angular distance between these stars, measured by an observer at Syene, is estimated as $24°$. The distance from one city to the other is 20,000 stadia. Can you estimate the circumference of the earth from this? Do you see a resemblance between this and Eratosthenes' model?

8 Suppose we send a satellite up to an altitude of h miles and snap a picture of the earth. The earth will appear as a circle, whose radius k' can be measured in inches. From our knowledge of the camera used and the laws of optics, the scale of the photograph can be found. In this way we can find k, the number of miles represented by the k' inches (see Figure 7). Let x be the radius of the earth. Write down three equations, each resulting from an application of the pythagorean theorem. Eliminating variables between these equations, find an equation involving x and the known quantity h.

9 Show that the equation for x which you found in Exercise 8 has a unique positive solution.

10 A ship whose mast is h feet tall sails in a straight line away from shore at the rate of v feet per

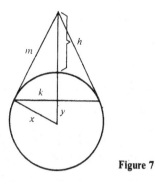

Figure 7

minute. An observer at exactly sea level measures the time, say m minutes, till the top of the mast disappears. Assume that the mast is perpendicular to the surface of the earth at all times. Show that the circumference of the earth C satisfies

$$\cos \frac{2\pi v m}{C} = \frac{1}{2\pi h C + 1}$$

11 Can you determine whether or not the equation of Exercise 10 determines a unique value of C (assuming v, m, and h are known)?

12 Is the list of sources of inaccuracy in Aristarchus' model complete? If not, what would you add?

13 Suppose the moon were only one-quarter (90° of arc) illuminated by the sun. Draw a figure which shows the three bodies in a right-triangle configuration, analogous to Figure 4. How much (measured in degrees of arc on the moon) of the moon is visible from the earth?

● **14** Suppose we take into account the real state of affairs in Aristarchus' model, namely that more than one-half of the moon is illuminated. (Why is this so?) Thus in Figure 4 the illuminated arc extends around the circle beyond A and B. Then where will the earthly observer be when a half-moon is seen? Between E_1 and E_2 or between E_2 and E_3?

● **15** Figure 8 shows a solar eclipse, a configuration in which an earthly observer sees the sun exactly covered up by the moon. (What a remarkable coincidence that the distances and sizes allow this to happen!) The portion of the moon illuminated is the larger arc from A to B, whose size is clearly

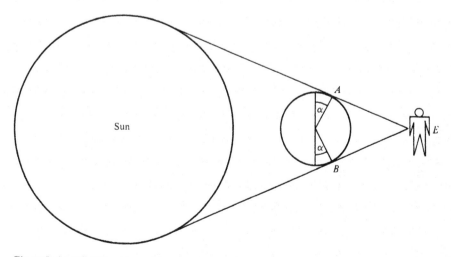

Figure 8 An eclipse (not to scale).

$180° + 2\alpha$. Thus 2α is the discrepancy between the truth and the assumption that half the moon is illuminated. How would you determine α from measurements made at point E on earth? (*Hint*: About the only thing you are capable of measuring is angle AEB.) Aristarchus estimated $\alpha = 2°$; a more correct answer is $\frac{1}{2}°$.

16 Suppose the moon was too small to blot out the whole sun in an eclipse but could only manage a small portion. Thus, in an eclipse the sun would appear to have a dark circular hole at the center. Draw a figure analogous to Figure 8 which shows such a state of affairs. Could you still find a way to estimate how much more than half the moon is illuminated by the sun?

17 Suppose we abandon the incorrect assumption that the moon has constant distance from the sun. Then the amount of the moon illuminated is variable. Show that it is greatest at the time of an eclipse by showing that at all other times the moon is farther from the sun. You may continue to assume that the earth has constant distance from the sun and the moon has constant distance from the earth.

18 Figure 9 shows the actual versus assumed portion of the moon that is illuminated. The difference is represented by arcs AC and BD. In Exercises 15 and 17 you showed how the maximum size of arc BD (measured by central angle α) could be estimated. But this doesn't tell us what angle would be subtended by this arc (angle θ in the figure) by a measuring instrument on earth. Estimate θ:

(a) Use the law of sines and the fact that $\beta \approx \sin \beta$ when β is small to show that $\alpha/\theta > (GF - OG)/OG$.

(b) Show $\theta < \alpha(\cot \phi - 1)^{-1}$.

(c) Compare what the inequality in (b) says for Aristarchus' estimate of $\alpha = \phi = 2°$ and the modern estimate of $\alpha = \phi = \frac{1}{2}°$.

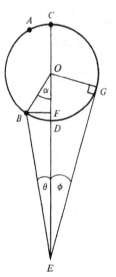

Figure 9 Illuminated part is the larger arc from A to B (sun's rays coming horizontally from right). Earth is at E.

● **19** Let $z = f(y)$ and $y = f(x)$. Can you write a formula for $S(z, x)$ in terms of $S(z, y)$ and $S(y, x)$?

● **20** Show algebraically that Equation (3) has the exact value $-1 - \Delta\theta/\theta_0$. Use this formula to carry out the calculation in Example 3.

21 Suppose we wish to compute a sensitivity which is not based on percentage errors, but on actual errors. What do you think the appropriate formula would be?

22 Carry out the analogue of Example 3 for Aristarchus' model; i.e., compute

$$\frac{\Delta r}{\Delta \theta} \frac{\theta_0 + \Delta\theta}{r_0 + \Delta r}$$

for $\theta_0 = 87$ and $\Delta\theta = 2.8$. Compare to the result of Example 3.

Computer exercises

23 In Example 3 concerning Eratosthenes' model, we find that, for $\Delta\theta = 3$, formula (3) and the small-error version computed in formula (4) are different. How close to 0 does $\Delta\theta$ have to be to give sensitivities which agree up to the first place after the decimal point? Investigate this with a computer program that computes formula (3) for a series of $\Delta\theta$ values, starting from $\Delta\theta = 3$ and approaching 0. Follow the method of Example 3 in computing formula (3). Could Exercise 20 simplify your work?

24 Work out Exercise 23 for Aristarchus' model. Is there anything like the formula in Exercise 20 that can simplify your work?

BIBLIOGRAPHY

Heath, Sir Thomas L.: "A Manual of Greek Mathematics," Dover Publications, New York. A good source for the work of Eratosthenes, Aristarchus, and others.

Penna, Michael A.: Surveying Outer Space, UMAP Module 580, available from COMAP, 271 Lincoln St., Lexington, MA 02173.

Woodcock, Alexander, and Monte Davis: "Catastrophe Theory," E. P. Dutton, New York, 1978. An elementary account, using pictures rather than equations. Good coverage of the applications and the controversy over them.

CHAPTER FOUR

OPTIMIZATION

1 CLASSICAL OPTIMIZATION

Abstract We begin with the concepts of feasible set and objective function and then proceed to review calculus techniques for optimization. Search methods make a brief appearance in the exercises. Emphasis is placed on the role of the boundary points as candidates for optimization.

Prerequisites Calculus of one and several variables.

Mathematical models are often used to help us make decisions. When we use a mathematical model to select the best alternative out of a large number of possibilities, we call this *optimization.*

Optimization is surely one of the most basic human activities. When a child is asked to choose 1 candy out of a handful of 10, the child faces a problem in optimization. Should the choice be the one with the largest volume, the heaviest, the one with the most chocolate? Having chosen a criterion, the child then needs to find the candy that measures up the best with respect to that criterion.

This example, simple though it is, illustrates two features which are also found in more complicated mathematical models: the feasible set and the objective function.

Definition

The *feasible set* is the set of all the possibilities from which we are allowed to choose.

In our example, the 10 candies form the feasible set.

Definition

The *objective function* is a function which assigns to each member of the feasible set a number which measures the desirability of that choice.

For example, suppose the child decided on the candy with the largest volume. Then volume would be the objective function. Each candy is assigned a number equal to its volume. The candy with the largest number wins. A child would not normally measure the volumes precisely—an eyeball estimate would be more likely—but in mathematical models we usually have a formula for the objective function.

Optimization sometimes involves minimization instead of maximization. For example, the child might wish to minimize the amount of nuts.

What makes our candy-choosing example so simple is that there are only 10 possibilities in the feasible set. In such a case, we can always, in principle, evaluate the objective function for each possibility and choose the best. Mathematical optimization problems are ones where the feasible set is very large, possibly infinite. In such problems, checking all the possibilities is impossible. Mathematical optimization may be thought of as a set of techniques that enable us to avoid enumerating all the possibilities in the feasible set.

Example 1

A traveling salesman must visit each of the five cities A, B, C, D, and E in Figure 1 exactly once. The numbers on the figure indicate air distances between the cities. The salesman's home is at A; so he wants to begin and end his trip at A. In what order should he visit the cities to minimize the total distance traveled?

In this problem, the feasible set consists of the various orderings of the five cities in which each city appears once except for A, which begins and ends the list. Two such orderings, along with their total distances are shown in the figure. There are $4! = 24$ orderings altogether making up the feasible set. It's possible to list them all by hand and calculate the travel distances for each. You could also write a computer program to generate all the orderings and calculate the distances.

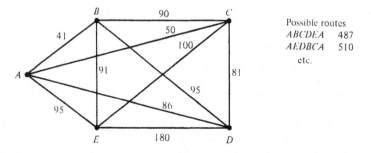

Possible routes
ABCDEA   487
AEDBCA   510
etc.

Figure 1 Air distances between five cities (not drawn to scale).

But what if you had more cities? With six cities the number of orderings is $5! = 120$ and working by hand isn't reasonable. However, computer enumeration is still possible. There are real-life problems of this kind with 50 or more cities, and here even computer enumeration is impossible because the number of orderings is too large (for example, $49!$ is over 6×10^{62}). It is necessary to use some mathematical theory to narrow the search down to a small part of the feasible region. Section 4 discusses this problem.

The previous example showed that, even with a finite feasible set, optimization can be difficult. One might suppose that, with an infinite feasible region, things might be even harder. But it turns out that there are large categories of problems with infinite feasible regions which are quite manageable. In the rest of this section we review some such problems where calculus tools apply. In the following sections, we discuss problems where linear programming and its relatives apply.

Example 2

The power P supplied by a certain battery depends on the external resistance x in the circuit according to the formula

$$P = x\left(\frac{10}{100 + x}\right)^2 \tag{1}$$

If x can be any positive number, how should we choose it to get the most power?

Here the feasible set is infinite, the interval $(0, \infty)$. The objective function is given by Equation (1). Obviously we cannot evaluate P for each number in the feasible set. Luckily calculus comes to the rescue. It tells us that:

Theorem 1

If $f(x)$ is a differentiable function on a certain feasible interval, then its maximum (minimum) either

1. Doesn't exist
2. Occurs at a critical point, i.e., a point where

$$\frac{df}{dx} = 0$$

3. Occurs at a point of the feasible set which is an endpoint of the feasible interval

In Example 2 the feasible region has only one endpoint, 0, and this is not in the feasible region so we can ignore possibility 3. We pursue possibility 2 by calculating that the derivative is

$$\frac{dP}{dx} = \frac{100(100 - x)}{(100 + x)^3}$$

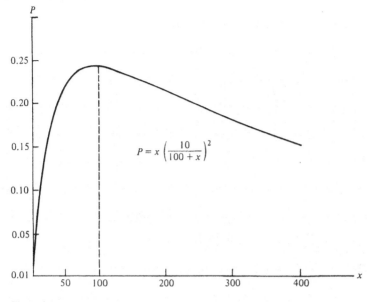

Figure 2 Power supplied by a battery as a function of external resistance.

The only critical value is, therefore, $x = 100$. To show that this is a maximum, we note that, for $x > 100$, $dP/dx < 0$, which means the curve decreases to the right of 100, while, for $x < 100$, $dP/dx > 0$, which means the curve is increasing when $x < 100$ (see Figure 2). Thus we have a maximum, and possibility 1 is ruled out.

Problems of the type discussed in Example 2 often have additional wrinkles: first, there may be a number of critical points; second, the maximum (or minimum) may occur at an endpoint. This last possibility would occur, for example, if the feasible set in Example 2 were $[200, \infty)$. The most common kind of feasible set encountered in practical optimization problems involving one variable is a closed interval $[a, b]$, and in such cases there are two endpoints to check. If, in addition, $f(x)$ is continuous, the following result rules out possibility 1.

Theorem 2

If f is a continuous function defined on a feasible set which is a closed interval $[a, b]$, then there is a maximum and a minimum for f on the interval $[a, b]$.

Example 3

Suppose $f(x) = cx + d$, a linear function, and we seek the maximum (or minimum) on the feasible set $[a, b]$. Since we know that there is a maximum (linear functions are continuous), we only need to find the critical points, evaluate f at these points and at the endpoints, and pick the largest.

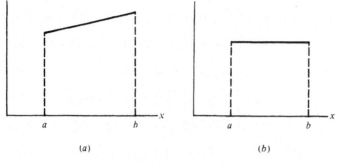

(a)   (b)

Figure 3

We have $f'(x) = c$ and, if $c \neq 0$, there are no critical points (Figure 3a). However, if $c = 0$, all x values are critical points (Figure 3b). In either case, one can restrict attention to the endpoints in looking for the maximum. In other words, for linear functions, calculus tells us that calculus is not needed.

A great many optimization problems involve a feasible region with dimension higher than one, as in the following problem.

Example 4

A refrigerated compartment is to be built in the shape of a box (Figure 4) and with a capacity of 8000 cubic feet. To save energy costs, we wish to find the dimensions that will minimize the amount of heat entering from outside.
Here are the rates, for some unspecified unit of time, at which heat will flow into the box:

Through the top, 1 unit per square foot
Through the bottom, 3 units per square foot
Through the sides, 2 units per square foot

If we let x and y represent length and width, then the height is $8000/xy$. The top

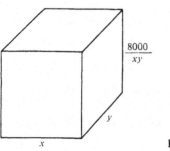

Figure 4

and bottom each have an area of xy. Two of the sides each have area $8000/x$, while the other two each have area $8000/y$. Therefore, the total heat-flow rate H is

$$H = 1xy + 3xy + \frac{2 \cdot 2 \cdot 8000}{x} + \frac{2 \cdot 2 \cdot 8000}{y} \tag{2}$$

The only restrictions on x and y are $x > 0$ and $y > 0$. Thus the feasible region consists of the interior of the first quadrant of R^2. We need to find which point in this region minimizes H.

In many-variable problems like Example 4, calculus again makes a contribution via the following theorem. (Compare this theorem with Theorem 1.)

Theorem 3

If $f(x_1, x_2, \ldots, x_n)$ has partial derivatives throughout a suitable* feasible set in R^n, then its maximum (minimum) either:

1. Doesn't exist
2. Occurs at a critical point, i.e., a point where $\partial f/\partial x_i = 0$ for $i = 1, 2, \ldots, n$
3. Occurs at a point of the feasible set which is on the boundary of the feasible set

In Example 4 possibility 3 doesn't occur. The boundary of the feasible set consists of the nonnegative portions of the x and y axes, and these points don't belong to the feasible set. Next we determine the critical points.

$$\frac{\partial H}{\partial x} = 4y - 32,000x^{-2} = 0$$

$$\frac{\partial H}{\partial y} = 4x - 32,000y^{-2} = 0$$

We solve for y in the first equation and substitute in the second to obtain $x^3 = 8000$, and therefore $x = 20$. Substituting in the first (or second) equation gives $y = 20$. Then the height is $8000/(20)(20) = 20$. For these dimensions the heat-flow rate H is 4800 units. This turns out to be the minimum because possibility 1 (no minimum) can be ruled out. However, doing this is not entirely routine.

In general, for functions of more than one variable, it can be a tricky matter to show that a minimum (maximum) does exist unless the feasible set has the following two characteristics:

1. Boundedness (This means you can find an n-dimensional box big enough to enclose the set.)
2. Contains all its boundary points

---

* A suitable feasible set in R^n is an open set, possibly augmented by some or all of its boundary.

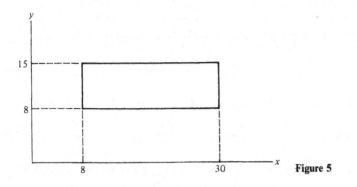

Figure 5

Theorem 4

If f is a continuous function defined on a feasible set which is bounded and contains all its boundary points, then there is a maximum and a minimum for f.

This is a useful theorem because feasible sets with these properties are common in optimization problems. In fact a close examination of Example 4 indicates that the feasible set described there is not realistic and a more practical one would be bounded and contain its boundary points.

The land on which we will build the refrigerated compartment will set upper bounds to the dimensions, say $x \leq 30$ and $y \leq 15$. Furthermore, there is a limit to how skinny a building can be built, and so there are lower limits also, say $x \geq 8$ and $y \geq 8$. Now Theorem 4 allows us to rule out possibility 1 of Theorem 3. But our new feasible region (Figure 5) is a mixed blessing because it contains four boundary segments, and possibility 3 of Theorem 3 rears its ugly head. These segments contain an infinite number of points, which means we can't evaluate f at all of them.

This is a drastically different situation from the one we face with one-dimensional feasible sets. For those 1-variable problems the feasible set is typically an interval of the form $[a, b]$, and there are only two endpoints to check, an easy chore. But with higher-dimensional optimization problems, dealing with the boundary is the hardest part of the problem.

Example 5

Maximize $f(x, y) = 2x + y$, subject to $0 \leq x \leq 10$ and $0 \leq y \leq 20$.

Since the feasible region (see Figure 6) satisfies the conditions of Theorem 4, the maximum does exist. When we check for critical points, we find $\partial f/\partial x = 2$ and $\partial f/\partial y = 1$. Since these partial derivatives can't be zero, there are no critical points. (Compare this to Example 3.) Therefore, the maximum occurs on one of the boundary segments.

We can easily find out where the maximum is in this example because the boundary and the objective function are fairly simple. Examine the objective

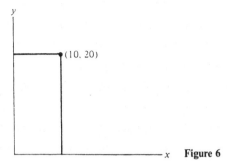

Figure 6

function and notice that, if x and y increase or one increases and the other stays the same, $f(x, y)$ will increase. Thus, any boundary point of the feasible set can be "improved" by moving up or to the right. The upper right corner is the only point where no further improvement is possible. So the maximum occurs at $(10, 20)$, and the maximum value of f is $f(10, 20) = 40$.

The problems involved in checking the boundary in Examples 4 and 5 are tame as these things go. (Exercise 9 shows a way to handle the boundary in Example 4 when Figure 5 is the feasible region.) It could be much harder if there are many boundary pieces or the feasible set has dimension higher than two, so that the boundary will have dimension greater than one.

Complications involving the boundary are only one of the two great stumbling blocks that limit the usefulness of calculus-based optimization methods. The other one is that not every objective function satisfies the conditions of Theorem 3 (Theorem 1 in the 1-variable case). For example $f(x) = |x^2 - \sin x|$ is not differentiable at points where $x^2 = \sin x$. The function $g(x) = [x^2 + 1]$ (where $[z]$ denotes the greatest integer less than or equal to z) is not even continuous, much less differentiable at points where $x^2 + 1$ is an integer. Exercises 14 through 16 describe some search methods that work for such functions.

Sidelight: Calculus of Variations

Feasible sets need not be sets of points in R^n. They can be sets of functions. A famous problem of this type is the brachistochrone problem proposed by Johan Bernoulli in 1696: what shape should a slide be if it is to connect two given points P and Q at different heights to minimize the time required for a ball to slide from P to Q?

The shape of a slide can be described by a function $f(x)$, where $f(x_a) = A$ and $f(x_b) = B$. The feasible set is a set of such functions. Each function, together with the laws of gravity, determines a time of descent. This time is the objective function. Problems of this type belong to the calculus of variations and are quite hard.

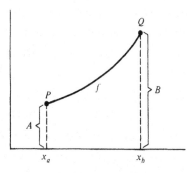

Another example is to find the shape of a string which is anchored at its ends along a given straight line and encloses the most area between itself and the line. This is often called Dido's problem after Queen Dido, who, according to legend, founded the city of Carthage in North Africa through a combination of trickery and mathematics. She persuaded a local chieftain to sell her as much land as an oxhide could contain. She then cut the oxhide into thin strips and tied them together to make a cord $2\frac{1}{2}$ miles long. She laid this cord down in a semi-circle whose ends touched the sea coast. This turns out to enclose the most area.

EXERCISES

● **1** The power formula in Example 2 is a special case of the formula $P = x[E/(r + x)]^2$, where E is the electromotive force and r is the internal resistance of the battery. E and r are constant for a particular battery. Show that the maximum power always occurs when the external resistance x equals the internal resistance, regardless of the value of E.

2 The intensity of microwaves at a distance d from a source whose intensity is I is I/d^2. If two sources have intensities I_1 and I_2 and are located at $x = 0$ and $x = 1$, find the location on $[0, 1]$ where the combined intensity is least. Does the minimum ever occur at an endpoint?

3 A ball is thrown into the air, and its height at any time t is $h = 64t - 16t^2$. At what time does it reach its highest point?

● **4** A certain chemical reaction consists of the conversion of chemical X to chemical Y. The rate of conversion at any given time is $R = 0.01xy$, where x and y are the amounts of X and Y present at that time. Initially there are 99 units of X and 1 of Y. Y arises only by conversion from X; so $y = 100 - x$.

    (a) How much X and Y are present when the conversion rate is highest?

    (b) Would your answer change if there were 40 units of X at the start?

5 A factory manager discovers that the cost of producing x items is $C(x) = 2500 + 3.7x - 0.009x^2 + 0.00001x^3$. Each item produced can be sold for \$10 (regardless of how many are produced). Find a formula for the profit in terms of x. Find the value of x that maximizes profit. Take into account that x must be an integer.

● **6** The cost of operating a certain truck is $0.01(20 + s/10)$ dollars per mile, when it is driven at s miles/hour. A truck driver earns \$6 per hour.

    (a) What is the most economical speed at which to operate the truck for a 600-mile trip?

    (b) Show that you get the same answer for any other length of trip.

    (c) Does your answer change if the speed limit of 55 miles/hour must be observed?

    (d) If truck drivers' wages go up, will the trucking company be more or less happy about observing the 55-miles/hour speed limit? Assume other costs stay fixed.

7 Show that, among all rectangular boxes having a surface area of 6 square inches, the cube has maximal volume. You may assume that a maximum exists.

● **8** A clothing store sells two kinds of overcoats which are similar but are made by different manufacturers. The cost to the store of the first kind is \$40, and the cost of the second kind is \$50. It has been determined by experience that, if the selling price of the first kind is x dollars $(x > 0)$ and the selling price of the second kind is y dollars $(y > 0)$, then the total monthly sales of the first kind is $(3200 - 50x + 25y)$ overcoats and the total monthly sales of the second kind is $25x - 25y$ coats. What selling prices should be chosen to maximize total profits? You may assume that a maximum exists.

9 Suppose we replace the original feasible region of Example 4 with that of Figure 5. To find the minimum when H is restricted to the boundary, we calculate the minimum for each of the four boundary segments separately and compare. To cope with segment AB, observe that on AB $y = 8$ and $x \in [8, 30]$. Substituting $y = 8$ in Equation (2), we obtain $H = 32x + 32{,}000/x + 4000$. Minimize this for $x \in [8, 30]$. Now do similar tricks for the other segments.

10 A manufacturing plant has two classifications for its workers, X and Y. Class-X workers earn \$14 per run, and Class-B workers earn \$13 per run. For a certain run the materials cost will be $y^3 + x^2 - 8xy + 600$ if x Class-X workers and y Class-Y workers are used. How many workers of each class should be used so that the cost of the run (including labor and materials) is a minimum if at least three workers of each class are required for a run? You may assume the minimum exists. (Problem 9 shows a method of handling the boundary.)

11 If f is the linear function $f(x_1, \ldots, x_n) = a_1 x_1 + a_2 x_2 + \cdots + a_n x_n + b$ and the feasible region is determined by the inequalities $c_i \le x_i \le d_i$, then show that the maximum and the minimum of f both occur on the boundary.

● **12** Suppose in Example 1 we are interested in the cost of a trip instead of the distance. Furthermore suppose airlines compute fares by this formula: \$0.50 per mile plus \$25 for each takeoff and \$25 for each landing. Assume all trips are nonstop (one takeoff and one landing). Can you show, without numerically solving, that the order that optimizes this objective function is the same as the order that optimizes the objective function based on distance?

Computer exercises

13 (a) Enumerate all 24 members of the feasible set of Example 1.

  (b) Write a computer program to enumerate feasible sets for problems like Example 1. Try it on Example 1 and compare the results with the results of part (a).

  (c) Add to your program the capability of evaluating the length of each member of the feasible set and then solve Example 1.

14 The exhaustive search procedure finds the maximum of $f(x)$ on an interval, say $[0, 1]$, like this: Choose an integer N and divide the interval into N equal parts by inserting partition points $1/N$, $2/N, \ldots, (N - 1)/N$. Now compute f at each of these and at 0 and 1. The largest of these values is your estimate of the maximum. It's only approximate, but, if N is large enough, it should be close. Try this for the function $f(x) = [10 \sin 4x]$ with feasible set $0 \le x \le 1$. ($[z]$ stands for the greatest integer less than or equal to z.)

15 The random search procedure finds the maximum of $f(x)$ on $[0, 1]$ like this: Choose an integer N; then choose N numbers randomly on $[0, 1]$ and evaluate f at these points and at 0 and 1. The largest of these values is your estimate of the maximum. If N is large enough, the approximation should be good. Try this for the same function as in Exercise 14 and compare results. Which method works better?

16 Suppose we know that $f(x)$ is monotonically increasing from $x = 0$ to a maximum at $x = s$ and then monotonically decreasing from $x = s$ to $x = 1$. But we don't know the value of s, and we want to find it. Dichotomous search is a method to find s, and it works like this: Choose points $x_1 < y_1$ and evaluate $f(x_1)$ and $f(y_1)$. If $f(x_1) < f(y_1)$, then s can't lie on $[0, x_1]$ (do you see why?); so we have narrowed our search to the interval $[x_1, 1]$. If $f(x_1) > f(y_1)$, then s must lie on $[0, x_1]$. Now we repeat the procedure on whichever subinterval we have narrowed down to; pick points $x_2 < y_2$ in the subinterval, evaluate $f(x_2)$ and $f(y_2)$, and so on. After many repetitions of this narrowing-down procedure, we get a small interval in which s must lie. Either endpoint of this small interval is a good estimate for s. Try this for $f(x) = 1 - |x^3 - 1|$. You can pick the x_i and y_i arbitrarily, but you might think about the most efficient way to choose them.

BIBLIOGRAPHY

Consult any standard calculus text for most of the material in this section. But for search methods (Exercises 14 through 16) see:

Cooper, Leon, and David Steinberg: "Introduction to Methods of Optimization," W. B. Saunders, Philadelphia, 1970.

Nevison, Christopher H.: Differentiation, Curve Sketching, and Cost Functions, UMAP Module 376, available from COMAP, 271 Lincoln St., Lexington, MA 02173.

Whitely, W. Thurmon: Five Applications of Max-Min Theory from Calculus, UMAP Module 341, available from COMAP, 271 Lincoln St., Lexington, MA 02173.

Wilde, Carroll O.: Calculus of Variations with Applications in Mechanics, UMAP Module 468, available from COMAP, 271 Lincoln St., Lexington, MA 02173.

2 LINEAR PROGRAMMING—FORMULATION AND GRAPHICAL SOLUTION

Abstract A surprising number of optimization problems fall into a category of mathematical problem called *linear programming*. Translating a word problem into the equations and inequalities of a linear programming problem is a key skill in mathematical modeling and one main objective of this section. The other is to give a geometric interpretation of what a linear programming problem is.

Prerequisites The analytic geometry of two dimensions, linear equations and inequalities.

Linear programming problems arise in many different fields. But we begin by describing the mathematical structure of linear programming problems, saving examples for later.

Definition

A linear programming problem is one where:

1. The feasible region is a subset of the nonnegative portion of R^n, defined by linear equations and inequalities.
2. The objective function to be maximized or minimized is linear, i.e., of the form $P = a_1 x_1 + a_2 x_2 + \cdots + a_n x_n$.

Example 1

Maximize $p = 1000x + 500y$ over the feasible region defined by

$$x \geq 0 \qquad y \geq 0$$
$$4x + y \leq 10$$
$$18x + 15y \leq 66$$

Example 2

Maximize $p = 60x + 45y + 30z$ subject to

$$x, y, z \geq 0$$
$$9x + 9y + 3z \leq 48$$
$$54x + 36y + 27z \leq 540$$
$$x + y + z = 18$$

Example 3

Minimize $p = 3x + 2y$ subject to

$$x, y \geq 0$$
$$5x + 7y \geq 35$$
$$10x + 4y \geq 40$$

In Examples 1 and 3 the feasible region is in R^2, and we can draw it and examine it visually. In Example 2 this is a little harder, but still possible, since we are in 3-space. But most linear programming problems have high-dimensional feasible regions; so graphing them is out of the question. Despite this, it will help us understand linear programming if we graph some low-dimensional examples.

Example 1 (Graphical Approach)

The inequality $x \geq 0$ has as its solution set the closed half-plane to the right of the y axis (Figure 1a). The inequality $y \geq 0$ specifies the closed half-plane above the x axis (Figure 1b). The solution set of the inequality $4x + y \leq 10$ is the half-plane on and below the line $4x + y = 10$ (Figure 1c). Finally, the solution set of the inequality $18x + 15y \leq 66$ is the half-plane on and below the line $18x + 15y = 66$ (Figure 1d). Points of the feasible region must satisfy all of these inequalities; so the feasible region is the intersection of all these half-planes (Figure 1e).

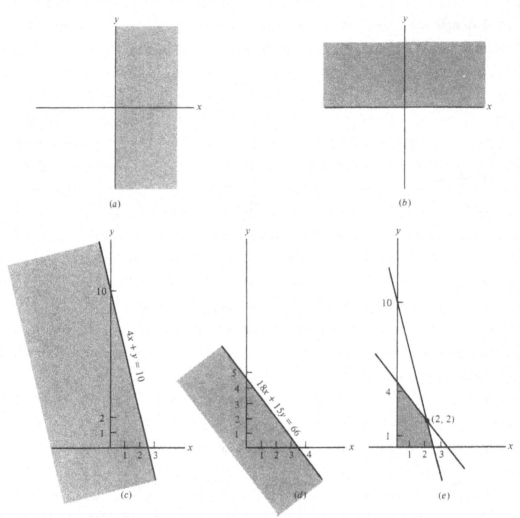

Figure 1

Example 3 (Graphical Approach)

The inequalities $x \geq 0$ and $y \geq 0$ specify the positive quadrant of R^2. The inequality $5x + 7y \geq 35$ specifies all points above (why not below?) the line $5x + 7y = 35$, while the inequality $10x + 4y \geq 40$ gives all points above the line $10x + 4y = 28$. The feasible region is shown in Figure 2.

Now how do we maximize or minimize an objective function for such a feasible region? First observe that calculus is no help. If we try calculus, in our first example we get

$$\frac{\partial p}{\partial x} = 1000 \qquad \frac{\partial p}{\partial y} = 500$$

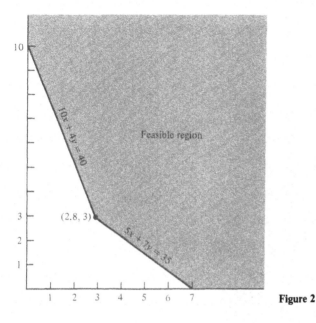

Figure 2

and there is no point of the feasible region where these partial derivatives are zero. There is no critical point in the interior of the feasible region of this linear programming problem (or any other nontrivial linear programming problem!). The maximum (or minimum) of an objective function for a linear programming problem either doesn't exist or lies on the boundary of the feasible region. In fact, there is a stronger theorem:

Theorem 1

The maximum (or minimum) in a linear programming problem either:

1. Doesn't exist (then we call the problem *unbounded*)
2. Occurs at a corner point of the feasible region

Instead of proving this theorem, we give an illustration in the case of Example 3. We begin by asking ourselves if it would be possible to achieve a value of the objective function as low as 24 (remember we want to minimize the objective function). In other words, are there points of the feasible region where

$$3x + 2y = 24$$

We plot this line (Figure 3) and discover that it crosses the feasible region; so the answer is yes. Now we become more ambitious and lower the value to 6. We plot

$$3x + 2y = 6$$

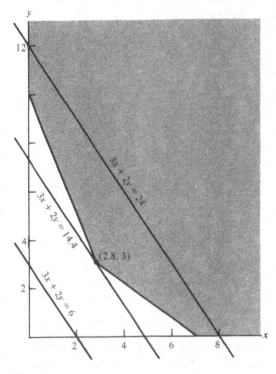

Figure 3

and discover it doesn't cross the feasible region. That is, 6 is not achievable. We call the lines we have plotted *constant-objective lines* because on each line the value of the objective function is constant. They are parallel, and this is no accident since the x and y coefficients are the same in both equations.

Our numerical experiments with the two constant-objective lines for $p = 24$ and $p = 6$ are meant to suggest that, as we lower our value from 24 to 6, the constant-objective lines slide smoothly down and to the left. We are looking for the lowest constant-objective line. This must be the one which passes through the corner point (2.8, 3).

To calculate the value of the objective function at this point, we substitute the coordinates of the corner point into the objective function and obtain, in this case,

$$p = 3(2.8) + 2(3) = 14.4$$

This is the minimum value of the objective function. This completes the illustration of part 2 of Theorem 1.

Part 1 of Theorem 1 can be illustrated with the same feasible region and objective function, but in the case where we wish to maximize instead of minimize.

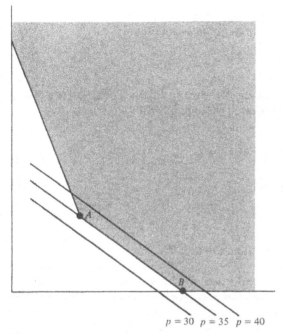

$p = 30 \quad p = 35 \quad p = 40$

Figure 4

It is plain to see that, for any value of $p \geq 14.4$, the constant-objective lines do cross the feasible region. Thus we can get as large a p as we like. There is no maximum p. This situation is rare in practical problems, and so, in this text, you can assume that part 1 of Theorem 1 doesn't occur.

Two clarifications should be made about the statement of Theorem 1. First, part 2 doesn't mean that the maximum (or minimum) occurs just at a single corner point. For example, it may occur at two corner points and along the whole edge connecting them. This is illustrated by Figure 4, in which we use the same feasible region as in Figure 3, but replace the objective function with $p = 5x + 7y$. Now the constant-objective lines are parallel to edge AB. The minimum value of p is 35 and is attained at each point of AB.

A second clarification: the word "unbounded" in part 1 applies to the entire problem and not just to the feasible region. As Figure 3 shows, a feasible region can be unbounded without the problem being unbounded. For a minimization problem to be unbounded, the feasible region must be unbounded in a direction in which the objective function is decreasing. For a maximization problem to be unbounded, the feasible region must be unbounded in a direction in which the objective function is increasing.

Theorem 1 suggests a way to do linear programming problems graphically when the feasible region is in R^2:

Step 1. Draw the feasible region accurately.
Step 2. Compute coordinates of all corner points.

Step 3. Evaluate the objective function at each corner point.
Step 4. Pick out the optimum value (maximum or minimum).

Example 4

We now solve Example 1. The drawing is shown in Figure 1. The corner points and the values of the objective function at these points are:

Corner point	Value of $1000x + 500y$
(0, 0)	0
(2.5, 0)	2500
(0, 4.4)	2200
(2, 2)	3000

The maximum value of the objective function is 3000.

Unfortunately the four-step method described above can get very difficult when there are many constraints. For example, if we add the additional constraint $3x + 9y \leq 27$ to the other constraints in Example 1, this has the effect of adding the line $3x + 9y = 27$ to Figure 1. What does the new figure look like, Figure 5a or b? It makes a difference because the corner points differ in the two cases. A careful plot would give the answer.

There is another way to tell, which does not require care in plotting. The question can be reduced to determining whether the point of intersection of $3x + 9y = 27$ and $4x + y = 10$ is in the feasible region or not. So calculate its coordinates and see if they satisfy all the constraints of the feasible region. This strategy works and can be applied repeatedly if there are many constraints (many doubtful corners), but it is tiresome.

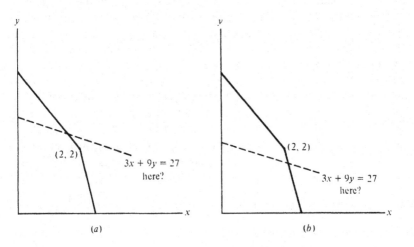

(a) (b)

Figure 5

There is an enormous variety of practical problems which can be formulated as linear programming problems. So many that it may be an interesting puzzle for historians of mathematics to determine why it took until the middle of the twentieth century to discover the usefulness of linear programming. The development of the computer is surely part of the explanation, but perhaps not all of it. The first computer solution of a linear programming problem occurred in 1952. By that time many important types of linear programming problems had been proposed and examples solved (e.g., the transportation problem by Hitchcock in 1941 and the diet problem by Stigler in 1945) and the simplex method of solution had already been devised (by Dantzig in 1947).

In any case, a primary tool of a mathematical modeler is skill at formulating linear programming problems. (Solving them is less crucial; there are canned computer programs for that.) A good way to begin acquiring this skill is to learn (by practice!) to recognize certain special and common types of linear programming problems. There are many, but in this text we confine ourselves to three main types:

1. Resource-allocation problems
2. Diet problems
3. Transportation problems

I. Resource-Allocation Problems

In a resource-allocation problem one has limited supplies of various resources, which can be combined in different proportions to make different products. For example, different kinds of grains can be blended to make different breakfast cereals. Resources can also include land, capital, and labor. Likewise, "product" needs to be interpreted loosely. In Example 6 below, the products are various agricultural crops.

Example 5 (Resource-Allocation Problem)

A fertilizer manufacturer uses two materials, nitrates and phosphates, to make two different varieties of fertilizer. A single batch of Sod-King needs 4 units of phosphate and 18 of nitrate, while a single batch of Gro-Turf uses 1 unit of phosphate and 15 units of nitrate. The profit on a batch of Sod-King is $1000, and the profit on a batch of Gro-Turf is $500. The company has 10 units of phosphate and 66 units of nitrate on hand. How many batches of each type should the company make from the available supplies if it wishes to earn the most profit?

Our unknowns are

$$x = \text{the number of Sod-King batches}$$

$$y = \text{the number of Gro-Turf batches}$$

Therefore the total number of units of phosphate used will be $4x + y$ and the total number of units of nitrate will be $18x + 15y$. Since these can't exceed available supplies,

$$4x + y \leq 10$$
$$18x + 15y \leq 66 \tag{1}$$

We can't make negative quantities; so

$$x \geq 0 \qquad y \geq 0 \tag{2}$$

Inequalities (1) and (2) express the only constraints faced by the fertilizer company. Thus the feasible region is that portion of R^2 consisting of points (x, y) satisfying Equations (1) and (2). This is shown in Figure 1.

The formula for profit is

$$p(x, y) = 1000x + 500y$$

The problem just described is nothing more than Example 1 fleshed out with a story. As we have seen, the maximum profit is \$3000, and it is achieved by making two batches of each type ($x = 2$ and $y = 2$).

Example 6 (Resource-Allocation Problem)

A farmer has set aside 18 acres of land to be divided into three plots: one for grapes, one for potatoes, and one for lettuce. Each crop has its own particular requirements for labor and capital. Also, each crop has a different profit per acre. The data are shown in the following table. The farmer's resources, beside the 18 acres, are \$540 of capital and 48 hours of labor. How many acres should each plot be?

Plot	Labor, hours per acre	Capital, \$ per acre	Net profit, \$ per acre
Grapes	9	54	60
Potatoes	9	36	45
Lettuce	3	27	30

Let x, y, and z be the number of acres of grapes, potatoes, and lettuce. The figures in the first column of the table, together with the 48-hour limitation, give $9x + 9y + 3z \leq 48$. Likewise, capital and labor give similar inequalities. Finally, all variables must be nonnegative. The full list of constraints is

$$x, y, z \geq 0$$
$$9x + 9y + 3z \leq 48$$
$$54x + 36y + 27z \leq 540$$
$$x + y + z = 18$$

The objective function, which we wish to maximize, is

$$p(x, y, z) = 60x + 45y + 30z$$

In this problem the feasible region is a set in 3-space, and so it's a little hard to visualize and find its corner points. Exercise 5 shows how to reduce this to a two-dimensional problem.

II. Diet Problems

In a standard diet problem one has a number of foods available, each of which has a different combination of nutrients. The problem is to devise a mix of these foods that will meet minimum requirements for each nutrient and have the least cost for all such mixes. In nonstandard diet problems the objective function need not involve cost; it could measure the amount of some impurity in the mix, and we might prefer to minimize this instead of cost. Example 9 shows that there are problems which don't deal with food at all, but which have the same underlying structure and are therefore considered diet problems.

Example 7 (Diet Problem)

The nutrition director of a college wishes to blend a soup to serve students. The director has two commercial products available: onion soup at 3 cents per ounce and chicken stock at 2 cents per ounce. Each ounce of onion soup has 5 units of protein and 10 units of iron, while an ounce of chicken stock has 7 units of protein and 4 units of iron. The nutrition director decides that a serving of soup ought to have at least 35 units of protein and 40 units of iron and doesn't care how many ounces are in a serving. How many ounces of each soup should be mixed for a serving to achieve or exceed the nutritional goals at the least cost?

Let $x =$ the number of ounces of onion soup per serving and $y =$ the number of ounces of chicken stock per serving. Then the number of units of protein will be $5x + 7y$, and the number of units of iron will be $10x + 4y$. Therefore, our constraints are

$$x, y \geq 0$$
$$5x + 7y \geq 35$$
$$10x + 4y \geq 40$$

The objective function, which we wish to minimize, is

$$p(x, y) = 3x + 2y$$

Example 3 shows the feasible region for this problem, and the discussion following Theorem 1 shows the solution: $x = 2.8$ and $y = 3$.

Example 8 (Diet Problem)

This example is virtually identical to the previous one, except that the nutrition director is concerned about minimizing contamination by DDT. Suppose each ounce of onion soup has 6 units of DDT and each ounce of chicken stock has 14 units. The director still needs to meet or exceed the same iron and protein minimums.

   The feasible region is the same as in the previous example. The only change is in the objective function. Now we wish to minimize

$$p(x, y) = 6x + 14y$$

Example 9 (Diet Problem)

Here is a diet problem that doesn't deal with food. Suppose that the U.S. Army gives each recruit a battery of tests and assigns him or her effectiveness ratings in two categories: combat and support services. The average male has a combat rating of 8 and support-services rating of 5, while the average female has a combat rating of 4 and a support-services rating of 9. It costs \$12,000 per year to maintain a male soldier and \$13,000 to maintain a female. It is necessary to have a total combat effectiveness (summed over all soldiers) of at least 7 million and a total support effectiveness of at least 6 million. How many soldiers of each sex should the army have to achieve the effectiveness levels at the least yearly cost?

   The analogy between this example and the previous two is that the two sexes play the role of the soups. Each sex is considered a combination of two kinds of effectiveness, just as each soup is considered a combination of two kinds of nutrition, iron and protein.

   Let $x =$ the number of female soldiers and $y =$ the number of male soldiers. The feasible region is described by

$$x, y \geq 0$$
$$4x + 8y \geq 7,000,000$$
$$9x + 5y \geq 6,000,000$$

We wish to minimize the objective function

$$p(x, y) = 13,000x + 12,000y$$

III. Transportation Problems

In a typical transportation problem one needs to ship goods of some type from various sources of supply, e.g., factories, to various destinations. At each destination there is a specified demand, while at each source the supply is limited. To ship from the ith to the jth destination costs c_{ij} per unit. The problem is to meet the demands at minimum total cost. Figure 6 shows a graphic view. The points on the

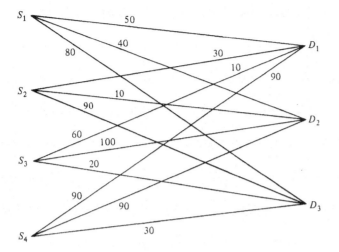

Figure 6

left are the sources, the points on the right are the destinations, and the connecting lines are the available transportation links, each labeled with its cost.

Our last example of a linear programming problem is the assignment problem. It is superficially quite different from the transportation problem because nothing is being moved and the objective function is sometimes to be maximized instead of minimized. But we will see in Example 11 that it can be thought of as a transportation problem after all.

Example 10 (Transportation Problem)

A truck rental company sometimes rents trucks for one-way trips and, as a result, now has too many trucks in some cities and too few in others. Cities S_1, S_2, S_3, and S_4 have 4, 3, 6, and 2 extra trucks, respectively. Cities D_1, D_2, and D_3 have 5, 3, and 6 too few, respectively. The table below shows this information, along with mileages between cities. How should the trucks be redistributed to minimize total mileage?

Source	Destination			Number available
	D_1	D_2	D_3	
S_1	50	40	80	4
S_2	30	10	90	3
S_3	60	100	20	6
S_4	90	90	30	2
Number required	5	3	6	

Let x_{ij} be the number of trucks to be shipped from S_i to D_j. Meeting the demand at D_1 means $x_{11} + x_{21} + x_{31} + x_{41} = 5$. Similarly each of the other two demands imposes an equality constraint (see below). But we can't exceed the supplies available. At S_1 this means $x_{11} + x_{12} + x_{13} \leq 4$. The entire list of constraints defining the feasible region is

$$x_{11} + x_{21} + x_{31} + x_{41} = 5$$
$$x_{12} + x_{22} + x_{32} + x_{42} = 3$$
$$x_{13} + x_{23} + x_{33} + x_{43} = 6$$
$$\text{Each } x_{ij} \geq 0$$
$$x_{11} + x_{12} + x_{13} \leq 4$$
$$x_{21} + x_{22} + x_{23} \leq 3$$
$$x_{31} + x_{32} + x_{33} \leq 6$$
$$x_{41} + x_{42} + x_{43} \leq 2$$

The total mileage, which we want to minimize, is

$$p = 50x_{11} + 40x_{12} + 80x_{13}$$
$$+ 30x_{21} + 10x_{22} + 90x_{23}$$
$$+ 60x_{31} + 100x_{32} + 20x_{33}$$
$$+ 90x_{41} + 90x_{42} + 30x_{43}$$

Figure 6 shows the situation graphically. The problem can be viewed as one of choosing a set of connecting links from among the ones shown and a quantity to flow along each link in such a way as to obey the supply and demand limitations.

Example 11 (Assigment Problem)

Suppose n employees $P_1, P_2, \ldots, P_n$ are being considered for assignment to n jobs $J_1, J_2, \ldots, J_n$. Based on past performance and aptitude tests, we determine that if the ith person is assigned to the jth job, the positive benefit to the company is given by a number c_{ij}. Each employee will be given one job, and each job will be filled by one employee. How should the assignment be made to maximize total benefit?

The analogy between this problem and the transportation problem of Example 10 is that the employees play the role of supply sources and the jobs play the role of destinations. If we draw a figure with dots on the left for employees, dots on the right for jobs, and connecting links representing possible assignments, then we arrive at the same kind of thinking as in the transportation problem: choose a set of connecting links so that each person is assigned and each job is filled. In an assignment problem the flow along each chosen link will be 1.

$$\text{Let } x_{ij} = \begin{cases} 0 & \text{if employee } i \text{ is not assigned to job } j \\ 1 & \text{if employee } i \text{ is assigned to job } j \end{cases}$$

The feasible set is defined by the following requirements:

$$x_{ij} = 0 \text{ or } 1 \qquad \text{for each } x_{ij}$$

$$\sum_j x_{ij} = 1 \qquad \text{for each } i$$

$$\sum_i x_{ij} = 1 \qquad \text{for each } j$$

The last equation asserts that for each given job j the total number of employees assigned to it is one. Can you interpret the equation before it? The objective function to be maximized is

$$p = \sum_{i,j} c_{ij} x_{ij}$$

EXERCISES

● 1 Find the minimum of $p = x + 3y$ on the feasible region defined by

$$x, y \geq 0$$

$$3x + y \geq 9$$

$$x + 2y \geq 8$$

2 In Example 3 (Figure 2) suppose the objective function is changed to $p = 10x + 14y$. Show that there are two corner points where the objective function is minimum. Draw a figure with a constant-objective line touching at these points.

● 3 In Example 1 (Figure 1) suppose we add the following constraint: $3x + 9y \leq 27$. Find the corner points of the new feasible region. Is Figure 5a or b a correct drawing of the feasible region?

4 In Example 3 (Figure 3) suppose we change the objective function to $p = 3x + 2y + 10$. Could you still be sure that the minimum value of p occurs at a corner point? Why?

5 Transform Example 2 into a 2-variable problem by using the constraint $x + y + z = 18$ to eliminate a variable from the other constraints and from the objective function. Solve the problem you obtain.

Problems 6 through 18 fall into one of the categories described in this section. Formulate each problem algebraically. Your formulation must include (1) a description of what your variables represent, (2) a listing of all constraints, and (3) an objective function. If the problem involves two variables, solve it graphically.

6 (Malkevitch, Meyer) The Little Varmint Undergarment Company makes two products, children's undershirts and children's underpants. Making a shirt requires just $\frac{1}{2}$ yard of cloth, while a pair of underpants needs $\frac{1}{4}$ yard of cloth and 18 inches of elastic. The company has on hand 200 yards of cloth and 1440 inches of elastic. If the profit on each type of garment is \$1, how many of each should be made?

● **7** (Lial, Miller) A biologist must make a nutrient for his algae. The nutrient must contain the three basic elements D, E, and F, and must contain at least 10 kilograms of D, 12 kilograms of E, and 20 kilograms of F. The nutrient is made from three ingredients, I, II, and III. The quantities of the basic elements in these ingredients is as given in the following table.

Ingredient, 1 unit	Nutrient, kilograms per unit of ingredient			Cost, $ per unit of ingredient
	D	E	F	
I	4	3	0	4
II	1	2	4	7
III	10	1	5	5

How many units of each ingredient are required to meet the biologist's needs at minimum cost?

● **8** (Kolman, Beck) A manufacturer of sheet polyethylene has two plants, one in Salt Lake City and the other in Denver. There are three distributing warehouses, one in Los Angeles, one in Chicago, and one in New York City. The Salt Lake City plant can supply 120 tons of the product per week, whereas the Denver plant can supply 140 tons per week. The Los Angeles warehouse needs 100 tons per week, the Chicago warehouse needs 60 tons per week, and the New York City warehouse needs 80 tons per week. Shipping costs per ton are in the table below.

Source	Destination		
	Los Angeles	Chicago	New York
Salt Lake City	5	7	9
Denver	6	7	10

How many tons should be shipped from each plant to each warehouse to meet demands and minimize costs?

9 (Gass) Lockburne Air Force Base (AFB) at Columbus, Ohio, has been testing a large item of equipment, weighing a ton, for a certain airplane. It is now desired that this equipment be tested at other bases. The bases to be used for testing and the number of the item required by each base are shown at the top and bottom of the columns in the table below. The item is available at the three locations corresponding to the rows in the table. The table itself gives air-mile distances between the sources of supply and the bases requiring the items. How many items should be shipped from the various sources of supply to the various bases requiring the item to minimize the total ton-miles required to do the shipping? (If t tons are sent for m miles, this comes to tm ton-miles.)

| City | Number of items available | Air-mile distance | | | | |
		MacDill AFB	March AFB	Davis Northern AFB	McConnell AFB	Pinecastle AFB
Oklahoma City	8	938	1030	824	136·	995
Macon	5	346	1818	1416	806	296
Columbus	8	905	1795	1590	716	854
Number of items required		3	5	5	5	3

● **10** (Maki, Thompson) A psychologist wishes to conduct two types of studies, designated type I and II. Each experiment requires white, gray, and black rats in the following numbers. One experiment of type I requires 5 white, 1 gray, and 2 black rats, while one experiment of type II requires 2 white, 3 gray, and 2 black rats. The psychologist has available 100 white, 60 gray, and 50 black rats. Also, the psychologist has decided that each experiment of type II has twice the value of an experiment of type I. How many of each type of experiment should be done to maximize the value?

11 (Levin, Kirkpatrick) The King Concrete Company manufactures bags of concrete from beach and river sand. Each pound of beach sand costs 6 cents and contains 4 units of fine sand, 3 units of coarse sand, and 5 units of gravel. Each pound of river sand costs 10 cents and contains 3 units of fine sand, 6 units of coarse sand, and 12 units of gravel. Each bag of concrete must contain at least 12 units of fine sand, 12 units of coarse sand, and 10 units of gravel. Graphically, find the best combination of beach and river sand which will meet the minimum requirements of fine sand, coarse sand, and gravel at the least cost, and indicate the cost per pound.

12 (Levin, Kirkpatrick) Linville Laboratories has just been notified that it has received three govern- ment research grants. The laboratory administrator must now assign a research director to each of these projects. There are four researchers available now who are relatively free from other duties. The time required to complete the necessary research activities will be a function of the experience and ability of the research director who is assigned to the project. The laboratory administrator has this estimate of project completion times (in weeks) for each director-grant combination:

| Research director | Grant | | |
	1	2	3
Louis Gump	60	90	54
Anne Aitken	54	108	30
Mary Albritton	36	84	18
Ned Powell	72	96	48

Since the three grants have about the same priority, the laboratory administrator would like to assign research directors in a way that would minimize the total time (in weeks) necessary to complete all three grant projects. What assignments should be made?

● **13** (Kolman, Beck) The Savory Potato Chip Company makes pizza- and chili-flavored potato chips. These chips must go through three main processes: frying, flavoring, and packing. Each kilogram of pizza-flavored chips takes 3 minutes to fry, 5 minutes to flavor, and 2 minutes to pack. Each kilogram of chili-flavored chips takes 3 minutes to fry, 4 minutes to flavor, and 3 minutes to pack The net profit on each kilogram of pizza chips is $0.12, while the net profit on each kilogram of chili chips is $0.10. The fryer is available 4 hours each day, the flavorer is available 8 hours each day, and the packer is available 6 hours each day. Maximize the net profit with your model.

14 (Levin, Kirkpatrick) A sausage company has several types of sausage—Super-Hot, Hot Special, Country Best, Superlean, and so forth—each representing a particular mix of eight very specifically described U.S. government grades of beef and pork with appropriate spices. These grades and their characteristics are summarized in the following table.

Grade and symbol	Percent lean	Cost per pound, $
Imported beef, IB	95	1.16
Boneless chuck beef, BCB	80	0.99
Boneless carcass beef, BKB	65	0.97
Boneless pork butts, BPB	85	0.98
Boneless pork picnics, BPP	70	0.91
Boneless pork trimmings (A quality), BPTA	50	0.82
Boneless pork trimmings (B quality), BPTB	30	0.61
Boneless pork fat, BPF	0	0.12

The company wants to figure out the least expensive way to mix up a batch (its standard batch of sausage is 1000 pounds) of its Half-n-Half brand. The label on this brand guarantees that it has a total lean content of 70 percent (that 70 percent of its total weight, exclusive of spices and additives, is either lean pork or lean beef). The label also promises that half the total weight, exclusive of spices and additives, is pork and the other half beef. The company generally always has several thousand pounds of each of the eight grades of meat on hand; so constraints on raw material are not necessary.

15 (Lial, Miller) An office manager is considering purchasing filing cabinets. Type-A cabinets cost $10 each, require 6 square feet of floor space, and hold 24 cubic feet of files. Type-B cabinets cost $20, require 8 square feet of floor space, and hold 32 cubic feet of files. Owing to budgetary limitations the manager cannot spend more than $140 and, owing to limitations of space, cannot devote more than 72 square feet to filing cabinets. (Cabinets may not be stacked on top of one another.) How many of each type should the manager buy in order to maximize the amount of storage space obtained?

16 (Olinick) The admissions director of a small college is faced with the task of admitting a freshman class of at most 500 students. The typical male applicant can be expected to have a combined SAT score of 1200, contribute $8000 to the college as an alumnus, cause $200 damage to dormitory buildings and classrooms, and cost $2400 per year to teach. The typical female applicant can be expected to have combined SAT scores of 1300, contribute $3000 as an alumna, and cause $100 in damages. Because of different course selections, she can be educated at a cost of $2000 per year.

The college president demands a freshman class that will eventually contribute at least $2.5 million to the college, the faculty insists that the average SAT score be 1250 or higher, and the maintenance department can handle up to $85,000 in damages. The college treasurer wants to educate the class at the lowest possible cost.

Set up a linear programming problem whose solution will tell the admissions director how many men and how many women to admit. Identify the variables and write down the constraints and the function to be optimized. Solve the problem geometrically.

● **17** (Lial, Miller) Seven patients require blood transfusions. Suppose there are four types of blood available: A, B, O, and AB. Blood type O can be given to a patient with any blood type at all, and a patient with blood type AB, called a universal recipient, can take any type of blood. With these exceptions, the types of donor and recipient must match up exactly. The following tables show the supplies on hand, the amounts required by the patients, and the cost per pint of the various types of blood.

Blood type	Supply, pints	Cost, $
A	7	1
B	4	4
AB	6	2
O	5	5

Patient	Blood type	Requirement, pints
1	A	2
2	AB	3
3	B	1
4	O	2
5	A	3
6	B	2
7	AB	1

Is it possible to give each patient what he needs? If so, what is the cheapest way to do so?

18 (Olinick) Mary Muttoni is the chairperson of the history department at a small university. One of her duties is to make up the teaching schedule. The catalog of the university promises that the department will offer four large lecture courses for the freshmen next term. These are:

History A: A Survey of American History
History B: Revolutions and Counterrevolutions
History C: European Intellectual History
History D: China and Japan

There are four professors in the department who can teach any of the four courses. Because of their different backgrounds, expertise, and enthusiasms, they will attract different numbers of students in each course. Muttoni estimates the student appeal of each instructor in each course and derives a set of enrollment estimates. These are displayed in the table below.

Each professor will be assigned to only one course, and each course is to be taught by only one faculty member. The chairperson wishes to maximize the total enrollment in the four courses by assigning the available professors to different courses. The number of students each professor will attract in each course is given below.

Professor	Course A	B	C	D
Doggoff	310	260	270	290
Josephs	270	330	250	210
Reapingwillst	210	230	190	280
Cragdodge	240	210	220	200

Problems 19 through 22 can be formulated as linear programming problems, but they do *not* fall into the categories discussed in the text. Formulate each as a linear programming problem; i.e., describe the variables and state the constraints and the objective function. Solve graphically if possible.

19 (Levin, Kirkpatrick) The owner of the Neighborhood Hamburger Stand has decided to operate on a 24-hour basis. Based upon estimates of trade throughout this period, the owner feels the need for at least the following number of employees during the given time periods:

Time period	Minimum number of employees required
0:01–4:00	3
4:01–8:00	5
8:01–12:00[†]	13
12:01–16:00	8
16:01–20:00	19
20:01–24:00[†]	10

[†] 12:00 is noon and 24:00 is midnight.

The employees may report for work at midnight, 4 a.m., 8 a.m., noon, 4 p.m., or 8 p.m. Once employees report in, however, they must stay continuously for an 8-hour shift. Set up the objective function and constraint equations which would generate a solution to the problem. You should determine the numbers of employees reporting at each of the six possible reporting times if the total overall number of personnel is to be held to the minimum.

20 (Kolman, Beck) Suppose that the financial advisor of a university's endowment fund must invest exactly $100,000 in two types of securities: bond AAA, paying a dividend of 7 percent, and stock BB, paying a dividend of 9 percent. The advisor has been told that no more than $30,000 can be invested in stock BB, while the amount invested in bond AAA must be at least twice the amount invested in stock BB. How much should be invested in each security to maximize the university's return?

21 (Hadley) An oil refinery blends five raw stocks to produce two grades of motor fuel, A and B. The number of barrels per day (bbl/day) of each raw stock available, the octane numbers, and the cost per barrel are given in the table below.

Stock	Octane	Bbl/day	Cost/bbl, $
1	70	2000	0.80
2	80	4000	0.90
3	85	4000	0.95
4	90	5000	1.15
5	99	3000	2.00

The octane number of motor fuel A must be at least 95, and that of fuel B at least 85. Assume that a contract requires that at least 8000 barrels/day of B must be blended. However, the refinery can sell its entire output of fuels A and B, whatever amounts are made. Motor fuel A is sold to distributors at $3.75 per barrel and motor fuel B at $2.85 per barrel. All raw stocks not blended into fuel and with an octane number of 90 or more are sold for use in aviation gasolines at $2.75 per barrel, and those of octane number 85 or less are sold at $1.25 per barrel for use in fuel oils. In order to maximize daily profits, how much of each motor fuel should be made and how should the raw stocks be blended? Is it necessary to know the cost per barrel of each raw stock?

22 (Gass) A caterer has the task of supplying the Mad Hatter's tea parties with napkins. Here is the schedule of napkins required each day of the week:

Monday	5
Tuesday	6
Wednesday	7
Thursday	8
Friday	7
Saturday	9
Sunday	10

Napkins cost 25 cents each to buy. The King's Laundry takes 2 days (e.g., napkins from Monday's party are back in time for Wednesday's party) and charges 15 cents per napkin. The Queen's Laundry takes 3 days and costs 10 cents per napkin. The caterer has no napkins now. He wants to find the cheapest schedule of purchases and launderings that satisfies the condition that there is no advance buying: each napkin purchased is used the same day. For each day of the week it is desired to determine how many napkins to purchase, how many of the dirty ones to leave dirty in the laundry room, and how many to send to each of the two laundries.

Computer exercises

23 Write a computer program to find the corners of a feasible region defined by any number of inequalities of the form $a_i x + b_i y \leq c_i$. (*Hint*: The hard part is finding the corner points. Read the discussion after Example 4.)

24 Write a computer program to solve the general 2×2 linear programming problem of maximizing $P = c_1 x + c_2 y$ on the feasible region defined by

$$x, y \geq 0$$

$$a_1 x + b_1 y \leq d_1$$

$$a_2 x + b_2 y \leq d_2$$

Consider the possibility that the lines $a_1 x + b_1 y = d_1$ and $a_2 x + b_2 y = d_2$ may not intersect in the feasible region. Apply your program to Example 1.

25 (Sensitivity Analysis) Use the program of Exercise 24 to study how the maximum for Example 1 changes as a result of small changes in the coefficients of the objective function. Make 1-percent increases and decreases in each coefficient separately and see which change has the greatest effect on the maximum.

BIBLIOGRAPHY

Gale, David: The Optimal Assignment Problem, UMAP Module 317, available from COMAP, 271 Lincoln St., Lexington, MA 02173.

Gass, Saul: "An Illustrated Guide to Linear Programming." McGraw-Hill Book Company, New York, 1970.

Hadley, George: "Linear Programming." Addison Wesley, Reading, Mass., 1962.

Kolman, Bernard, and Robert E. Beck: "Elementary Linear Programming with Applications," Academic Press, New York, 1980.

Levin, Richard I., and Charles C. Kirkpatrick: "Quantitative Approaches to Management," McGraw-Hill Book Company, New York, 1978.

Lial, Margaret, and Charles D. Miller: "Finite Mathematics: With Applications in Business, Biology and Behavioral Sciences," Scott, Foresman and Company, Glenview, Ill., 1977.

Maki, Daniel P., and Maynard Thompson: "Mathematical Models and Applications," Prentice-Hall, Englewood Cliffs, N.J., 1973.

Malkevitch, Joseph, and Walter Meyer: "Graphs, Models and Finite Mathematics," Prentice-Hall, Englewood Cliffs, N.J., 1974.

Olinick, Michael: "An Introduction to Mathematical Models in the Social and Life Sciences," Addison Wesley, Reading, Mass., 1978.

Rosenberg, Nancy: Linear Programming in Two Dimensions: I and II, UMAP Modules 453 and 454, available from COMAP, 271 Lincoln St., Lexington, MA 02173.

3 AN OUTLINE OF THE SIMPLEX METHOD

Abstract The simplex method is the main solution procedure used for linear programming problems. It is generally carried out by standard and widely available computer programs; so there is little need for modelers to do it by hand. Nor is there much need for modelers to tinker with the theory behind the simplex method. Nevertheless, it is useful to know something about how the method works, if only to be able to use a program and interpret its results intelligently. This section aims for this level of understanding. In particular, most of the terminology concerning the simplex method is explained. Most of the theory is left for the exercises.

Prerequisites Graphical solution of linear programming problems (as in Section 2); algebraic manipulations on linear equations.

Most linear programming problems are too large for the kind of graphical solution described in the previous section. The method most commonly used is the simplex method. There are also special methods that work just for the transportation problem or the assignment problem. In practice, the implementation of the simplex method or any of the special purpose methods is carried out by a "canned" computer program. The purpose of this section is to provide a rough guide to the simplex method that will be useful to someone whose main contact with it will be through a computer program.

We shall illustrate our discussion with the example: maximize $p = 5x_1 + 6x_2$ subject to

$$x_1, x_2 \geq 0$$
$$2x_1 + 4x_2 \leq 24 \tag{1}$$
$$6x_1 + 3x_2 \leq 30$$

Slack Variables

Each inequality constraint of the form

$$a_1 x_1 + \cdots + a_n x_n \leq b$$

is replaced by the pair of conditions

$$a_1 x_1 + \cdots + a_n x_n + s = b$$
$$s \geq 0$$

Here s is a new variable, invented on the spot. It is called a *slack variable*. We create separate slack variables for each inequality. For an inequality with a $\geq$ sign, we subtract a nonnegative slack variable.

Example 1 The constraints for the feasible region of Equation (1) become

$$x_1, x_2, s_1, s_2 \geq 0$$
$$2x_1 + 4x_2 + s_1 \qquad = 24 \qquad (2)$$
$$6x_1 + 3x_2 \qquad + s_2 = 30$$

Most computer programs for the simplex method will not require you to input the problem with slack variables. The program will create the slack variables as its first step. However, the output of the program may make reference to slack variables.

Feasible Solutions and Basic Feasible Solutions

After introducing slack variables, we have a system of simultaneous linear equations and are looking for a solution, with all variables nonnegative, which maximizes p. Any nonnegative solution of this system is called a *feasible solution*. For example, $x_1 = 1, x_2 = 1, s_1 = 18$, and $s_2 = 21$ is a feasible solution. A feasible solution corresponds to a point of the feasible region. Specifically, if we discard the coordinates of the slack variables, the remaining coordinates belong to a point of the feasible region. For example, $(1, 1)$ is the point of the feasible region corresponding to $x_1 = 1, x_2 = 1, s_1 = 18$, and $s_2 = 21$.

Normally there are infinitely many feasible solutions, but the simplex method concentrates on a particular type of feasible solution called a *basic feasible solution*.

Definition

In a linear programming problem with n variables and m equations (after adding slack variables), if we set some $n - m$ variables equal to 0 (leaving m equations in m unknowns) and then obtain a unique nonnegative solution for the remaining variables, we call this set of values a basic feasible solution (BFS). The variables set equal to 0 are called *nonbasic variables* and the others are called *basic variables* for this particular BFS.

It can be shown that each BFS corresponds to a corner point of the feasible region and conversely.

Example 2 If we set x_1 and $x_2 = 0$ in Equation (2), we get the unique positive solution $s_1 = 24$ and $s_2 = 30$ for the remaining variables. Thus $x_1 = x_2 = 0$, $s_1 = 24$, and $s_2 = 30$ is a BFS. [The corner point it corresponds to is $(0, 0)$.] This can easily be read off Tableau (4), which is described below. For this BFS the value of the objective function is 0, which is the number in the upper right corner of the tableau.

Tableaux

In addition to Equation (2) describing the feasible region, we also are interested in the objective function; so we add it to the equations like this:

$$
\begin{aligned}
-5x_1 - 6x_2 \qquad\qquad + p &= 0 \\
2x_1 + 4x_2 + s_1 \qquad\qquad &= 24 \\
6x_1 + 3x_2 \qquad + s_2 \qquad &= 30
\end{aligned}
\tag{3}
$$

The manipulations we have to do on these equations concern only the coefficients; so it is common to detach them and form a tableau like this:

x_1	x_2	s_1	s_2	p	
-5	-6	0	0	1	0
2	4	1	0	0	24
6	3	0	1	0	30

$$\tag{4}$$

The operations of the simplex method can be carried out either on the system of Equation (3) or Tableau (4). The tableau is generally used in practice.

Pivot Exchanges and Gaussian Elimination

By picking x_1 and x_2 to set equal to 0 in Example 2, we made things easy for ourselves because we could read off the other values. If we had wanted to set $x_1 = 0$ and $s_1 = 0$, it would have been slightly harder; we would need a little algebra instead of just reading off the solution. One way to do this algebra is to carry out an operation called a *pivot exchange*. A pivot exchange is nothing more than a method of putting our system of equations into a form where a BFS can be found by reading off values after certain variables have been set equal to 0. The simplex method consists of a carefully chosen series of pivot exchanges.

Definition

In a system of simultaneous linear equations, a pivot exchange consists of selecting a particular variable and eliminating it from all equations except for one previously selected one. (This is also called *gaussian elimination*.)

Example 3 Suppose we wish to eliminate x_2 from all but the second equation of Equation (3). We can achieve this by first dividing the second equation by the coefficient of x_2 and getting

$$\tfrac{1}{2}x_1 + x_2 + \tfrac{1}{4}s_1 = 6$$

Now subtract 3 times this equation from the third and add 6 times it to the first.

$$-2x_1 + \tfrac{3}{2}s_1 \qquad + p = 36$$
$$\tfrac{1}{2}x_1 + x_2 + \tfrac{1}{4}s_1 \qquad = 6$$
$$\tfrac{9}{2}x_1 - \tfrac{3}{4}s_1 + s_2 \qquad = 12$$

Here is the value of this: if we want a BFS in which x_1 and s_1 are set equal to 0, we can read off the values of x_2 and s_2; $x_2 = 6$ and $s_2 = 12$. Furthermore, from the first equation we find that, for this BFS, the value of p is 36.

This pivot exchange is usually carried out directly on Tableau (4) and produces the following tableau.

x_1	x_2	s_1	s_2	p	
-2	0	$\tfrac{3}{2}$	0	1	36
$\tfrac{1}{2}$	1	$\tfrac{1}{4}$	0	0	6
$\tfrac{9}{2}$	0	$-\tfrac{3}{4}$	1	0	12

(5)

Notice that the value of the objective function is once again the number in the upper right. This is no accident; in every tableau produced by the simplex method, the number in the upper right is the value of the objective function for the BFS that can be read off from the tableau. (Some simplex-method programs put the objective-function equation at the bottom of the tableau. In that case the value of the objective function is at the lower right.)

If a pivot exchange is not carefully chosen, it will give a tableau where the solution we can read off has some variables negative. Therefore, this is not a BFS and the pivot exchange is a poor one. The next example illustrates this. The simplex method has built-in safeguards that avoid this kind of pivot exchange (see Exercise 7).

Example 4 In Tableau (4) suppose we attempt to remove x_2 from all the equations but the lst. To do this, first divide the last row by 3, then subtract 4 times it from the second, and add 6 times it to the first.

x_1	x_2	s_1	s_2	p	
7	0	0	2	1	60
-6	0	1	$-\frac{4}{3}$	0	-16
2	1	0	$\frac{1}{3}$	0	10

The solution we read off is $x_1 = 0$, $x_2 = 10$, $s_1 = 0$, and $s_1 = -16$. But the value -16 is not permissible for s_1; so we do not have a BFS. The pivot exchange was poorly chosen.

How the Simplex Method Works

Step 1. Find an initial tableau from which a BFS and an associated value of the objective function can be read off.

Step 2. Do a pivot exchange to get a new BFS with a value of the objective function which is no worse than before (and usually better).

Step 3, 4, ... Repeat step 2 until no further improvement in the objective function is possible.

Not every pivot exchange will improve the value of the objective function in step 2. There are rules to follow to select the variable to eliminate and which equations to eliminate it from. It is possible to prove that, when you get to a tableau where the rules no longer work, then the value of the objective function for this tableau is generally optimum.[*]

Condensed Tableaux

Often computer programs report condensed tableaux instead of the kind described above. In a condensed tableau the columns of the basic variables are left out, as is the p column. There is little lost this way because these columns are all 0 except for a single 1. We could reconstruct the missing column if we merely knew in which row the 1 was. The position of the 1 is indicated in the condensed tableau by simply labeling that row with the name of the removed variable.

[*] There are a few exceptional cases described later in the section titled "Additional Wrinkles."

Example 4 The condensed version of Tableau (4) is

	x_1	x_2	
p	-5	-6	0
s_1	2	4	24
s_2	6	3	30

(6)

The label s_1 on the second row takes the place of the column labeled s_1 in Tableau (4). Labels p and s_2 also take the place of removed columns.

Example 5 Create the full simplex tableau for the following condensed tableau:

	x	y	s_1	
p	2	4	-6	4
z	3	4	1	6
s_2	-2	1	5	8
s_3	6	2	-3	10

The full tableau is

x	y	z	s_1	s_2	s_3	p	
2	4	0	-6	0	0	1	4
3	4	1	1	0	0	0	6
-2	1	0	5	1	0	0	8
6	2	0	-3	0	1	0	10

In the case of uncondensed tableaux we proceeded from one tableau to another by pivot exchanges. When working with condensed tableaux, we need a slightly different arithmetic procedure to achieve the same effect. Even though the arithmetic is a little different, it is usually still called a pivot exchange.

Geometric Interpretation

BFS's correspond to corner points of the feasible region and vice versa. A medium-sized problem involving 40 variables and 60 constraints might have over a billion corners. The simplex method finds the optimum solution by checking a very tiny proportion of the corner points.

Suppose the polyhedron of Figure 1 represents the feasible region and the planes are constant-objective planes, with the objective function increasing in the direction of the arrow. We want to maximize the objective function. Step 1 of the simplex method is to find a corner to start with. This is called an *initial BFS*.

The method then finds a corner point which neighbors the initial one but which lies on a higher constant-objective plane (or, in certain cases, on the same constant-objective plane). This process of trading in one BFS for one that is usually better continues till we get to the "top" of the polyhedron.

Additional Wrinkles

1. *Cycling.* It is remotely possible for the simplex method to fall into a never-ending pattern of pivot exchanges that cycle repeatedly through the same tableaux and give the same value of the objective function each time. Examples have been specially devised to show that this can happen, but no examples have ever occurred in real applications. In any case, there are simplex-method programs that can handle this problem.
2. *No optimum solutions.* If the feasible region is unbounded, then it is possible that the objective function may not have an optimum value. (This is illustrated in the discussion following Theorem 1 in Section 2.) This can easily be detected in the simplex method, and your program will do the detection for you.
3. *Finding a starting BFS.* Step 1 of the simplex method calls for finding a tableau where a BFS can be read off. In many problems the tableau you supply to the

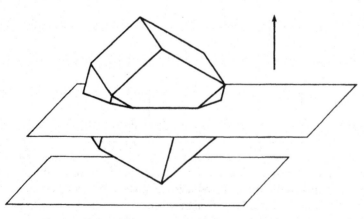

Figure 1

computer program will be a suitable starting one. The case where this is not so can be divided into two categories:

   a. The feasible region is empty. In this case there are no BFS's at all, there is no suitable starting tableau, and, since there is no solution at all, there certainly is no optimal one. The program will detect this situation for you.

   b. The feasible region is not empty. In this case there exists a suitable starting tableau, even though the one supplied to the program as input may not be it. The program will generally be capable of finding a suitable tableau on its own. One technique for doing this involves *artificial variables*, which are temporarily added to the problem. Your program will probably do this without letting you know about it. In the event that the program output does make reference to artificial variables, you can safely ignore it. These variables always are eventually set equal to 0 and have no meaning in the application giving rise to the linear programming problem you are trying to solve. In particular, they are not to be confused with slack variables.

EXERCISES

1 Use slack variables to describe the feasible region of Example 5 in Section 2 in terms of equations involving nonnegative variables. Can you give a physical interpretation of what these slack variables mean in that example?

2 In Equation (3) suppose we want to eliminate x_1 from all equations but the third. (Then we would say 6 is the pivot.)

   (*a*) Carry out this pivot exchange.

   (*b*) Can you do it directly on Tableau (4)?

   (*c*) What is the BFS you obtain after pivoting?

   (*d*) What is the corresponding value of the objective function for this new BFS?

   (*e*) Is this pivot exchange better or worse than the pivot exchange, carried out in Example 3 on this same system, which produced Tableau (5)?

3 Find the condensed version of Tableau (5).

4 Determine the rules for pivot exchange when dealing with condensed tableaux. In particular, show how the pivot exchange that transforms Tableau (4) into Tableau (5) can be carried out on Tableau (6), which is the condensed version of Tableau (4). Your result should be the condensed version of Tableau (5).

5 Carry out a graphic solution of the illustrative problem discussed in this section. Find the corner points that correspond to the BFS's found in Examples 2 and 3.

6 Can you prove that there are only finitely many BFS's? Can you give a formula for an upper bound to the number of BFS's there can be in a problem with n variables and m constraints?

7 We didn't describe how the simplex method picks the best pivot exchange to do for a given uncondensed tableau. In the case of maximizing the objective function, here's how:

   1. Find the most-negative entry of the objective-function row (leaving aside the last two entries). The column of this entry is called the *pivot column*. This column indicates the variable to be eliminated from all equations except one (which one is determined by rule 2). If there are no negative entries in the objective-function row, you have reached the maximum solution.

   2. Divide each positive entry in the pivot column, except the top one, into the last entry of its row. Find the entry in the pivot column that gives the smallest ratio. This is called the *pivot*. The row of this pivot is the one we don't eliminate the variable from. If there is no positive entry in the

pivot column, then there is no maximum. The objective function can be made as large as you wish.

(*a*) Apply these rules to the tableau below and carry out the pivot exchange.

x	y	s_1	s_2	p	
-2	-1	0	0	1	0
4	1	1	0	0	10
18	15	0	1	0	60

(*b*) Using this tableau, show that, if rule 1 is followed but rule 2 is not, then the solution read off of the next tableau will not have all variables nonnegative.

(*c*) Using this tableau, show that, if the pivot column is not chosen by rule 1 but rule 2 is followed, then the improvement in the objective function is not as great as if rule 1 were followed.

8 In a linear programming problem involving maximizing (as opposed to minimizing), suppose there is a tableau in which there is a negative entry in the objective-function row but no positive entries elsewhere in the column of that negative entry. This means that there is no maximum. Can you explain why? Consider the system of equations corresponding to that tableau.

Computer exercise

9 Go to your computer center and use a canned simplex program to solve the problem discussed in this section.

BIBLIOGRAPHY

Glicksman, Abraham M.: "An Introduction to Linear Programming and the Theory of Games," John Wiley & Sons, Inc., New York, 1963. One of the simplest expositions of the simplex method and the theory behind it.

Kolman, Bernard, and Robert E. Beck: "Elementary Linear Programming with Applications," Academic Press, New York, 1980. A comprehensive account in the language of linear algebra.

4 INTEGER PROGRAMMING—THE KNAPSACK AND TRAVELING SALESMAN PROBLEMS

Abstract In many linear programming problems, answers must be integers to make sense. Two of the most important types of such problems are described. The simplex method may not work for such problems, even if we round off the answers.

Prerequisites Formulating linear programming problems.

There are many linear programming problems where answers that are not integers are not acceptable. For example, when we solved the diet problem (Example 7) of

Section 2, we discovered that one serving of soup included 2.8 ounces of onion soup. But what if the cook can't or won't measure out soup in fractions of an ounce? The same problem can arise whenever we deal with something that comes in indivisible units: people (Example 9 of Section 2), trucks (Example 10 of Section 2), etc.

Such problems are called *integer programming problems*. If we are looking for an integer solution to a linear programming problem, we should specify this when describing the feasible region. Here's how our soup problem is described this way. (Compare Example 3 of Section 2.)

Example 1

Minimize $p = 3x + 2y$ subject to

$$x, y \geq 0$$
$$x, y = \text{integers}$$
$$5x + 7y \geq 35$$
$$10x + 4y \geq 40$$

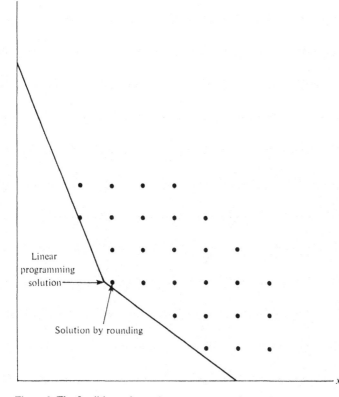

Figure 1 The feasible set for an integer programming problem.

The feasible set for this problem is the infinite collection of *lattice points* with integer coordinates that lie above or on the boundary lines in Figure 1. (Compare Figure 2 of Section 2.)

One approach to solving this integer programming problem is to take the answer given by ordinary linear programming and round the coordinates up to the nearest integer; in this case we round $(2.8, 3)$ up to $(3, 3)$. This turns out to be the minimum (among integer points); so rounding up has served us well.

However, there are two difficulties with rounding:

1. There are many ways to round. Each coordinate can be either rounded up or rounded down. How do we decide?
2. Even if we make the best decisions about rounding, we may not get the optimum.

The next example illustrates these problems.

Example 2

Maximize $p = 3.8x + 2.4y$ subject to

$$x, y \geq 0$$

$$x, y = \text{integers}$$

$$3.8x + 2.2y \leq 15.2$$

$$y \leq 3.8$$

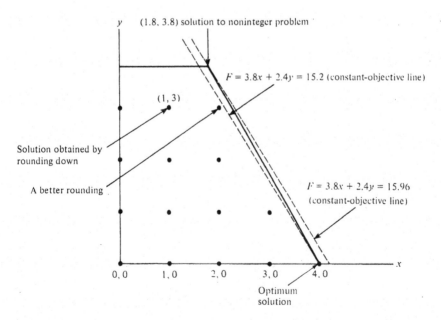

Figure 2

The feasible region for the noninteger version of this problem is the trapezoid bounded by solid lines in Figure 2. The heavy dots constitute the feasible set for the integer version. The solution to the noninteger problem is (1.8, 3.8) and gives a value to the objective function of $(3.8)(1.8) + (2.4)(3.8) = 15.96$. If we rounded both coordinates up to (2, 4), we are now outside the feasible region. If we round down to (1, 3), the objective function value is $(3.8)(1) + (2.4)(3) = 11$. A better rounding would be to round x up and y down (but how would we know this if we didn't look at the picture?) to get (2, 3). Then the objective function value is $(3.8)(2) + (2.4)(3) = 14.8$. The best integer point turns out to be (4, 0), where the objective function has the value 15.2. This point cannot be obtained from (1.8, 3.8) by any sort of rounding. (The fact that it is a corner of the feasible region is just an accident.)

There is a special category of integer programming problem in which the variables do not represent quantities in the ordinary sense but are so-called *decision variables*, as in the following examples.

Example 3 (Knapsack Problem)

A total of m items, whose weights are $a_1, a_2, \ldots, a_m$, are available for packing a knapsack. The total weight to be packed cannot exceed b. The objective is to pack as many items as possible.

$$\text{Let } x_i = \begin{cases} 0 & \text{if item } i \text{ is not to be packed} \\ 1 & \text{if item } i \text{ is to be packed} \end{cases}$$

We wish to maximize $p = x_1 + x_2 + \cdots + x_m$ subject to

$$x_i = 0 \text{ or } 1$$

$$a_1 x_1 + a_2 x_2 + \cdots + a_m x_m \leq b$$

Example 4 (Traveling Salesman Problem)

The traveling salesman problem is probably the most famous and thoroughly studied integer programming problem. In this problem there are n cities and a salesman must arrange a tour that visits each one exactly once and then returns to the starting city in a minimum total time. The time required to go from city i to city j is denoted t_{ij}.

The data of a traveling salesman problem are often depicted with a graph (Figure 3), in which the vertices represent the cities and the edges connecting the cities are labeled with the times.

$$\text{Let } x_{ij} = \begin{cases} 0 & \text{if edge } ij \text{ is not part of the tour} \\ 1 & \text{if edge } ij \text{ is part of the tour} \end{cases}$$

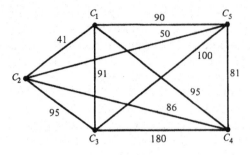

Figure 3 Travel times between five cities.

As an illustration, the following assignment of variables represents the tour that goes around the outer border of Figure 3.

$$x_{12} = x_{23} = x_{34} = x_{45} = x_{51} = 1$$

All other $x_{ij} = 0$

This is a connected tour that visits every city exactly once (except the start which is visited twice).

As we will now see, not every assignment of 0s and 1s to the variables x_{ij} will be a suitable tour; so some conditions will have to be imposed on the x_{ij}. (These constraints will define the feasible set.) The assignment

$$x_{12} = x_{23} = x_{34} = x_{35} = 1$$

All other $x_{ij} = 0$

is unsuitable because, after visiting x_3, there are two cities to go to next rather than one. This could be ruled out if we insist that only one edge leaves city 3. In algebra, this means

$$x_{31} + x_{32} + x_{34} + x_{35} = 1$$

Similar conditions must hold at the other cities. In general,

$$\sum_j x_{ij} = 1 \qquad \text{for each } i \tag{1}$$

We also need a condition that ensures that each city is actually visited. The requirement that we visit city 3 is

$$x_{13} + x_{23} + x_{43} + x_{53} = 1$$

Similar conditions hold for the other cities. In general,

$$\sum_i x_{ij} = 1 \qquad \text{for each } j \tag{2}$$

Finally, we must ensure that we have one connected tour, instead of a disjoint collection of partial tours whose union covers the cities. The kind of thing we wish to rule out can be illustrated on Figure 3 by the assignment

$$x_{12} = x_{23} = x_{31} = 1$$
$$x_{54} = x_{45} = 1 \tag{3}$$
$$\text{All other } x_{ij} = 0$$

Notice that this satisfies all the requirements so far.

To rule this out, we begin by observing that, if s is a set of cities, then

$$\sum_{i,j} x_{ij} \qquad \text{for } i, j \in s \text{ and } i \neq j$$

is the number of edges in the proposed solution which join members of s to one another. If s is a set of cities covered by a tour (full or partial), then

$$\sum_{i,j} x_{ij} = |s| \qquad \text{for } i, j \in s \text{ and } i \neq j$$

where $|s|$ denotes the number of members of s. Thus one way to rule out partial tours is to insist that, for any proper subset s of the cities, we have

$$\sum_{i,j} x_{ij} \leq |s| - 1 \qquad \text{for } i, j \in s \text{ and } i \neq j \tag{4}$$

For example, for the subset $s = \{1, 2, 3\}$, we have

$$x_{12} + x_{21} + x_{13} + x_{31} + x_{23} + x_{32} \leq 2$$

This rules out solution (3).

The number of constraints of type (4) is large: for n cities it is $2^n - 1$. There are more devious ways of eliminating partial tours without adding so many constraints.

The full list of constraints consists of the equations and inequalities in Equations (1), (2), and (4), together with the requirement that each x_{ij} be 0 or 1.

Finally we come to an easy part. The objective function, which we wish to minimize, is

$$p = \sum t_{ij} x_{ij}$$

Generally speaking, solving integer programming problems is a lot harder than solving linear programming problems that have no integer constraints. There is a variety of methods that sometimes work (the most useful of these at present are "cutting plane" methods and "branch and bound" methods), but there is no single method that can be relied upon to work for any commonly occurring problem in a practical amount of time. Even in cases where one of the special methods will work, it is sometimes hard to tell in advance which one will work. For these reasons we will not study any solution methods for integer programming problems. Solving these problems is a task for specialists.

EXERCISES

● **1** For Exercises 6 through 22 of Section 2 determine which definitely require integer solutions. (If there are doubtful cases, list them separately.)

2 Suppose the constraints defining the feasible set do not include any equations and suppose each inequality is a $\leq$ inequality with all coefficients on the left side positive.

    (a) If we solve the problem without the integer constraints and then round all coordinates down, show that this gives a point in the feasible set.

    (b) Would this still be true if there were an equation among the constraints? Would it depend on the signs of the coefficients in the equation?

● **3** (a) Under what conditions would rounding up all coordinates of a noninteger solution of an integer programming problem give a point of the feasible set? (Try Exercise 2 for some clues.)

    (b) Can you formulate any conditions under which mixed rounding works, i.e., some coordinates are rounded up and others down?

4 Suppose you have a noninteger solution to an integer programming problem involving n variables. How many different roundings are possible if each coordinate can be either rounded up or down independently of the other coordinates? Answer the question first for $n = 2$, $n = 3$, and then give a formula in terms of n.

● **5** In the knapsack problem, suppose item i has a value of v_i and it is desired to maximize the total value of items packed (instead of the total number). How does the problem formulation change?

6 In the knapsack problem, suppose item i has volume s_i and the knapsack has total volume capacity c. If you need to stay under both the weight and volume limits, how does the problem formulation change?

● **7** Suppose we have a knapsack problem in which item i has value v_i and wish to maximize total value, subject to a weight constraint. Suppose further that each item has greater value than the sum of the values of all items with lower weight. In this case the following strategy suggests itself: at each stage of filling the knapsack, use the heaviest item you can without going over the weight limit. Does this strategy always give the optimal solution?

8 The strategy used in Exercise 7 will not always give an optimal solution if we do not have the condition that the value of an item exceeds the sum of the values of all lighter items. Create an example that demonstrates this.

9 Suppose a cashier must give a customer 87 cents in change, using standard American coins. The cashier wishes to know how many of each type of coin to use to have a minimum number of coins that add up to 87 cents.

    (a) Formulate this as an integer programming problem. (Is it a knapsack problem or similar to one?)

    (b) Why is it that cashiers routinely manage to solve this problem by common sense, without integer programming?

10 Show that the feasible set for a knapsack problem is always finite. Is this true for any integer programming problem? How about a traveling salesman problem?

11 In a traveling salesman problem, suppose you insist on starting with C_1 and ending with C_n (there is no return to the starting city). Write out the constraints for this new problem. Try it first for Figure 4 below.

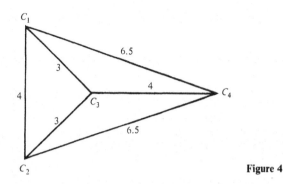

Figure 4

● **12** In the traveling salesman problem of Figure 4, suppose you insist on starting at C_1 and having the last leg of your tour being from C_4 to C_1. Write out the constraints for this problem.

13 In a traveling salesman problem, suppose that certain links are unavailable. For example, there may be no direct route from C_1 to C_4, as in Figure 5. How would we change the integer programming formulation of the problem? Do this first for Figure 5 where only the available links have been drawn.

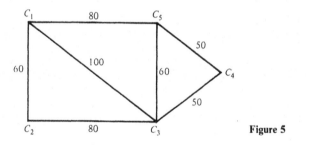

Figure 5

14 Suppose we have a graph, such as in Figure 5, and are looking for the longest tour in which we visit no city more than once except for the starting city, which we return to at the end. Show how to formulate this as an integer programming problem. For inspiration, read the traveling salesman problem and Exercise 13. (*Note*: A tour of the type described which goes through every city is called a *Hamilton circuit*.)

Computer exercises

15 Write a computer program that makes change (see Exercise 9).

16 Modify the change-making program (Exercise 15) so that, when the cash register gets low on a certain type of coin (a condition you specify by some kind of input to the program), no coins of that type are used in making change.

17 (*a*) Invent your own random traveling salesman problem like this:

1. Use the random-number generator to obtain 10 points (x_i, y_i) in the plane.
2. Use the Euclidean-distance formula to compute the distances between pairs of these points.

(*b*) Generate 100 random tours that cover every city once and return to their start. Compute the length of each random tour. Find the average.

(c) Plot the points obtained in part (a) on graph paper and try finding a good tour by trial and error. You will probably do better than most of the random tours in part (b). Why do you suppose that is?

(d) Using the insights gained in part (c), try to modify the random-tour program of part (b) to incorporate some "artificial intelligence" in place of the completely random-choice procedure.

BIBLIOGRAPHY

Cooper, Leon, and David Steinberg: "Introduction to Methods of Optimization," W. B. Saunders, Philadelphia, 1970.

Hoffman, A. J., E. L. Johnson, P. Wolfe, and M. Held: Aspects of the Traveling Salesman Problem, IBM Research Report, available from IBM Thomas Watson Research Center, Yorktown Heights, NY 10598. A good survey of the state of the art for this problem as of 1981.

Kolman, Bernard, and Robert E. Beck: "Elementary Linear Programming with Applications," Academic Press, New York, 1980.

Maynard, James M.: A Linear Programming Model for Scheduling Prison Guards, UMAP Module 272, available from COMAP, 271 Lincoln St., Lexington, MA, 02173.

5 THE TRANSPORTATION PROBLEM

Abstract The transportation problem is a special type of optimization problem which occurs very commonly. Although it is a linear programming problem and can, therefore, be solved by the simplex method, there is a special purpose algorithm that is more efficient and intuitive. This algorithm is the subject of this section.

Prerequisites No formal prerequisites.

A transportation problem involves finding the cheapest way of shipping goods from certain sources of supply to certain destinations.

Example 1

Suppose that two oil refineries S_1 and S_2 act as sources of gasoline for three cities $D_1, D_2,$ and D_3. In Figure 1, the numerical labels on the refineries are their daily production capacities and the numerical labels on the cities are their daily demands. The line segments represent shipment routes connecting the refineries and cities. Notice that any refinery can supply any city. The numerical label on a segment is the cost of shipping 1 unit of gasoline along that route.

In this example, the sum of the capacities is 38, which is also the sum of the demands. Therefore, there is no difficulty in satisfying the demands. There are many ways it can be done. Two possibilities are shown in Figure 2. In this figure, we have drawn in only the shipment routes actually used. The labels in circles on the routes show how much is sent along them. The solution in Figure 2a is cheaper

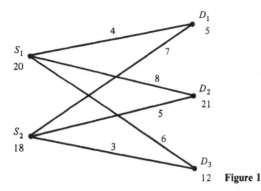

D_1
5

S_1
20

4

7

8

D_2
21

5

S_2
18

3

6

D_3
12   **Figure 1**

than the solution in Figure 2b. But since we haven't worked out all possible solutions, we don't know if it is the cheapest of all.

In general, a transportation problem involves m sources $S_1, S_2, \ldots, S_m$, having capacities (or supplies) $s_1, s_2, \ldots, s_m$, and n destinations $D_1, D_2, \ldots, D_n$, having demands $d_1, d_2, \ldots, d_n$. We assume that the total demand equals the total supply:

$$\sum_1^n d_i = \sum_1^n s_i$$

Many practical problems do not conform to this restriction. But there is a trick that easily transforms any transportation problem where supplies exceed demand to one where supplies equal demand. This will be described later; now we proceed with the case where total supply and demand are equal.

The cost of sending one unit from S_i to D_j is denoted c_{ij}. The s_i, d_j, and c_{ij} are all ≥ 0.

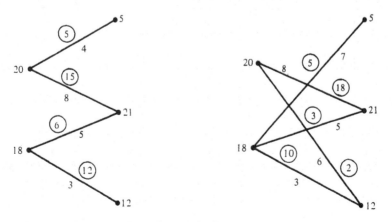

(a) Cost = $5 \cdot 4 + 15 \cdot 8 + 6 \cdot 5 + 12 \cdot 3 = 206$ ·   (b) Cost = $5 \cdot 7 + 3 \cdot 5 + 10 \cdot 3 + 18 \cdot 8 + 2 \cdot 6 = 236$

Figure 2 Two feasible solutions to the transportation problem of Figure 1.

Table 1

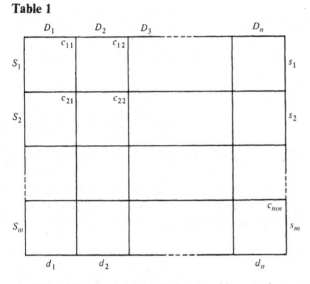

We will solve transportation problems using a tableau containing the costs, supplies, and demands. The tableau contains one row for each source S_i and one column for each destination D_j (see Table 1). The number in the (i, j) square (i.e., the ith row and jth column) is c_{ij}. We place each cost in the upper right corner of its box to make room for other numbers which may appear in the boxes during the course of the algorithm. The supplies s_i and demands d_j are placed around the rim: s_i on the ith row and d_j on the jth column.

Table 2 shows the tableau for Example 1.

Table 2

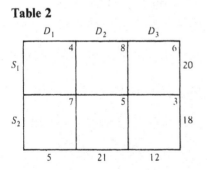

A solution is displayed on the tableau as follows: suppose x_{ij} is the amount sent from S_i to D_j; then we place the number x_{ij} inside a circle in the (i, j) square. If $x_{ij} = 0$, we leave the square blank. The squares with circled amounts are said to be *in the solution.*

Example 2

Table 3a displays the solution shown in Figure 2a. Table 3b displays the solution shown in Figure 2b. Notice that in each row the sum of the circled numbers equals

Table 3

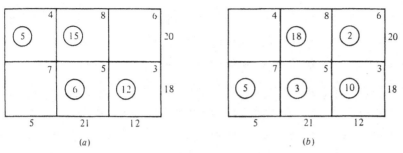

(a) (b)

the supply for that row. For each column, the sum of the circled numbers equals the demand for that column. These equalities are called *rim conditions*.

Definition

A *circuit* among the squares of a transportation tableau is a sequence of squares such that:

1. The first square is the same as the last.
2. Each square can be connected to the next square in the sequence by a horizontal or vertical line in the tableau.

Example 3

Table 4a and b shows examples of circuits.

Table 4

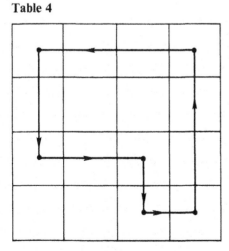
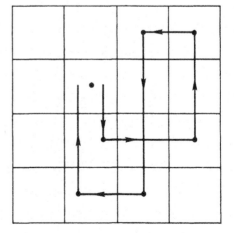

(a) Squares in circuit:
 (1, 1)–(3, 1)–(3, 3)–(4, 3)–
 (4, 4)–(1, 4)–(1, 1)

(b) Squares in circuit:
 (2, 2)–(3, 2)–(3, 4)–(1, 4)–
 (1, 3)–(4, 3)–(4, 2)–(2, 2)

Definition

A solution is called a *basic feasible solution* (BFS) if it has these two characteristics:

1. It involves $n + m - 1$ squares with circles.
2. There are no circuits among the squares in the solution.

Table 3*a* shows a BFS. (Note that $m = 2$ and $n = 3$.) However, the solution in Table 3*b* is not a BFS. The reason we are interested in BFS's is:

Theorem 1

The lowest cost in a transportation problem is always achieved by some BFS.

Proof We omit the proof of this theorem.

Figure 3 shows a flowchart that summarizes the transportation algorithm. The first step is to find an initial BFS. Any BFS will do, and any method—including trial and error—that finds one is acceptable. One of the simplest is to use the *Northwest Corner Rule*, which we will describe in detail.

A key aspect of the transportation algorithm is a method for testing a given BFS to see whether it is the lowest-cost solution. This test involves the concept of the improvement index of a square. If we apply this test to the initial BFS produced by the Northwest Corner Rule and discover that we do not have the optimal solution, then we must find a new BFS to test.

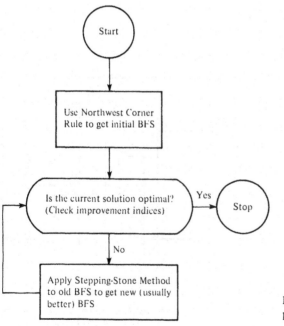

Figure 3 A flowchart for the transportation algorithm.

A new BFS is found by the Stepping-Stone Method. When this method is applied to one BFS to produce a second, the cost of the second BFS is never greater than that of the first. In most cases, the cost is strictly less. Consequently, if we repeatedly apply the Stepping-Stone Method, we will almost always reach the optimum solution eventually.

The few cases where this does not occur are due to the phenomenon called *cycling*, are quite rare, and will be disregarded here. (The interested reader should read about cycling on page 39 of the book by Donald P. Gaver and Gerald L. Thompson, cited in the Bibliography, to see how these exceptions arise and how they can be dealt with.)

The Northwest Corner Rule

1. Choose the entry in the northwest (upper left) corner of the matrix. This represents the shipment route from S_1 to D_1.
2. Use this route to satisfy as much of the demand at D_1 as possible from the supply at S_1. Record the shipment with a circled number in the upper left square.
3. If the supply at S_1 is not used up, use the remaining supply to fill remaining demands $D_2, D_3, \ldots$, in order, until the supply at S_1 is used up. Record shipments with circled numbers.
4. When one supply is used up, go to the next supply and start filling demands in order with the first D_i where there is still some demand not yet fulfilled. Record all shipments.

Example 4

In Table 2 we begin by shipping 5 units from S_1 to D_1. There are still 15 units of supply at S_1; so we send them all to D_2 and have 6 units of demand left at D_2. Supply S_1 is now used up; so we proceed to S_2. We send 6 units from S_2 to fill the remaining demand at D_2. The remaining supply at S_2 is used to fill the demand at D_3. The solution obtained in this way is shown in Table 3a.

Notice that the Northwest Corner Rule pays no attention to costs.

The Northwest Corner Rule generally produces a BFS. Furthermore, the circled entries make a staircase pattern, stretching from the northwest corner to the southeast corner. There is a situation, called *degeneracy*, when the solution produced by the Northwest Corner Rule does not have these characteristics. This is illustrated in the next example.

Example 5

Let us apply the Northwest Corner Rule to the tableau in Table 5a. Before long we reach the stage shown in Table 5b, where supplies at S_1 and S_2 have been used up and they exactly fill the demands at D_1 and D_2. When we continue with the Northwest Corner Rule, we produce the solution in Table 5c. It is not a BFS since it only has four entries.

Table 5

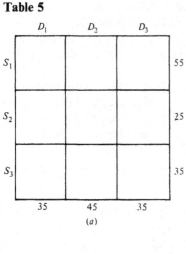

(a)

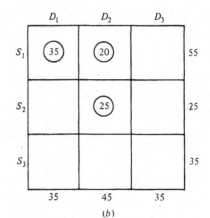

(b)

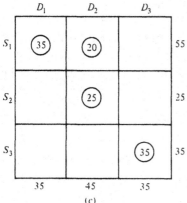

(c)

Definition

A solution to a transportation problem is called *degenerate* if the number of squares in the matrix with circled entries is less than $m + n - 1$. If a transportation problem has degenerate solutions, then the problem itself is called degenerate.

    Degeneracy occurs in the Northwest Corner Rule when a supply is used up and a demand fulfilled simultaneously. In the more common (nondegenerate) case, when a supply is used up there is still some unfulfilled demand at the destination you are working on.

    A degenerate solution may also arise when we apply the stepping-stone algorithm to try to produce a new BFS.

    If degeneracy does occur, usually the algorithm we will present here still works if we make the following modification. Choose some blank squares and add them to the solution by placing circled zeros in them. Choose the number and positions of these added squares so as to produce a BFS: there should be no cycles among the squares and there should be $n + m - 1$ squares altogether. In rare cases this modification is not sufficient due to cycling, as we mentioned earlier.

In this section, we shall only present problems where none of the solutions are degenerate.

The Stepping-Stone Improvement Method

The Stepping-Stone Method takes a given BFS and examines each shipping route (square of the matrix) not used by that BFS to see whether a cheaper cost could be obtained by diverting some material to the unused route from another route. This is explained in detail in the following example.

Example 6

In the tableau of Table 6 suppose we try to send 1 unit from S_1 to D_2. We symbolize this with a plus sign in the (1, 2) square. To accomplish this without violating the rim condition involving the first row, we need to divert 1 unit from the route connecting S_1 to D_1, thereby reducing the (1, 1) square from 56 to 55. This is symbolized by the minus sign in the (1, 1) square.

Table 6

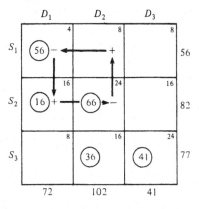

But now the rim condition involving the first column is no longer satisfied (i.e., the demand of 72 for D_1 is not fulfilled). To fix this, we increase the amount shipped from S_2 to D_1 from 16 to 17. This disturbs the rim condition originally satisfied by row 2. To fix this, we decrease the amount shipped from S_2 to D_2 from 66 to 65. One might expect this to disturb the rim condition involving the second column. However, this condition has already been disturbed by the first change, involving the (1, 2) square. Our last change merely compensates for this.

Notice that the cells with plus and minus signs form a circuit. Formation of such a circuit is always the first step in the Stepping-Stone Method. This circuit is called the *alteration circuit*.

Now we must determine whether the modification we have just described lowers the cost. This is done using the costs in the upper right corners of the squares of the alteration circuit (those squares with plus or minus signs). The plus sign in the $(1, 2)$ cell means an increase of 8 in the cost. The minus sign in the $(1, 1)$ square means a decrease of 4 in the cost. After checking all the signed squares, we find the net change in cost and call it the *improvement index* for the $(1, 2)$ square:

$$\text{Improvement index for } (1, 2) \text{ square} = 8 - 4 + 16 - 24$$
$$= -4$$

The fact that the improvement index is negative means that the modifications described would be worthwhile. The total cost has been reduced by 4.

So far we have contemplated sending only 1 unit along the $(1, 2)$ route and making compensating changes. But we can send more, as long as none of the required subtractions make a negative amount. In the present case we could divert 56 units to the route from S_1 to D_2. Thus each plus or minus sign in the table can be regarded as referring to a change of 56 units. Now the total improvement is 56 times the index of improvement, i.e., $56(-4) = -224$.

Table 7

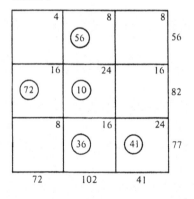

The general rule for the amount to divert to the previously unused square is to use the smallest circled amount appearing in any square with a minus sign.

Table 7 shows the new solution after these changes have been made. Notice that it is a BFS. (This is no accident.) The $(1, 1)$ square has been reduced to zero and is no longer in the BFS. Its place is taken by the previously unused $(1, 2)$ square.

Here's a summary of what we have done.

Rules for the Stepping-Stone Method

1. Begin with a square which is unused in the current BFS. Place a plus sign in it.
2. Trace a circuit, beginning and ending at this square, by moving horizontally and vertically only and changing direction only at squares of the current BFS. It can be shown that there is always exactly one such circuit.
3. Place plus and minus signs alternately at the squares where the circuit changes direction.
4. Calculate the improvement index by adding the costs in squares with plus signs and subtracting those whose squares have minus signs.
5. If the improvement index is nonnegative, choose another unused square and return to step 2. If there is no unused square with negative improvement index, your current BFS is optimal.
6. If the improvement index is negative, find the smallest circled amount in a negatively labeled square. Add this to the squares of the circuit with plus signs and subtract it from those with minus signs. Circle the amount in the previously unused square and delete the circle whose amount has become zero.

Here are two key facts that make the Stepping-Stone Method useful:

Theorem 2

In a transportation problem with no degenerate solutions, the Stepping-Stone Method always produces a solution which is a BFS.

Proof Omitted.

Theorem 3

1. If there is no square with a negative improvement index, then the current BFS is optimal.
2. If there are no degenerate solutions and there is a square with a negative improvement index, then the Stepping-Stone Method will produce a cheaper BFS.

Proof Omitted.

The next example illustrates part 1 of Theorem 3 and also shows that an alteration circuit may not be a simple four-cornered "square" circuit, as in Example 6.

Example 7

For the tableau and BFS shown in Table 8a, we show the alteration circuit for the unused $(1, 3)$ square. The improvement index is

$$8 - 4 + 16 - 24 + 16 - 24 = -12$$

Table 8

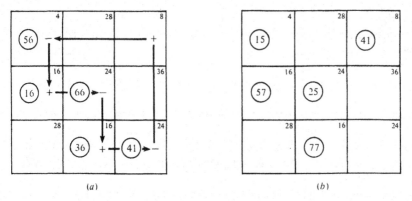

(a)   (b)

The smallest circled amount in a negatively labeled square is 41; so the amount that can be diverted to the (1, 3) square is 41. The new BFS is shown in Table 8b.

The improvement indices for the unused squares in Table 8b are:

(1, 2) square: $28 - 4 + 16 - 24 = 16$
(2, 3) square: $36 - 16 + 4 - 8 = 16$
(3, 1) square: $28 - 16 + 24 - 16 = 20$
(3, 3) square: $24 - 16 + 24 - 16 + 4 - 8 = 12$

None of these is negative; so Theorem 3 tells us that the BFS in Table 8b is the lowest-cost solution.

The only question which remains now is whether we can be sure that repeating the Stepping Stone Method eventually brings us to a BFS which is optimal, i.e., where no unused squares have negative improvement indices. How do we know that we don't get caught in an infinite loop in the flowchart of Figure 3?

Theorem 4

In a transportation problem with no degenerate BFS's, the algorithm of Figure 3 eventually produces a BFS which is optimal.

Proof At each application of the Stepping-Stone Method to a nondegenerate BFS and an unused square with negative improvement index, we obtain a new BFS, which has lower cost. Therefore, as we repeatedly apply the Stepping-Stone Method, we never produce the same BFS twice. It can be shown that there are only a finite number of BFS's, and so the process must eventually stop at the lowest-cost BFS.

The upshot of these theorems is that, for a nondegenerate problem, we repeatedly apply the Stepping-Stone Method. Eventually we reach a BFS where no unused square has a negative improvement index. At this point, the BFS is the optimal solution. The following example shows the complete solution of a transportation problem, using this method.

Example 8

Starting with the tableau in Table 9, we apply the Northwest Corner Rule and produce the tableaux in Tables 10 through 15.

Table 9

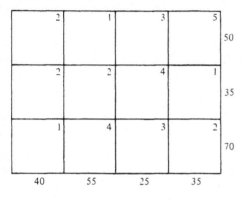

Our first application of the Stepping-Stone Method will be to the $(3, 1)$ square in Table 15. (This is not the only square in this table we could use.) The index of improvement is $1 - 2 + 1 - 4 = -4$. The amount we can divert to the $(3, 1)$ square is 10 (the minimum circled amount in a negative square). When this is done, the new tableau is shown in Table 16.

We now apply the Stepping-Stone Method to this BFS and the unused $(2, 4)$ square, whose index of improvement is $1 - 2 + 1 - 2 + 1 - 2 = -3$. The largest amount we can place in the new square is 30. After doing this and compensating at the corners of the alteration circuit, we obtain Table 17.

Table 10   **Table 11**

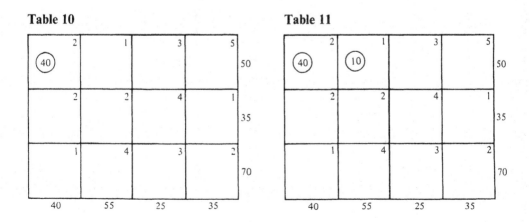

Table 12

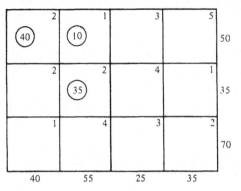

Table 13

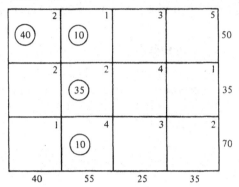

Table 14

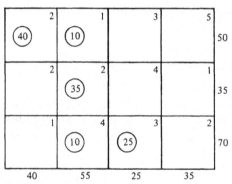

Table 15

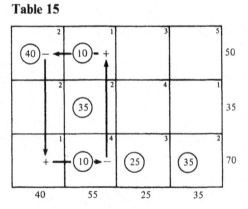

Table 16

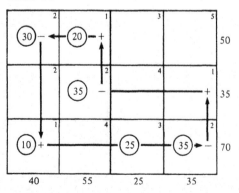

Table 17

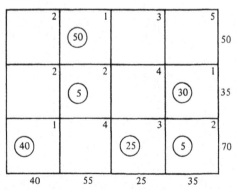

Each of the unused squares in Table 17 has an index of improvement that is positive. This means we have arrived at the optimal BFS. According to this BFS, the shipments we would make are:

50 units from S_1 to D_2
 5 units from S_2 to D_2
30 units from S_2 to D_4
40 units from S_3 to D_1
25 units from S_3 to D_3
 5 units from S_3 to D_4

So far we have dealt only with problems where total supply equals total demand. If total supply exceeds total demand, we first compute the excess supply, then create a new fictitious destination with this excess as its demand. The amount "shipped" to this destination is, in reality, left where it is. Therefore, the costs of shipping to this destination are 0. We now have a new problem where total supply equals total demand; so we can solve it, using the transportation algorithm described in this section.

In Example 10 of Section 2 of this chapter, we showed how one particular transportation problem could be formulated as a linear programming problem. In fact, every transportation problem can be so formulated. Consequently every transportation problem can be solved by the simplex method. However, even for small values of m and n the simplex method is more time consuming than the special-purpose transportation algorithm presented here.

Beside the transportation problem, there are many other types of optimization problems where a special-purpose algorithm is more effective than the simplex method. Many good examples of this may be found in the book by Eugene Lawler, cited in the Bibliography.

EXERCISES

● **1** Carry out the Northwest Corner Rule on the tableau in Exercise 5. Disregard the circled entries shown in Table 21.

● **2** For the tableau in Table 18, find the alteration cycle for each unused square. Circles indicate the BFS.

● **3** For the tableau in Table 19, find the improvement index for each unused square. Circles indicate the BFS.

● **4** For each unused square in Table 20 which has a negative improvement index, find how much could be diverted to the unused square.

5 For each unused square in Table 21, do the following:
   (*a*) Calculate the index of improvement.
   (*b*) Determine the maximum amount that can be diverted to the unused square.
   (*c*) Calculate the new BFS.

6 Carry out all steps of the transportation algorithm to find the solution for the tableau in Table 22.

Table 18

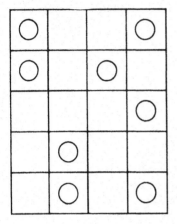

Table 19

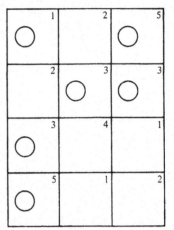

Table 20

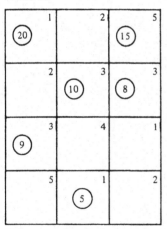

Table 21

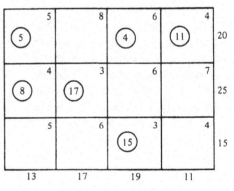

Table 22

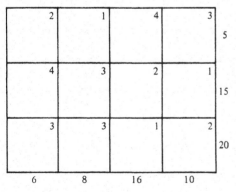

7 A trucking company has a contract to move 115 truckloads of sand per week between three sand-washing plants W, X, and Y and three destinations A, B, and C. Cost and volume information are given in the tables below. Compute the optimal transportation cost, using the transportation algorithm.

Project	Requirement per week truckloads	Plant	Available per week, truckloads
A	45	W	35
B	50	X	40
C	20	Y	40

	Transportation cost, $		
From	To project A	To project B	To project C
Plant W	5	10	10
Plant X	20	30	20
Plant Y	5	8	12

8 Find the lowest-cost solution to the transportation problem shown in Table 23. Notice that total supply exceeds total demand.

● **9** Table 24 shows a transportation problem in which the proper subset $\{S_1, S_2\}$ is capable of exactly supplying the demands of the proper destination subset $\{D_1, D_3\}$. Can you use this fact to demonstrate a degenerate solution?

10 The Northwest Corner Rule pays no attention to costs; so the BFS it produces may be a poor one. The *minimum-entry method* attempts to do better as follows:

1. Begin by finding the lowest-cost entry of the matrix.
2. Permute rows and columns to put the lowest-cost entry in the northwest corner.
3. Then apply the Northwest Corner Rule.

Table 23

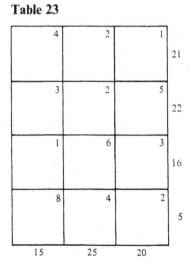

Table 24

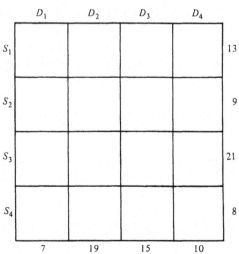

(a) Apply this method to the tableau of Table 2.

(b) Apply this method to the tableau of Example 7.

Computer exercises

11 Write a computer program to carry out the Northwest Corner Rule. Input to the program should be: the two-dimensional matrix of costs (c_{ij}), the one-dimensional array of supplies (s_i), and the one-dimensional array of demands (d_j). The output should be a two-dimensional array whose nonzero entries are the circled amounts in the BFS obtained by the Northwest Corner Rule.

12 Write a computer program to carry out the minimum-entry method of finding an initial BFS, which is explained in Exercise 10. Inputs and outputs should be as described in Exercise 11.

BIBLIOGRAPHY

Gale, David: The Optimal Assignment Problem, UMAP Module 317, available from COMAP, 271 Lincoln St., Lexington, MA 02173. This module discusses a problem which is closely related to the transportation problem.

Gaver, Donald P., and Gerald L. Thompson: "Programming and Probability Models in Operations Research," Brooks-Cole Publishing Co., Monterey, Calif., 1973. This book begins with a full account, including all the theory, of the transportation problem.

Lawler, Eugene: "Combinatorial Optimization: Networks and Matroids," Holt, Rinehart and Winston, New York, 1976.

Levin, Richard I., and Charles A. Kirkpatrick: "Quantitative Approaches to Management," 4th ed., McGraw-Hill Book Company, New York, 1978. Chapter 11 has a detailed, elementary account of the transportation problem, including many nice numerical exercises that show the scope of the applications.

6 COMBINATORIAL OPTIMIZATION—THE CHINESE POSTMAN PROBLEM

Abstract The Chinese postman problem is one of many optimization problems involving graphs—graphs of street networks in our examples. Many public services, such as mail delivery, garbage collection, etc., involve routes on such a network. Finding efficient routes is a spin-off on elementary Euler circuit theory (which is developed here).

Prerequisites No formal prerequisites.

In the year 1735 a popular puzzle no one could solve came to the attention of the Swiss mathematician Leonhard Euler, who solved the problem and then returned to more serious work. Until recently the puzzle stayed a puzzle, but now it has developed a second identity in applied mathematics.

The puzzle concerned the city of Koenigsberg and its seven bridges, which span the Pregel river and join different portions of the city (see Figure 1a).

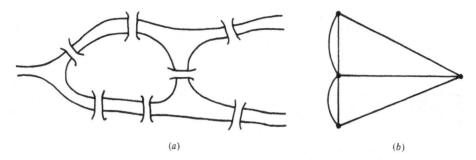

(a)   (b)

Figure 1 The seven bridges of Koenigsberg—the original Euler circuit problem.

The citizens of Koenigsberg wanted to know whether it was possible to take a walk through the city and cross each bridge exactly once and return to the starting point. Euler reasoned that, since walking on land didn't matter, he could replace each land mass by a point and let the bridges be represented by links connecting the points. The structure he obtains is shown in Figure 1b and is called a *graph*.

Definition

A graph is a finite set of points, called *vertices*, together with a finite set of connecting links, called *edges*. Each edge joins two different vertices.

If we ignore the walking done on land, we can visualize a walk in the city of Koenigsberg as a succession of edges (bridges) in the graph, eventually leading us back to the starting vertex.

Definitions

A *path* in a graph is a sequence of edges where consecutive edges have a vertex in common. The same edge may appear more than once in a path.

A *circuit* in a graph is a path which starts and ends at the same vertex.

An *Euler circuit* in a graph is a circuit in which each edge of the graph is used exactly once.

Using this terminology, we can state the problem this way: is there an Euler circuit in the graph of Figure 1b?

Modern applications involving Euler circuits usually involve graphs of street networks, as in Figure 2. For example, suppose a garbage collection truck needs to travel down every street to collect the garbage. We will suppose that each street is two-way; so travel in either direction is possible. Furthermore we suppose that it is only necessary to go down a street once to collect all the garbage from both sides. Therefore, the ideal arrangement would be to travel each street exactly once—

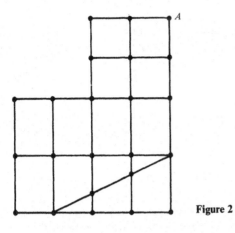

Figure 2

none left out and no duplications. The final requirement is that the truck ends its tour at the garage from which it starts, say at intersection *A*. Putting all these requirements together, it is clear that we are looking for an Euler circuit.

Unfortunately, neither the Koenigsberg graph nor the street graph of Figure 2 has an Euler circuit.

The conditions that determine whether a graph has an Euler circuit deal with the concepts of connectedness and valence.

Definitions

The *valence* of a vertex is the number of edges meeting at that vertex. A vertex of valence *n* is said to be *n-valent*.

A graph is *connected* if, given any two vertices, there is at least one path connecting them.

Technically, it is possible for a valence to be zero; such a vertex is said to be *isolated*. Graphs with isolated vertices are of no interest in the theory or applications of Euler circuits and won't appear in this section.

Theorem 1

In a graph without isolated vertices, the following conditions are necessary and sufficient for the existence of an Euler circuit:

1. The graph is connected.
2. Each valence is even.

To see that these conditions are necessary, first observe that an Euler circuit covers all edges and therefore all vertices too. For any two vertices we choose, we can get a path from one to the other by following the Euler path from one vertex till it gets to the other. Therefore the graph is connected.

To say that the valences must be even in order for there to be an Euler circuit is equivalent to saying that, if there is an odd valence, there can't be an Euler circuit. To see why this is so, consider the 5-valent vertex in Figure 1*b*. Let's follow an alleged Euler circuit around, starting from some other vertex. Each time we come through the 5-valent vertex, we use up two edges which haven't been used before. After two visits we have only one of the five edges left. When we come around next to use that edge, we will be stuck at that vertex. Thus we won't be able to get back to our start, and the circuit we are following can't be an Euler circuit.

A different style of proof is used to show that the conditions of Theorem 1 are necessary. First we give a method for finding Euler circuits. Then we prove that it works when the conditions of the theorem hold. The basic method is to build the circuit piecemeal. We wander around the graph at random, making the largest circuit we can that doesn't duplicate edges. If this results in an Euler circuit, we are done. If not, save this circuit and then generate a new circuit in the unused part of the graph. Build this into the first as a side trip. Here it is in more detail. (We skip the proof that this method really works.)

Algorithm for Constructing Euler Circuits

1. Choose any starting vertex
2. Begin tracing a path randomly, except for the restriction that you never use an edge twice. Continue this until you are at a vertex where all edges are already used. It can be proved that the only vertex where you can get stuck like this is the one you started from.
3. If the circuit generated in step 2 covers all edges, then you have an Euler circuit and you are done. If not, then this circuit is called an *approximate Euler circuit* and we proceed to step 4.
4. Choose a new starting vertex which is on the approximate Euler circuit generated so far but which also has at least one unused edge touching it.
5. Begin tracing a path randomly, except avoid using edges already used on this path or on the approximate Euler circuit generated so far. Continue till you get stuck. This will happen at the vertex you started with in step 4.
6. Create a new circuit by taking the old approximate Euler circuit and adding the new one from step 5 as a side trip as follows:
   a. Trace the old approximate Euler circuit till you get to the vertex where the side trip found in step 5 starts.
   b. Trace out the side trip completely, returning to its start.
   c. Finish tracing the old approximate Euler circuit.
7. If the new circuit is an Euler circuit, stop. Otherwise we call this circuit a new approximate Euler circuit and go to step 4 again.

Example 1

We use the above algorithm to get an Euler circuit for the graph in Figure 3a.

Step 1. Choose x_1.

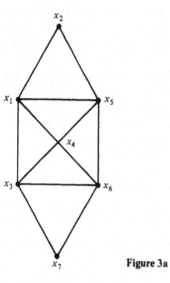

Figure 3a

Step 2. We generate the approximate Euler circuit (boldface edges)

$$x_1 x_2 x_5 x_4 x_1 x_3 x_6 x_5 x_1$$

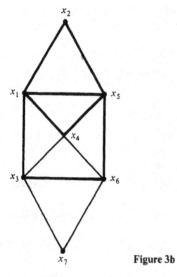

Figure 3b

Step 3. We are not done; so proceed to step 4.
Step 4. Choose x_3 (x_4 or x_6 would also be correct).

Step 5. The side trip (hatched edges) is

$$x_3 x_7 x_6 x_4 x_3$$

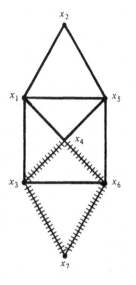

Figure 3c

Step 6. $x_1 x_2 x_5 x_4 x_1 x_3 x_6 x_5 x_1$

$$+ \overbrace{x_3 x_7 x_6 x_4 x_3}$$

$$= x_1 x_2 x_5 x_4 x_1 x_3 x_7 x_6 x_4 x_3 x_6 x_5 x_1$$

Step 7. We observe that this circuit is an Euler circuit.

Now what if a graph has no Euler circuit? In the case of the Koenigsberg puzzle that was the end of it. But in our garbage collection problem this is only the beginning. The garbage has to be collected somehow and the quicker the better. In most cities the routes are worked out without mathematical analysis and involve lots of duplication (using edges already covered). This duplication is called *deadheading*. The fact that actual routes involve so much deadheading suggests the following mathematical problem:

Find a circuit that covers all edges of a given graph with the minimum number of duplications.

This problem (and a variant soon to be described) is sometimes called the Chinese postman problem.

The Chinese postman problem can be analyzed using Euler circuit theory. Suppose we have a minimum deadheading circuit, such as $x_1 x_3 x_4 x_2 x_3 x_2 x_1$ in the graph of Figure 4a. Let's draw the circuit so that, when an edge is duplicated,

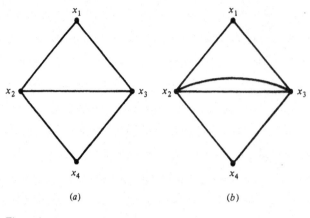

(a)   (b)

Figure 4

we add an extra edge to the graph for the circuit to use (Figure 4*b*). By doing this, we create a new graph where the circuit can be traced without duplication, using the new edge when needed. On this new graph our circuit is an Euler circuit.

The solution to the Chinese postman problem is obtained by working this idea in reverse. We take our given graph, which has no Euler circuit, and add edges, duplicating existing ones, in such a way that the graph becomes connected and even-valent. We call this *Eulerizing* the graph. Now we trace an Euler circuit on the Eulerized graph. Finally, we trace out this circuit in the original graph, duplicating an edge each time the Euler circuit used an added edge. The result of this procedure is a circuit that covers all edges of the original graph. But it may do so with more deadheading than necessary.

Example 2

The graph of Figure 4*a* can be Eulerized by adding a pair of edges, as in Figure 5, or a single edge, as in Figure 4*b*.

In order to get a minimum-deadheading circuit, we might add to our procedure one more trick: find the minimum number of edges that need to be added

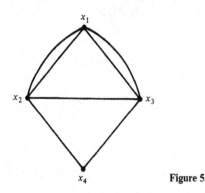

Figure 5

to Eulerize the graph. But this isn't the best formulation. Since blocks are not all the same length, it is not enough to count edges and minimize them. In actual applications each edge is weighted with the length (or travel time) of the corresponding block. Now when we add an edge paralleling an already existing one with length a_i, we are letting ourselves in for an extra a_i units in the circuit we eventually obtain. Therefore, we pose the Chinese postman problem in the following more general (and much more useful) form:

Eulerize a given weighted graph so that the total weight of added edges is a minimum.

Example 3

In the three-by-three-block rectangular street network in Figure 6, each block has length 1 or 4. The eight added edges, shown in Figure 6a, are a better Eulerization than the four added edges in Figure 6b. Even though twice as many edges are used, the total length is 8, as opposed to 10.

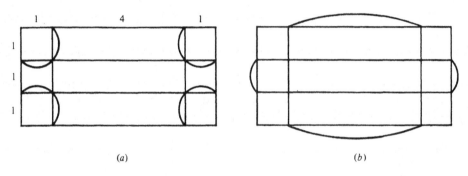

(a)   (b)

Figure 6 Two Eulerizations for the same graph.

If the street network is at all large or complicated, it may be hard to see exactly how to add the new edges in a minimal way. There is a fairly efficient procedure that can be followed to get the best Eulerization (see Beltrami, cited in the Bibliography), but it is a bit complex; so we will be content to solve these problems by ingenuity and trial and error. When using this "seat of the pants" approach, it is sometimes useful to keep in mind that one added edge can convert at most two odd-valent vertices to even. Therefore,

Theorem 2

If v_0 is the number of odd-valent vertices, then no Eulerization can involve fewer than $v_0/2$ added edges.

Sometimes one can get away with only $v_0/2$ added edges, as in Figure 7a. But there are graphs, such as in Figure 7b, where more than $v_0/2$ edges are needed.

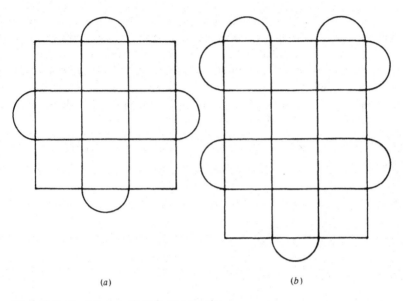

(a)   (b)

Figure 7 Optimal Eulerizations for two graphs.

Anyhow, one shouldn't be too preoccupied with the number of edges (unless all edges have equal or nearly equal weight) since it is the total weight of the added edges that counts.

The methods we have described for routing garbage trucks could apply in a variety of other situations, including:

Delivering mail
Snow plowing the streets
Salting icy roads
Railroad track inspections and maintenance
Reading electric meters in the houses along a set of streets

Mathematical analysis can yield major savings in these service and maintenance problems. For example, a pilot study carried out in New York City in the 1970s (see Kursh and Bodin, cited in the Bibliography) discovered much room for improvement in the way motorized sweepers clean the streets. In one sanitation district there were 88 miles of street to sweep. Existing routes had been worked out without the benefit of mathematical theory and involved 65 miles of deadheading. Mathematical analysis showed that deadheading could be reduced to 38 miles. This would have saved about $30,000 per year, just in that one district. (There are 57 districts altogether.)

Unfortunately, bureaucratic inertia, unions, and politics often stand in the way of this sort of improvement. The New York City pilot study was never implemented because of these obstacles. But there are also examples where improvements have been made. An electric company in Israel was able to cut 40 percent off the time required for its electric-meter readers to make their rounds (see Stern and Dror, cited in the Bibliography) by applying the sort of theory we have discussed in this section.

EXERCISES

● **1** Eulerize each of the street networks shown in Figure 8 in such a way that the total travel time of the added edges is a minimum. Assume all edges which are unlabeled have a travel time of 1.

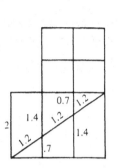

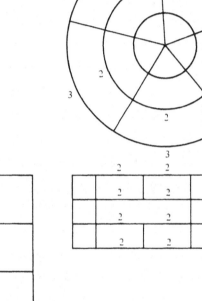

Figure 8

2 For a rectangular network of the type shown in Figure 9, $m \times n$ blocks, all of unit length, and where m and n are both odd, show a way to Eulerize the network using $m + n - 2$ edges. Do you think you could do better? What if m or n (or both) is even?

Figure 9

● **3** Find a minimum duplication circuit (minimum deadheading) on the street network shown in Figure 10. First Eulerize the network; then find an Euler circuit.

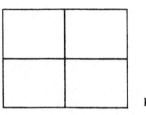

Figure 10

● **4** A *directed graph* is one in which each edge has an arrow and this arrow specifies the only direction that edge can be traversed. Suppose the network of Exercise 3 consists of one-way streets, as shown in the directed graph of Figure 11. Can you find a circuit that covers every block exactly once (no deadheading)? If not can you explain why not?

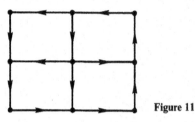

Figure 11

5 Suppose a street network consists entirely of two-way streets and each block must be covered at least once in each of the two directions. How would you model this situation with a directed graph (see Exercise 4 for definition)? Use the network of Exercise 3 as an example. Can you arrange a circuit that covers each block exactly once in each direction (no deadheading)?

6 Do the same as in Exercise 5 except use the networks of Exercise 1.

7 In a directed graph a necessary and sufficient condition for an Euler circuit is that both the following hold:

1. It is possible to find a directed path from any given starting vertex to any given end vertex.
2. At each vertex, the number of edges leaving equals the number of edges entering (invalence = outvalence).

How does this bear upon Exercises 5 and 6?

8 In each of the networks of Exercise 1, suppose two vehicles are to share the task of covering the edges. Each vehicle must start and end at the same vertex (you may decide where), and the vehicles are expected to do roughly equal amounts of driving. Try to find routes to minimize total deadheading.

● **9** Let e_i denote the edges of a network; let the length of e_i be l_i, as in the graph of Figure 12. Here's an attempt to describe the Chinese postman problem algebraically: let x_i be a variable that denotes

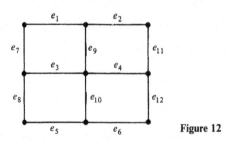

Figure 12

the number of edges to be added which duplicate e_i. There are certain algebraic conditions that need to be satisfied. For example, $x_1 + x_2 + x_9$ must be odd. Write the full list of such conditions. Then write out the function that needs to be minimized.

Computer exercise

10 Write a computer program to find an Euler circuit in a connected, even-valent graph, given as input to the program. [One way to specify the graph is with a two-dimensional array A, where $A(i, j) = 1$ if there is an edge from vertex i to j and $A(i, j) = 0$ otherwise.]

BIBLIOGRAPHY

Balakrishnan, V. K.: The Chinese Postman Problem, UMAP Module 582, available from COMAP, 271 Lincoln St., Lexington, MA 02173.

Beltrami, Edward J.: "Models for Public Systems Analysis," Academic Press, New York, 1977. A nice and not too technical survey of a wide variety of optimization models applicable to urban planning and public policy.

Edmonds, J., and E. Johnson: Matching Euler Tours, and the Chinese Postman Problem, *Math. Prog.* vol. 5, pp. 88–124, 1973.

Kursh, Samuel, and Lawrence Bodin: A Computer Assisted System for the Routing and Scheduling of Street Sweepers, *Oper. Res.*, vol. 26, no. 4, 1978.

Lawler, Eugene L.: "Combinatorial Optimization: Networks and Matroids," Holt, Rinehart and Winston, New York, 1976. Covers many types of general combinatorial optimization problems.

Stern, Hellman I., and Moshe Dror: Routing Electric Meter Readers, *Comps. and Oper. Res.*, vol., 6, pp. 209–223, 1979.

Tucker, Alan: "Applied Combinatorics," John Wiley & Sons, Inc., New York, 1980. A nicely written overview without too much technicality.

FIVE

CHOOSING THE MATHEMATICS FOR THE MODEL

1 SIMULATION VERSUS ANALYTIC SOLUTION—QUEUEING UP IN THE DOCTOR'S WAITING ROOM

Abstract Mathematics is often thought to consist of finding formulas for quantities of interest to us. In that approach, called the *analytic approach*, we rely heavily on mathematical theory. There is another approach, which might almost be called *experimental mathematics*, which can sometimes be used. One example of this is called *simulation*. It needs less theory but lots of patience and/or computer time. This section compares the analytic and simulation approaches for the problem of managing a doctor's waiting room. We study two examples, with different levels of complexity.

Prerequisites The nature of probability; the multiplication rule. Aspects of the binomial model make appearances, but previous knowledge of them is not assumed.

If you mix potassium nitrate, sulfur, and charcoal in proportions of 75, 10, and 15 percent, respectively, and light a match to the mixture it will burn or explode. In fact this is a recipe for gunpowder. You could try the experiment, but it might be safer to rely on chemical theory to predict the outcome. In other examples, an experiment might be better than theory. The point is that there are two great pillars upon which the experimental sciences rest, theory and experiment. By contrast, the mathematical manipulations in a mathematical model often seem to be based wholly on one pillar, that of theory. This is largely true, but not completely. In fact, the experimental approach finds some use when we do mathematics. (And it is commonplace when we evaluate a model, but that's another matter.)

One form of mathematical experiment is simulation. Our objective in this section is to describe some examples of simulation and to compare the simulation approach with the theoretical one (which is generally called the analytical approach).

To illustrate these ideas, we shall examine a problem in queueing theory, in ordinary language, the theory of waiting lines. Queues are distressingly common in everyday life: we often have to wait on line in supermarkets and banks and at theater box offices; airplanes queue up to land at busy airports; perhaps you place your homework assignments in a queue and make them wait till you get to them (do they mind waiting?). The particular examples we will study, however, all have to do with the waiting that goes on in a doctor's waiting room.

We shall consider two studies along these lines. The first is extremely simple and limited in scope. The second problem is a bit more interesting and a bit more complicated.

Our objective in each case is to compare two ways of attacking the problem: simulation and analytic formulation.

Problem 1

The doctor decides to study the strategy of not giving patients definite appointments but allowing them to show up in the waiting room whenever they please. In this strategy the waiting room would be open for 3 hours, beginning at 9 a.m. Since time keeping is never perfectly precise, we can suppose that patient arrivals are recorded by rounding the exact time down to the nearest minute. Thus patients can arrive at $9:00, 9:01, \ldots$, up until $11:59$ a.m. (The door closes at 12 sharp, and so there are no arrivals at 12.) This is a total of 180 points in time. In principle we could imagine two or more patients arriving at the same one of these time points, but this would gum up our model, and so, to keep the mathematics simple, we'll suppose that there is never more than one arrival per time point.

We now assume that, at any of these points in time, the probability of an arrival is $\lambda = 0.075$. (This assumption sweeps a number of questions under the rug: Isn't it possible that the probability should be higher near $9:00$ than later in the morning? How would you determine λ anyhow? We'll leave these and other questions under the rug.)

It may seem that, with arrivals being random, it is impossible to say anything about what would happen under these circumstances. Although we can't say definitely what would happen, we can say what the probabilities are for various desirable or undesirable outcomes. As a warm-up exercise, we'll calculate the probability that the first patient arrives at $9:00$ and the second arrives at $9:01$.

With this example in mind, it may be useful to pause and provide rough definitions of the concepts of simulation and analytic solution, which we wish to compare.

Definition

An *analytic solution* involves finding a formula that relates the quantity we are trying to estimate to other quantities known to us. In this instance we want the probability of arrivals at 9:00 and 9:01, expressed as a function of λ.

Definition

A *simulation solution* attempts to estimate the value of a quantity by mimicking (simulating) the dynamic behavior of the system involved.

Simulation Solution of Problem 1

We create a spinner (Figure 1) in which one sector has d degrees, where d is calculated from the formula

$$\frac{d}{360} = \lambda = 0.075 \tag{1}$$

The idea is that $d/360$ represents the probability that the spinner will land in the sector with d degrees. Therefore, when we choose d according to Equation (1) ($d = 27°$ in this case), the spinner experiment simulates (mimics) the experiment of determining whether an arrival has occurred at a given time point (9:00 say).

If we want to simulate the two time points 9:00 and 9:01, we spin the spinner twice. In order to find out the probability of arrivals at both 9:00 and 9:01, we could repeat this pair of spins 100 times (a total of 200 separate spins) and calculate the fraction of times both spins ended in the sector with d degrees. This fraction is our estimate of the probability of arrivals at both 9:00 and 9:01. Table 1 shows how

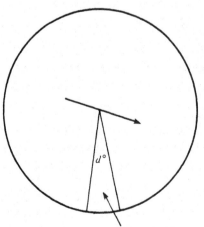

Spinner in this sector indicates arrival **Figure 1**

Table 1†

Trial	First spin	Second spin
1	Arrival	$\cdots$
2	$\cdots$	$\cdots$
3	$\cdots$	Arrival
4	$\cdots$	$\cdots$
5	$\cdots$	$\cdots$
6	Arrival	Arrival
7	$\cdots$	$\cdots$
8	$\cdots$	$\cdots$
9	$\cdots$	$\cdots$
10	$\cdots$	$\cdots$

† Dots indicate no arrival.

this might work out if we had done merely 10 pairs of spins. Actually it is not advisable to do such a small number of spins because the chance of a fluke result is high. 100 pairs of spins would be better; 1000 would be much better. (There is some theory that bears on the question of how many pairs of spins to do, but we won't go into it.) The conclusion we would reach from Table 1 is that the chance of arrivals at both 9:00 and 9:01 is 0.1. Actually this is a very bad answer. When we carried out a simulation involving 100 trials, we obtained the estimate 0, which is closer to the theoretically correct value. With 1000 trials we got 0.005, which is almost exactly the theoretically correct answer.

If you are wondering where we got the patience to do 2000 spins of a spinner, the fact is we didn't. We used a computerized version of a random-number table, which plays the role of the spinner. Table 2 shows a portion of a random-number table (the full table takes many pages). The word "random" refers not to the individual numbers of the table, but to their sequence, which is carefully designed to show no preference for any digit and no pattern in the digits. In addition, it means that, if we group the digits by two's (or three's or any other size), the resulting sequence of multidigit numbers is random. For example, if we start at the top left digit of the table and group by three's, we get the following random sequence: 044, 338, 067, 424, etc.

Here's how we would use our random-number table to replace the spinner:

1. Choose a digit in the table at random (close your eyes and jab your finger at the page).
2. Start reading off numbers in groups of three digits, proceeding from left to right down the page and onto additional pages till you have 2000 three-digit numbers,
3. Now check your list and identify which three-digit numbers are between 000 and (999)(0.075). The idea of this is that such three-digit numbers come up with probability 0.075, and so finding one of these three-digit numbers is like seeing that a patient has arrived.

Table 2 Random numbers[†]

04433	80674	24520	18222	10610	05794	37515
60298	47829	72648	37414	75755	04717	29899
67884	59651	67533	68123	17730	95862	08034
89512	32155	51906	61662	64130	16688	37275
32653	01895	12506	88535	36553	23757	34209
95913	15405	13772	76638	48423	25018	99041
55864	21694	13122	44115	01601	50541	00147
35334	49810	91601	40617	72876	33967	73830
57729	32196	76487	11622	96297	24160	09903
86648	13697	63677	70119	94739	25875	38829
30574	47609	07967	32422	76791	39725	53711
81307	43694	83580	79974	45929	85113	72268
02410	54905	79007	54939	21410	86980	91772
18969	75274	52233	62319	08598	09066	95288
87863	82384	66860	62297	80198	19347	73234
68397	71708	15438	62311	72844	60203	46412
28529	54447	58729	10854	99058	18260	38765
44285	06372	15867	70418	57012	72122	36634
86299	83430	33571	23309	57040	29285	67870
84842	68668	90894	61658	15001	94055	36308
56970	83609	52098	04184	54967	72938	56834
83125	71257	60490	44369	66130	72936	69848
55503	52423	02464	26141	68779	66388	75242
47019	76273	33203	29608	54553	25971	69573
84828	32592	79526	29554	84580	37859	28504
68921	08141	79227	05748	51276	57143	31926
36458	96045	30424	98420	72925	40729	22337
95752	59445	36847	87729	81679	59126	59437
26768	47323	58454	56958	20575	76746	49878
42613	37056	43636	58085	06766	60227	96414
95457	30566	65482	25596	02678	54592	63607
95276	17894	63564	95958	39750	64379	46059
66954	52324	64776	92345	95110	59448	77249
17457	18481	14113	62462	02798	54977	48349
03704	36872	83214	59337	01695	60666	97410
21538	86497	33210	60337	27976	70661	08250
57178	67619	98310	70348	11317	71623	55510
31048	97558	94953	55866	96283	46620	52087
69799	55380	16498	80733	96422	58078	99643
90595	61867	59231	17772	67831	33317	00520
33570	04981	98939	78784	09977	29398	93896
15340	93460	57477	13898	48431	72936	78160
64079	42483	36512	56186	99098	48850	72527
63491	05546	67118	62063	74958	20946	28147
92003	63868	41034	28260	79708	00770	88643
52360	46658	66511	04172	73085	11795	52594
74622	12142	68355	65635	21828	39539	18988
04157	50079	61343	64315	70836	82857	35335
86003	60070	66241	32836	27573	11479	94114
41268	80187	20351	09636	84668	42486	71303

[†] From John Freund, "Modern Elementary Statistics," Prentice-Hall, Inc., Englewood Cliffs, N.J., 1967, pp. 393–396.

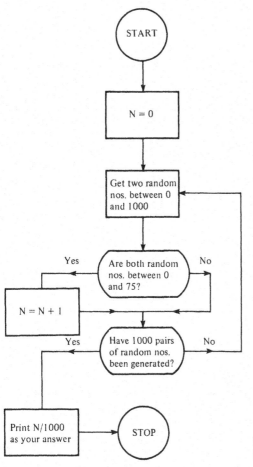

Figure 2 Flowchart to simulate arrivals in first two minutes.

If the number of trials we wish to do is not 1000, this method needs modification. For example, for 130 trials we choose 260 three-digit numbers in step 2.

Obtaining 2000 random three-digit numbers from a table in a book is very time consuming; so it is fortunate that random numbers can be generated by computer. Figure 2 shows a flowchart designed to carry out the simulation we performed earlier on the spinner.

For details about getting random numbers, consult your local computing center. These details vary from one computing center to another.

Analytic Solution of Problem 1

In this problem, even the simplest simulation solution is inconvenient compared with the approach we now describe.

In the language of probability, we are interested in the event

E = a patient arrives at 9:00 and a patient also arrives at 9:01

This can be regarded as $E_1 \cap E_2$, where

$$E_1 = \text{a patient arrives at 9:00}$$

$$E_2 = \text{a patient arrives at 9:01}$$

Since the patients are not in any sort of conspiracy, it seems reasonable to assume E_1 and E_2 are independent events. Therefore the multiplication rule applies and

$$P(E) = P(E_1 \cap E_2) = P(E_1)P(E_2)$$

$$= (0.075)(0.075)$$

$$= 0.005625$$

Presto! We are done. Compare this to the answers obtained by simulation.

Here's a problem that's a little harder than Problem 1.

Problem 2

The doctor wants to know how many patients will be in the waiting room at 11:59 when it closes its doors. This is one measure of how long past 11:59 the doctor must work. Since arrivals are probabilistic, we can't give a definite answer, but we can try to find an average value.

Our assumptions thus far are inadequate to answer the question because they say nothing about how long it takes the doctor to see patients. To take an extreme example, if it only takes the doctor 1 minute to see a patient, then no patient will ever have to wait, because we have assumed that patients can't arrive any faster than one per minute (and even that is unlikely because the arrival probability 0.075 is so low). In this extreme case, the doctor will have either no patients or one patient left to see at 11:59, depending on whether there was an arrival at 11:59. At the other extreme, if it takes the doctor 4 hours to see a patient, then it is very likely that the waiting room will be crowded at 11:59. We will make the following assumption:

Assumption Each patient takes exactly 15 minutes of the doctor's time for treatment, and the doctor treats only one patient at a time.

Analytic Solution of Problem 2

This problem is sufficiently complicated that we don't know of a formula that gives an exactly correct answer. Since we are not experts on probability models, there may actually be an analytic approach to this problem, but we decided not to spend the time searching for one. This is the sort of decision that often faces people who do mathematical modeling: is it better to settle for a "sure-thing" method, even if it is time consuming and lacks elegance, or to invest an uncertain amount of time looking for a quick and neat way?

There is, however, a way to obtain a lower bound by theoretical arguments concerning the binomial probability model.

The binomial model deals with the situation in which an experiment is repeated many times and each trial can have one of two possible outcomes.

Examples

1. A coin is tossed 5 times. The outcome of each trial is either H or T. For the combined experiment the outcomes are all the sequences of length 5 consisting of H's and/or T's. For example, $HTHTH$, $HHHHH$, etc.
2. We observe the doctor's waiting room at each of the 180 times an arrival may occur. A single trial has either of these two outcomes: "Arrival" or "No Arrival."

An additional assumption in the binomial model is that the trials are independent of one another. What this means in practice is that, if you know the outcome of one trial, this does not give any useful information about what will happen on another trial. In Examples 1 and 2, common sense and practical experience both suggest that the trials are independent. (An example of an experiment with non-independent trials is to offer your dog a bowl of beer twice and see if the dog likes it. The result of the first trial would give a definite clue about what would happen on the second.)

Theorem

Suppose n independent trials of an experiment are performed. Suppose the outcomes for a single trial are A and B, with probabilities p and $1 - p$. Then the expected (i.e., average) number of A outcomes out of the n trials is pn.

Examples

1. If we toss a fair coin 10 times, the expected number of heads is 5. This can be interpreted to mean that, if we repeat this 10-flip experiment many times and keep track of how many heads appear in each separate series of 10 flips, then the average of all these values will in the long run be 5.
2. In the doctor's waiting room, we observe whether or not an arrival occurs at each of the 180 different time points where it is possible to have an arrival. According to the theorem the expected number of arrivals is $(0.075)180 = 13.5$. We interpret this to mean that, if we averaged the number of arrivals on a long run of many different days, we would obtain 13.5 in the long run.

Part 2 of the last example is the key to our reasoning concerning the number of patients left over at 11:59. Suppose we could figure out how many patients the doctor sees up to 11:59 *on the average*. (This is not necessarily the same as the number he finishes with; he could be in the middle of seeing a patient at 11:59.) Call the average number seen k. Then

$$\text{Average number left over} = 13.5 - k \tag{2}$$

We won't try to find k; that seems like a tough problem. But it is not hard to see that

$$k \leq 12 \tag{3}$$

Here's why: the best the doctor can manage is to see one patient after another with no idle time in between spent waiting for the next patient to show up. But this won't always be possible. If patient arrivals are poorly timed (e.g., if no patient shows up till 10 o'clock), then the doctor will have idle time and not be able to see the maximum number of patients between 9 and 11:59. Assuming no idle time, the number of patients the doctor can see in the 179 minutes between 9 and 11:59 is 12. But there will often be idle time; so k, the average the doctor will see, must be less than the maximum of 12. Now that we have established Equation (3), we apply this result to Equation (2) to see that the average number left over is ≥ 1.5.

Unfortunately this theoretical argument gives no information about how much greater than 1.5 the average number of leftover patients is. Readers who have a taste for further theoretical assaults on this problem can try Exercises 13 through 15, which allow us to conclude that the average is really at least 2.4.

Now let's exercise the better part of valor and try a simulation approach.

Simulation Solution of Problem 2

As in Problem 1 we will simulate 1000 different days (by computer of course) and make a record for each day of the number of patients left at 11:59. At the end we average these values.

But the simulation of one particular day, called a *trial* in the language of simulation theory, is harder than the simulation of one trial in Problem 1. In Problem 1 a trial consisted of 2 minutes of the day (2 random numbers), and there was no need to keep track of when people got in to see the doctor (and thereby reduced the waiting room population). But now we do need to keep track of this, as well as keeping track of when people arrive in the waiting room. In addition, we will need to keep track of when patients are finished seeing the doctor, because that's when the next patient gets to leave the waiting room to enter the treatment room and see the doctor.

Here is the basic plan of the simulation:

1. We simulate 1000 trials (days).

2. A single trial is simulated on a minute-by-minute basis. We first examine what happens at 9:00, then what happens at 9:01, etc., until we get to 11:59—a total of 180 instants.
3. At each minute we determine whether any of the following events has occurred (and we check for these events in the order in which we now list them):
    a. An arrival to the waiting room.
    b. A patient is done seeing the doctor (i.e., leaves the treatment room).
    c. A patient leaves the waiting room for the treatment room to see the doctor.
4. These three types of events are "detected" in the simulation like this:
    a. To detect an arrival to the waiting room, the program uses the random-number generator (as in Problem 1).
    b. To detect when a patient is done with the doctor, the program must keep track of when that patient started to see the doctor and then add 15 minutes.
    c. To detect when a patient leaves the waiting room for the treatment room, we create a sort of an "alarm" to go off in the program 15 minutes after the previous patient started to see the doctor.

Here's how step 4c might be done in FORTRAN. Suppose a patient got in to see the doctor at time I1, and suppose I is the variable that keeps track of the current moment. Then the program should contain a statement like:

$$\text{IF (I. EQ. (I1 + 15)) } \langle \text{take needed action} \rangle \qquad (4)$$

The action taken depends on whether there is a patient in the waiting room. If there is, send the patient into the treatment room. Otherwise do nothing.

To carry this out we need to keep running track of the number of patients in the waiting room with a variable, say, NWAIT. It is also helpful to keep track of whether the doctor is busy (for example, with a variable BUSY, where BUSY = 0 means the doctor is not busy and BUSY = 1 means that the doctor is busy). Sending someone in to the treatment room is carried out by

$$\text{NWAIT} = \text{NWAIT} - 1$$
$$\text{BUSY} = 1$$
$$\text{I1} = \text{I}$$

Figure 3 shows a flowchart to carry out one trial of the simulation. For reliable results you should do at least 100 trials. In order to convert this flowchart to a computer program, you will have to create variables beside those shown on the flowchart. For example, the creation of the alarm device will require a variable like I1 in Equation (4). You may find need for others as well.

When we performed 1000 trials of the simulation described in the flowchart, we obtained an average of 2.98 patients "left over" at 11:59 in the waiting room. Table 3 and Figure 4 show more details about how the results of the individual trials broke down into categories. You will undoubtedly get results slightly different from these if you try this yourself, because your computer will probably generate a different sequence of random numbers.

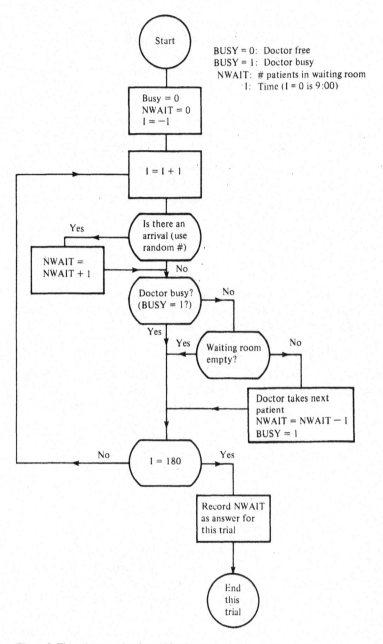

BUSY = 0: Doctor free
BUSY = 1: Doctor busy
NWAIT: # patients in waiting room
·I: Time (I = 0 is 9:00)

Figure 3 Flowchart to simulate 180 minutes in the doctor's waiting room.

Table 3

Number left at 11:59	Number of trials
0	183
1	159
2	148
3	147
4	125
5	76
6	62
7	39
8	26
9	18
10	6
11	7
12	2
13	0
14	2
	1000

Average number left 2.98

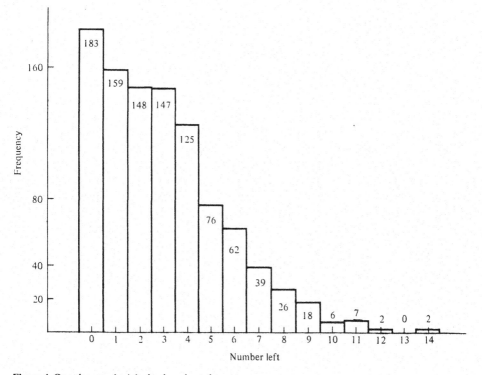

Figure 4 One thousand trials, broken down by outcome.

Sidelight: Nonmathematical Simulation

There is a less mathematical (but still useful) type of simulation in which we want to study the behavior of people faced with a certain real-world situation. For example, how will pilot trainees react if they fly into large flocks of birds? It may be awkward to study a situation like this "in the wild." Instead, pilots receive some of their training in a flight simulator. This is an exact copy of a cockpit, but constructed in a laboratory and outfitted with some type of viewing screen to replace the front windshield. A movie of the birds can be projected on this viewing screen.

Another example of simulation concerns the behavior of people living in flood plains. A flood plain is an area near a river which is very prone to being flooded. Not only do people willingly live and invest their money in flood-plain areas but they often refuse to buy flood insurance to protect their farms, homes, and businesses — even in some cases when the price of the insurance is artificially reduced to $\frac{1}{10}$ of its actuarially fair value.

In order to study this situation, researchers created a game, rather like Monopoly, which simulated life on a flood plain. Players had the opportunity to live, work, invest, buy insurance, etc. When the dice rolled a certain way, a flood occurred and created a certain amount of loss of life and property. By having people play this game, researchers obtained insights about the reasons why people behave the way they do. These insights would have been harder to obtain from real life, where one needs to wait years before floods actually occur.

Sidelight: Monte Carlo Methods

The simulations we are describing here are often referred to as Monte Carlo methods, because it is so easy to think of them in terms of gambling devices, such as the spinner of Figure 1 or dice or roulette wheels. The name was bestowed by John von Neumann, one of this century's greatest mathematicians, during his work on the Manhattan project during World War II.

One of the trickiest problems encountered in that project was the design of shielding and dampers for nuclear bombs. Direct experimentation was out of the question: applied mathematics would have to do the job. The problem could be reduced to trying to find out how far neutrons could penetrate various materials. Unfortunately it proved impossible to find a formula for the probability that a neutron would travel a given distance without absorption. The analytic approach had failed. What to do? As Daniel McCracken tells the story (*Scientific American*, p. 90, May 1955):

> At this crisis the mathematicians John von Neumann and Stanislas Ulam cut the Gordian knot with a remarkably simple stroke. They suggested a solution which in effect amounts to submitting the problem to a roulette wheel. Step by step the probabilities of the separate events are merged into a composite picture which gives an approximate but workable answer to the problem.

EXERCISES

Computer exercises

1 Write a computer program to implement the algorithm for solving Problem 1, which is shown in Figure 2.

2 If the probability of an arrival is not constant but has value 0.07 up till 9:59, 0.075 from 10:00 till 10:59, and 0.08 from 11:00 till 11:59, show how you would modify the flowchart in Figure 3.

● **3** Show how to modify the flowchart of Figure 2 if you want to do 500 trials instead of 1000.

● **4** Suppose your random-number generator returns numbers in the interval 0 to 9999, that is, it gives one of the four-digit numbers 0000, 0001, . . . , 9999. How would you modify the flowchart of Figure 2?

● **5** Show how the flowchart of Figure 3 can be expanded to show, in detail, how to keep track, on a minute-to-minute basis, of when the doctor is busy and when the doctor gets done seeing a patient. (As presently written, the flowchart doesn't show when or how the variable BUSY changes value from 1 to 0.)

6 Modify the flowchart of Figure 3 to study the following question: what is the average number of minutes past 11:59 which the doctor must work?

7 Suppose that approximately half the doctor's patients require 15 minutes for treatment, while the other half need 20 minutes. How would you modify the flowchart of Figure 3 to cope with this extra complication?

8 Write and run a computer program to carry out the simulation for Problem 1 with 1000 trials. After each 50 trials have the program print the fraction f of trials which had patients arriving at both 9:00 and 9:01. Make a table of values and a graph of f versus the number of trials.

● **9** In 1709 Nicolaus Bernouilli was interested in the following problem in demography:

If n men of equal age die within t years, what is the mean duration of life for the last survivor?

He transformed this to the following:

If n points are chosen randomly on an interval of length t, what is the average distance of the largest value from the origin?

This problem can be solved by simulation. One way is to throw n darts at a wall of length t and then repeat this experiment 10 times or so. Can you find a better way to do the simulation with a random-number generator? Use the values $n = 10$ and $t = 20$.

● **10** What are the probabilities of the following events? You may use simulation or the analytic approach to determine your answer.
    (*a*) A patient arrives at 9:00, but no patient arrives at 9:01.
    (*b*) The first patient arrives at 9:02.
    (*c*) No patients arrive all day.

11 Could you modify the analytic argument concerning Problem 1 of the text to take into account the nonconstant probabilities described in Exercise 2?

● **12** In our text description of the plan of the simulation for Problem 2, we distinguished three types of events to detect at each minute: type *a*, *b*, and *c*. Explain by example why it would not be a good model of a real doctor's office if we performed the detections in the order *a*, *c*, *b* instead of *a*, *b*, *c*.

Analytic exercises

Exercises 13 through 15 are all related and should be done in order.

● **13** Let the probability of an arrival at any particular time instant be p. (Using p in place of 0.075 will make things easier.)

(*a*) Find, as a function of p, the probability of the event $E_1 = $ the first arrival occurs at 9:01. (*Hint*: This can be described as "no arrival at 9:00, but an arrival at 9:01." It's very much like Problem 1 in the text.)

(*b*) Generalize part (*a*) to find the probability of $E_k = $ the first arrival is k minutes after 9:00.

(*c*) Find the probability that the waiting room closes before any patients arrive.

14 Exercise 13 implies that the mean waiting time till either a patient arrives or the waiting room closes is

$$f(p) = 0p + 1(1 - p)p + 2(1 - p)^2p + \cdots + 179(1 - p)^{179}p + 180(1 - p)^{180}$$

(Do you see why?) Now we do some old-fashioned algebra. Set

$$k(p) = (1 - p)p + (1 - p)^2p + \cdots + (1 - p)^{179}p + 180(1 - p)^{180}$$

and show that

$$f(p) - k(p) = (1 - p)[f(p) - 180(1 - p)^{180} - 179p(1 - p)^{179}]$$

Use this to show that

$$f(p) = \frac{k(p) - (1 - p)[179p(1 - p)^{179} + 180(1 - p)^{180}]}{p}$$

Use the formula for a finite geometric series to find a formula for $k(p)$ and then show that

$$f(p) = \frac{(1 - p)[1 - (1 - p)^{180}] + p(1 - p)^{180}}{p}$$

Since $(1 - p)^{180}$ is near zero for $p = 0.075$ (check this), we get the approximate formula

$$f(p) \approx \frac{1 - p}{p}$$

15 Use the formula of Exercise 14 to show that the doctor's idle time (time not seeing a patient) must have a mean value of at least 12.33 minutes. Consequently the mean number of patients seen can't be more than

$$\frac{179 - 12.33}{15} \approx 11.1$$

What does this imply about the mean number of patients left over at 11:59 o'clock?

16 Could the complication described in Exercise 7 be handled analytically?

BIBLIOGRAPHY

Maki, Daniel P., and Maynard Thompson: "Mathematical Models and Applications," Prentice-Hall, Inc., Englewood Cliffs, N.J., 1973. See the discussion of simulation in section 10.3.2.

Olinick, Michael: "An Introduction to Mathematical Models in the Social and Life Sciences," Addison-Wesley Publishing Company, Reading, Mass., 1978. Chapter 14 has an extended example of a computer simulation of the use of operating and recovery rooms in a hospital.

2 DISCRETE VERSUS CONTINUOUS MODELS

Abstract Models involving changes over time are often expressed through differential equations, the derivative being the instantaneous rate of change.

But sometimes it makes as much sense to build discrete models around noninstantaneous rates of change. Then difference equations are in order. Our objective in this section is to introduce difference equations and compare them to differential equations.

The main example used in the subsections of this section is population: first the exponential growth model, then logistic growth. Commodity price cycles in economics are discussed in Section 2E.

A key point of this section is that differential equations and difference equations have very different stability behavior. The general lesson is that difference equations and differential equations are not interchangeable.

Prerequisites Calculus, some elementary ideas about differential equations. No knowledge of difference equations is required.

Included subsections

2A. Introduction to Difference Equations
2B. Exponential Population Growth—Difference and Differential Equations
2C. Logistic Population Growth—Difference and Differential Equations
2D. Equilibrium, Stable and Unstable
2E. Stability of Linear Difference Equations—Commodity Prices and Cobweb Analysis
2F. Stability of some Nonlinear Difference Equations—Logistic Population Growth

2A INTRODUCTION TO DIFFERENCE EQUATIONS

A good deal of applied mathematics is dedicated to the problem of trying to describe how something changes over time. One particular example we will study in later sections concerns population. Populations of people, animals, or plants increase and sometimes decrease as time goes by. It would be nice to be able to predict the future size of a population, on the basis of past and present data.

There are two different styles of handling this sort of problem: a discrete model involving difference equations and a continuous model involving differential equations. A very rough way to compare these methods is illustrated in Figure 1. Figure 1a shows a population graph, based upon counts of the population at the time values $t = 0, 1, 2, \ldots$. The method of difference equations attempts to predict the positions of the data points in the future, on the basis of knowledge of the present $(t = 0)$ and some information about the population changes from one time instant to the next.

By contrast, the method of differential equations attempts to predict a smooth and continuous curve of future population values (the solid curve in Figure 1b), often by finding a formula for it.

The two methods differ not only in their objectives, a series of discrete values in one case and a continuous curve in the other, but also in their mathematical

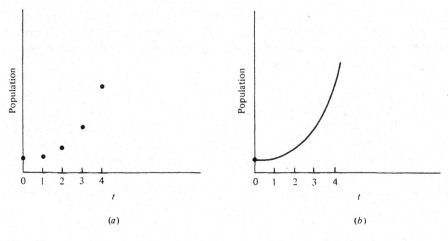

(a)

(b)

Figure 1

means of achieving these objectives. Differential equations require a knowledge of calculus, while difference equations can be approached with more elementary mathematics. Despite the differences between the two approaches, they often give similar results. However, there are ways in which the predictions of the two methods may differ. Our main goal is to compare the two methods and show the similarities and differences.

We begin with difference equations and the related problem of finding a pattern in a sequence of numbers. Here are three number sequences. Can you spot the patterns in them?

1. 1, 2, 3, 4, 5, . . .
2. 4, 6, 8, 10, 12, . . .
3. 1, 2, 4, 7, 11, . . .

In each case the pattern in the sequence can be described by a difference equation. If we let $x(0)$, $x(1)$, $x(2)$, . . . denote the members of the sequence, then here are the descriptions:

1.* $x(t + 1) - x(t) = 1$ for $t = 0, 1, 2, \ldots$.
2.* $x(t + 1) - x(t) = 2$ for $t = 0, 1, 2, \ldots$.
3.* $x(t + 1) - x(t) = t + 1$ for $t = 0, 1, 2, \ldots$.

Difference equation 1* merely states that the difference between one term of the sequence and the next is always 1. In difference equation 2* this difference is always 2. In equation 3* the difference is not constant but follows the given formula.

In addition to equations 1*, 2*, and 3* there are many other kinds of difference equations, and it is difficult to frame a satisfactory definition that encompasses them all without making things quite confusing. However, it is easy enough to describe the kinds of difference equations which are of interest to us. These are equations of the form

$$x(t + 1) - x(t) = F(x(t), t) \qquad t = 0, 1, 2, \ldots \qquad (1)$$

where F is some function of two variables. In any particular case, the function F may not actually involve $x(t)$, t, or either. Example 1* and 2* do not involve $x(t)$ or t, while example 3* does not involve $x(t)$. Here are some additional examples of difference equations:

$$
\begin{array}{ll}
x(t + 1) - x(t) = 3x(t) & t = 0, 1, 2, \ldots \\
x(t + 1) - x(t) = x(t) + t + 2 & t = 0, 1, 2, \ldots \\
x(t + 1) - x(t) = [x(t)]^2 + 2t^3 & t = 0, 1, 2, \ldots \\
x(t + 1) - x(t) = \sin [x(t) + 2t] & t = 0, 1, 2, \ldots
\end{array}
$$

A difference equation may be regarded as an infinite set of equations involving the variables $x(0)$, $x(1)$, ... For example, equation 1* yields these equations when we substitute the t values 0, 1, 2, ... :

$$
\begin{array}{l}
x(1) - x(0) = 1 \\
x(2) - x(1) = 1 \\
\vdots
\end{array}
$$

Definition

A sequence of numbers $a_0, a_1, a_2, \ldots$ is a solution to difference Equation (1) if, when we set $x(0) = a_0$, $x(1) = a_1$, etc., these values satisfy each of the equations resulting from Equation (1) by substituting the values 0, 1, 2, ... for t.

It turns out that a difference equation can have many solutions. Here are some additional solutions to 1*, 2*, and 3*:

1.′ 4.5, 5.5, 6.5, 7.5, 8.5, ...
2.′ 7, 9, 11, 13, 15, ...
3.′ 3, 4, 6, 9, 13, ...

If we specify, as an *initial condition*, some value for the first term of the sequence, say $x(0) = x_0$, then there is a single definite solution for a difference equation of type (1). For example, if we specify the initial condition $x(0) = 1$ for difference equation 1*, then sequence 1′ is no longer a solution, while sequence 1 still is.

Theorem 1

There is one and only one solution to a difference equation of the form of Equation (1) which satisfies the initial condition $x(0) = x_0$.

Proof The first term, $x(0)$, of any solution is required to be x_0. But, substituting $t = 0$ in Equation (1), we get

$$x(1) = x(0) + F(x(0), 0)$$

$$= x_0 + F(x_0, 0)$$

which means that the value of $x(1)$ is determined since x_0 is known and F is a known function. Substituting $t = 1$ in Equation (1), we get

$$x(2) = x(1) + F(x(1), 1)$$

This tells us that $x(2)$ is determined. Clearly we can continue this way indefinitely and determine the values of $x(t)$ for $t = 0, 1, 2, \ldots$.

The method used to prove this theorem can be used in practice to compute the first few values of the solution of a difference equation for a specified initial condition. Here is an example:

Example 1

Find the first four terms of the solution to

$$x(t + 1) - x(t) = -\frac{[x(t)]^2 - 4x(t) + 4}{x(t)}$$

which satisfies the initial condition $x(0) = 4$.

Solution Substituting $t = 0$ and transposing gives

$$x(1) = x(0) - \frac{[x(0)]^2 - 4x(0) + 4}{x(0)}$$

$$= 4 - \frac{4^2 - 4(4) + 4}{4} = 3$$

Likewise,
$$x(2) = 3 - \frac{3^2 - 4(3) + 4}{3} = \frac{8}{3}$$

The next two terms are $x(3) = \frac{10}{4}$ and $x(4) = \frac{12}{5}$.

Unfortunately one is often not satisfied with finding a solution in this way. At best we can compute only a finite number of terms in the solution, and there are other shortcomings as well. What is preferable, but not always possible, is to find a formula which describes the sequence. For example, the solution to 1* with initial condition $x(0) = 1$, which is the sequence $1, 2, 3, \ldots$, can be described by the function $x(t) = t$, defined on the domain consisting of the nonnegative integers. Likewise, the solution to 2*, with initial condition $x(0) = 7$, is $x(t) = 2t + 5$ for $t = 0, 1, 2, \ldots$.

Verifying that a certain formula satisfies a difference equation plus initial condition is usually done by direct substitution, as in the following examples.

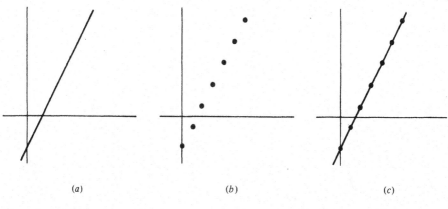

(a)   (b)   (c)

Figure 2

Example 2

Verify that $x(t) = 2t + 3$ satisfies the difference equation $x(t + 1) - x(t) = 2$, with initial condition $x(0) = 3$.

Solution First we verify the initial condition by substituting $t = 0$ into the proposed solution. $x(0) = 2(0) + 3 = 3$ as required. Now, for the difference equation: by substituting $t + 1$ in the proposed solution we obtain $x(t + 1) = 2(t + 1) + 3$, so that

$$x(t + 1) - x(t) = 2(t + 1) + 3 - (2t + 3) = 2$$

which means that the difference equation is also satisfied.

We are accustomed to thinking of a function like $x(t) = 2t + 3$ as being defined for all real values of t. In other words, we could draw its graph as shown in Figure 2a. However, it is important to realize that the only values of t for which the difference equation and initial condition imply an x value are $t = 0$ (the initial condition gives us the x value here) and $t = 1, 2, \ldots$. It is meaningless to ask what the value of $x(\frac{1}{2})$ is. Consequently, when we draw the graph of the solution we should strictly speaking draw a figure like Figure 2b. To make a picture more appealing to the eye we may occasionally connect the dots, as in Figure 2c.

Example 3

The difference equation plus initial condition of Example 1 has $x(t) = 2(t + 2)/(t + 1)$ as its solution. Verify this.

Solution First we observe that the difference equation of Example 1 can be more conveniently written (after simple algebra):

$$x(t + 1) = 4 - \frac{4}{x(t)} \tag{2}$$

If $x(t) = 2(t + 2)/(t + 1)$, then $x(t + 1) = 2(t + 3)/(t + 2)$. In addition, $4 - 4/x(t) = 4 - 2(t + 1)/(t + 2)$. Therefore, in order for Equation (2) to be satisfied, we must have

$$\frac{2(t + 3)}{t + 2} = 4 - \frac{2(t + 1)}{t + 2}$$

This can easily be verified. Finally, we note that $x(0) = 2(0 + 2)/(0 + 1) = 4$, in accordance with the initial condition.

So far we have introduced two main problems involving difference equations:

1. Given a difference equation plus initial condition, find the first few terms of the solution sequence.
2. Given a difference equation plus initial condition and given a function, check whether the given function is a solution of the difference equation plus initial condition.

Both of these problems are fairly easy: 1 is arithmetic and 2 is algebra. There is, however, another problem which is often harder:

3. Given a difference equation with initial condition, find a formula for the solution.

There is no general theory of how to do this, although there are tricks and theories for special types of equations. These special methods are not our main interest, but we will encounter some of them in the exercises and in the next sections.

EXERCISES

● **1** Find the first four terms of the solution sequence in each of the following cases:
  (a) $x(t + 1) - x(t) = x(t)^2 + t; x(0) = 1$.
  (b) $x(t + 1) - x(t) = t^2 + t + 1; x(0) = -1$.
  (c) $x(t + 1) - x(t) = t/x(t); x(0) = 1$.
● **2** Show that the difference equation $x(t + 1) - x(t) = a$, with initial condition $x(0) = b$, has the solution $x(t) = at + b$.

3 (a) Plot the solutions to $x(t + 1) - x(t) = 1$ for these three initial conditions: $x(0) = -1, x(0) = 0$, $x(0) = 1$. Plot them on the same coordinate axes for ease of comparison. What do the solutions have in common?
  (b) Plot the solutions to $x(t + 1) - x(t) = a, x(0) = 1$, for three values of $a: a = -1, a = 0, a = 1$. What do the solutions have in common?

● **4** Show that $x(t) = t^2$ is not a solution to $x(t + 1) - x(t) = 2t$ for any initial condition.

5 (a) Find a formula for the solution to the difference equation $x(t + 1) - x(t) = -2x(t)$ with initial condition $x(0) = 3$. (*Hint*: Follow the scheme of the proof of Theorem 1.) Verify your solution by substitution.
  (b) Do the same for the equation $x(t + 1) - x(t) = kx(t)$ with initial condition $x(0) = a$.

6 Show that the solution to $x(t + 1) - x(t) = x(t)^2 + 1$, with $x(0) = a_0$ and $a_0 > 0$, has all terms positive.

Computer exercise

7 In many cases the solution of a difference equation is said to "depend continuously on the initial condition." For example, consider the difference equation $x(t + 1) - x(t) = t + x(t)$. Consider also the infinite family of initial conditions:

1. $x(0) = (\frac{1}{2})^1$.
2. $x(0) = (\frac{1}{2})^2$.
$\vdots$
i. $x(0) = (\frac{1}{2})^i$.
$\vdots$

As you can see, these initial conditions approach $x(0) = 0$ in the limit. Now let $x_i(t)$ be the solution corresponding to initial condition i and let $x^*(t)$ be the solution corresponding to the limiting initial condition $x^*(0) = 0$. Continuous dependence on the initial conditions means that, for each t, $x_i(t) \to x^*(t)$ as $i \to \infty$. Use a computer to compute $x^*(5)$ and $x_i(5)$ for $i = 1, \ldots, 10$ and see if it appears that convergence occurs. Do the same for $t = 6, 7,$ and 8.

BIBLIOGRAPHY

Goldberg, Samuel: "Introduction to Difference Equations," John Wiley & Sons, Inc., New York, 1958. A standard text.

2B EXPONENTIAL POPULATION GROWTH—DIFFERENCE AND DIFFERENTIAL EQUATIONS

Abstract The simplest growth model of population can be described with difference equations or differential equations with somewhat different results. Neither is completely realistic since population is a step function, defined on all time values.

Prerequisites Section 2A.

If a small number of yeast cells are placed in a test tube of some nutrient at time 0 ($t = 0$), the growth of this population with time might look like Figure 3, at least at the outset.

The numbers tabulated in Figure 3 were obtained by Raymond Pearl from data collected by a German investigator, T. Carlson. Carlson studied the growth of yeast by periodically centrifuging a culture to separate the yeast and then determining the volume and mass of the residue. The fractions in the table result from this indirect way of "counting."

Examination of the data shows that the ratios of population sizes for successive t values seem to lie between about 2 and 3. Suppose we denote by $x(t)$ the

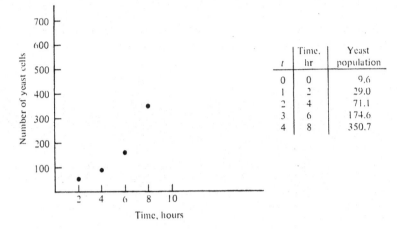

Time, t	hr	Yeast population
0	0	9.6
1	2	29.0
2	4	71.1
3	6	174.6
4	8	350.7

Figure 3 Growth of a colony of yeast cells.

number of cells at time t, where the time unit is 2 hours. Then, we might take, as a simplified model,

$$x(t + 1) = kx(t) \qquad t = 0, 1, 2, \ldots \tag{3}$$

for some suitable k value between 2 and 3, say $k = 2.5$. The initial condition would be $x(0) = 9.6$.

So far the argument in favor of this equation is based entirely on looking for a pattern in a set of numbers. Another strategy for arriving at a mathematical model is to derive it from the underlying mechanism which gives rise to the numbers. In this case we could proceed from knowledge of how yeast reproduce.

Yeast cells reproduce by splitting in two. After a certain time, which depends on the species of yeast and the environmental circumstances, each of the "baby" yeast cells splits again. Naturally the many cells in a culture do not all split simultaneously. A reasonable way to describe the splitting rate would be to observe the culture for some short period of time and see what percentage of the cells split. Suppose we observe that 36 percent of the cells split in a certain 40-minute period—one-third of a 2-hour time unit. We make the assumption that in every 40-minute period the same percentage will split. (How reliable do you think this is?) This assumption yields the following equations:

$$
\begin{aligned}
x(t + \tfrac{1}{3}) &= 1.36x(t) & \text{for } t = 0, 1, 2, \ldots \\
x(t + \tfrac{2}{3}) &= 1.36x(t + \tfrac{1}{3}) \\
&= (1.36)^2 x(t) & \text{for } t = 0, 1, 2, \ldots \\
x(t + 1) &= 1.36x(t + \tfrac{2}{3}) \\
&= (1.36)^3 x(t) & \text{for } t = 0, 1, 2, \ldots \\
&\approx 2.5x(t)
\end{aligned}
\tag{4}
$$

The last equation is exactly the type we conjectured by merely examining the numbers. Compare it with Equation (3).

The following theorem shows the solution to an equation of this type.

Theorem 2

If $x(t)$ is a function which satisfies

$$x(t + 1) = kx(t) \qquad t = 0, 1, 2, \ldots \tag{5}$$

then for these t values $x(t)$ can be expressed by the formula

$$x(t) = k^t x(0)$$

Proof Successively substituting $t = 0, 1, 2, \ldots$ in Equation (5) gives

$$x(1) = kx(0)$$
$$x(2) = kx(1) = kkx(0) = k^2 x(0)$$
$$x(3) = kx(2) = kk^2 x(0) = k^3 x(0)$$
$$\vdots$$

Continuing in this way, we obtain

$$x(t) = k^t x(0)$$

Example 4

When we apply this theorem to the equation $x(t + 1) = 2.5x(t)$, which describes the data in Figure 3, we obtain the solution

$$x(t) = 9.6(2.5)^t$$

It is important to realize that the solution shown in Theorem 2, like Equation (5) itself, holds only for $t = 0, 1, 2, \ldots$.

Equation (3) can be rewritten by simple algebra as a difference equation.

$$x(t + 1) - x(t) = (k - 1)x(t) \qquad t = 0, 1, 2, \ldots$$

or

$$x(t + 1) - x(t) = rx(t) \qquad t = 0, 1, 2, \ldots \tag{6}$$

where we have set $r = k - 1$. In this form, the solution given by Theorem 2 is $x(t) = x(0)(r + 1)^t$.

The difference equation form for $x(t + 1) = 2.5x(t)$ is

$$x(t + 1) - x(t) = 1.5x(t) \qquad t = 0, 1, 2, \ldots$$

The left side of Equation (6) can be regarded as a difference quotient used to approximate the derivative of $x(t)$. Recall that by definition

$$\frac{dx}{dt} = \lim_{h \to 0} \frac{x(t + h) - x(t)}{h}$$

If we simply set $h = 1$ instead of letting $h \to 0$, then, on the right side, we have $x(t + 1) - x(t)$ as an approximation to dx/dt. In other words, the left side of Equation (6) is nearly a derivative. This suggests that the population growth function could also be described by a suitable differential equation. This is partly true and partly false, as we shall see.

The differential equation that is often considered equivalent to Equation (6) is

$$\frac{dx}{dt} = (\log k)x \qquad t \geq 0 \tag{7}$$

For the yeast growth data, where $k = 2.5$, this becomes

$$\frac{dx}{dt} = (\log 2.5)x \qquad t \geq 0$$

Notice that we assume this holds for all $t \geq 0$, not just integer values. This is necessary so that we can solve the differential equation. But there is no justification for this assumption in terms of anything we have observed in the test tube.

Theorem 3

The solution to the differential equation

$$\frac{dx}{dt} = sx \qquad t \geq 0 \tag{8}$$

is $\qquad x(t) = x(0)e^{st}$

Proof

$$\frac{dx}{x} = s \, dt$$

$$\int \frac{dx}{x} = \int s \, dt$$

$$\log x = st + c$$

$$x(t) = e^{st + c}$$

$$= e^c e^{st}$$

$$= c'e^{st}$$

Substituting $t = 0$ gives $x(0) = c'e^0 = c'$; so we can write the solution

$$x(t) = x(0)e^{st}$$

When we apply this theorem to Equation (7), we see that it has the solution

$$x(t) = x(0)e^{(\log k)t}$$
$$= x(0)k^t \qquad t \geq 0$$

which is the same formula as we had for the solution of the difference equation. See Theorem 2.

The fact that Equations (6) and (7) have the same formula for their solutions is what leads to the idea that difference Equation (6) and differential Equation (7) are similar. But similar is not the same as identical. The key difference is that the difference equation only holds for $t = 0, 1, 2, \ldots$, and so its solution only holds for these integer t values. On the other hand, the differential equation is assumed to hold for all $t \geq 0$, and the solution therefore holds for all $t \geq 0$. Figure 4 displays the distinction graphically.

The difference equation gives us no information about what's going on between the integer time values. We have, therefore, no information about how to connect the dots in Figure 4a. There are many possibilities. If we assume the same formula $x(t) = x(0)k^t$ holds for noninteger t values as well, then we obtain Figure 4b, which is also the solution of the differential equation. On the other hand, as the next example shows, we could have a solution like the one in Figure 4c.

Example 5

The function

$$x(t) = x(0)k^t + a \sin 2\pi t \tag{9}$$

satisfies the difference Equation (5) for any constant a. This can be verified by straightforward substitution. Note that, when $t = 0, 1, 2, \ldots$, we have $\sin 2\pi t = 0$, and so for these values $x(t)$ reduces to the solution given by Theorem 2. Figure 4c shows a solution of this type with $a = 10$.

Many other examples of this type could be constructed. The idea is to take the function $x(t) = x(0)k^t$, which satisfies both the difference equation and the differential equation, and add to it a term which leaves the value unchanged at $t = 0, 1, 2, \ldots$, but which changes the function at other t values. Other terms which do this when added include $\sin^2 2\pi t$ and $1 - \cos 2\pi t$.

Of the two models discussed here, the one most commonly found is the one involving differential equations. However, in our view the difference equation is both simpler and more natural, while the differential equation is best thought of as an approximation.

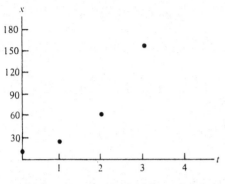

(a) Points on $x(t) = 9.6(2.5)^t$ for $t = 0, 1, 2, 3$

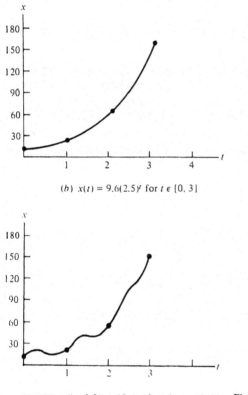

(b) $x(t) = 9.6(2.5)^t$ for $t \in [0, 3]$

(c) $x(t) = 9.6(2.5)^t + 10 \sin(2\pi t)$ for $t \in [0, 3]$ **Figure 4**

The popularity of the differential equations model is due to at least the following important factors:

1. While not based on impeccable logic, it nevertheless gives good results. (As we have seen, it can be made to give exactly the same formula as the difference equation.)

2. Analysis of differential equations uses the well-known tools of the calculus. By contrast the theory of difference equations is less well known.

To put the comparison between continuous (i.e., differential equations) models and discrete (i.e., difference equations) models in better perspective, it is useful to lay aside mathematical machinery for a moment and try to imagine what the actual graph of yeast population growth might look like. Figure 5 shows what the general appearance might be. Flat spots on the graph, such as *AB*, correspond to intervals of time when no births (fissions) occur. At times such as t_1, one or more (5 in the case of t_1) additions are being made to the population. The height of the jump at a discontinuity indicates how many splittings are taking place at that instant. The number of splittings and the times when they take place undoubtedly have a somewhat random character. A more realistic description would require probability theory.

The biologist takes observations at a finite number of instants, separated in time, say $t = 0, 1, 2, 3, 4$, and obtains the values indicated by heavy dots. The rest

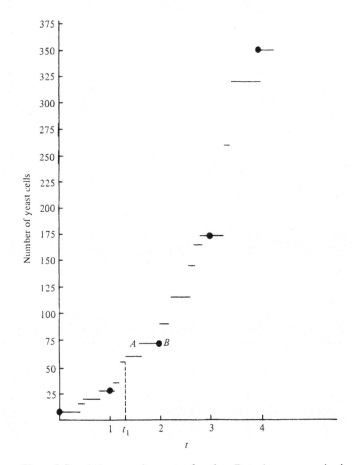

Figure 5 Population growth as a step function. Dots show measured values.

of the graph aside from these dots, the curve connecting them, is unknown. If a difference-equation model is built, the biologist is trying to find an equation that will predict a set of five dots that fit the actual data points (as in Figure 4a). This model purposely ignores what happens between the dots. If instead a differential-equation model is built, the biologist finds a continuous curve, defined for all t values (see Figure 4b). This model is incorrect in that it does not show the discontinuities and flat spots which clearly exist in the actual graph of population size.

EXERCISES

1 Suppose 20 percent of a yeast population splits in any 15-minute interval. If 1 time unit designates 2 hours, what formula connects $x(t + 1) - x(t)$ to $x(t)$? Follow the method of Equation (4).

2 Presumably our nutrient-filled test tube can only hold a certain maximum number of yeast. Is this consistent with either the continuous or discrete models presented here? In particular, what do these models predict for $\lim_{t \to \infty} x(t)$?

3 Suppose we have a strain of bacteria which all split simultaneously and the time between splits is 1 unit. Write the equation relating $x(t + 1)$, the population at $t + 1$, to $x(t)$, the population at t. If there are 100 bacteria to start with, what is the formula for $x(t)$? What differential equation would you use to describe the growth of the population?

4 Suppose we have two types of yeast in a culture, type x and type y. In 1 time unit 10 percent of the type x's split and give rise to new type x's, while 20 percent of the type x's split into type y's. In one time unit 15 percent of the type y's split into two type x's, while 5 percent of the type y's split into two type y's. If $x(t)$ and $y(t)$ denote the numbers of type x and y yeast at time t, respectively, find formulas for $x(t + 1)$ and $y(t + 1)$ in terms of $x(t)$ and $y(t)$.

5 Rework Exercise 4 under this revised assumption: In 2 time units 15 percent of the type y's split into two type x's, while 5 percent of the type y's split into two type y's. Now find formulas for $x(t + 2)$ and $y(t + 2)$ in terms of $x(t)$, $x(t + 1)$, $y(t)$, and $y(t + 1)$.

6 Suppose 40 percent of the bacteria of a certain type split in 1 time unit. How many will split in $\frac{1}{2}$ time unit? Do you think 20 percent? If so, you would be wrong, for, suppose 20 percent is correct, then $x(t + \frac{1}{2}) = 1.20x(t)$ and $x(t + 1) = 1.20x(t + \frac{1}{2})$. These equations imply that $x(t + 1) = (1.20)^2 x(t)$.
  (a) What does this last equation imply about the percent splitting in 1 unit of time?
  (b) Can you find the correct fraction s which splits in $\frac{1}{2}$ time unit?

7 Generalize Exercise 6 by finding the formula for the fraction splitting in a time interval of $1/n$, assuming still that 40 percent split in any 1 time unit.

8 Suppose a bank gives 5 percent interest per year. The first interest payment of 5 percent is added to your account one year after your deposit. After each additional year there is another addition of 5 percent of the amount on deposit at the start of that year. Suppose you start with $1000, make no withdrawals, and leave all interest in the account. If $M(t)$ denotes the money in your account after t years,
  (a) Write an equation relating $M(t + 1)$ to $M(t)$.
  (b) Find the formula for $M(t)$ in terms of t.

9 Suppose the bank in Exercise 8 gives interest n times a year, but in the amount of $5/n$ percent for each of the n periods.
  (a) Find formulas relating $M(t + 1/n)$ to $M(t)$, $M(t + 2/n)$ to $M(t)$, etc.
  (b) Using the formulas in part (a), relate $M(t + 1)$ to $M(t)$.
  (c) Find the formula for $M(t)$ in terms of t and n. (Assume a starting amount of $1000.)
  (d) Find the limit as $n \to \infty$ of the function of n and t you found in part (c) to represent $M(t)$.
[*Hint*: $\lim_{n \to \infty} (1 + 1/n)^n = e$.]

10 (*a*) Show that the function of Example 5 satisfies the following difference equation (for any value of *a*):

$$x(t + 1) = kx(t) \qquad t = \tfrac{1}{2}, 1\tfrac{1}{2}, 2\tfrac{1}{2}, \ldots$$

(*b*) Can you generalize this result?

11 (*a*) Show that $x(t) = x(0)k^t \cos 2\pi t$ satisfies difference Equation (5). (See Example 5.)

(*b*) The function in part (*a*) is not a reasonable one to describe a yeast population (even though it satisfies the right difference equation). Why?

Computer exercise

12 Since the situation in Exercise 4 is close to the situation described in this section, we might guess that $x(t)$ and $y(t)$ were exponential functions. Calculate (successively) $x(1)$, $y(1)$, $x(2)$, $y(2)$, ..., $x(100)$, $y(100)$, starting from the assumption that $x(0) = 1000$ and $y(0) = 1000$. Do $x(t)$ and $y(t)$ look like exponential functions? One way to tell is to compute the ratios of successive x values and the ratios of successive y values.

2C LOGISTIC POPULATION GROWTH—DIFFERENCE AND DIFFERENTIAL EQUATIONS

Abstract Exponential growth can't go on forever because resources run out. We introduce density-dependent growth rates (logistic equations) to reflect this. In the continuous case, we can solve the differential equation. This is not possible for the difference equation.

Prerequisites Solving an elementary differential equation (integration by partial fractions).

In the last section we developed two models for yeast growth: a continuous model, involving the differential equation $dx/dt = (\log 2.5)x(t)$, and a discrete model, involving the difference equation $x(t + 1) - x(t) = 1.5x(t)$. Both models led to the same equation for $x(t)$, namely $x(t) = x(0)(2.5)^t$. But both models have a serious shortcoming: as $t \to \infty$, $x(0)(2.5)^t \to \infty$, an impossibility in the real world. Since the resources of the test tube are finite, there must surely be some level, say K yeast cells, beyond which the culture cannot grow.

It turns out that the data presented in Section 2B is only part of the data collected by Pearl. The complete set is shown in Figure 6. Notice that, for larger t values, there is a definite leveling off, and the formula $9.6(2.5)^t$ doesn't fit the data set very well for these large t's. In this section we tinker with both our continuous and discrete models to fit all the data.

Before proceeding, let us be somewhat more general. In the equation $x(t + 1) - x(t) = 1.5x(t)$, let us replace the specific growth rate 1.5 with the general constant r. The analogous replacement in the differential equation would

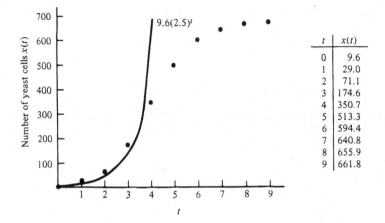

t	$x(t)$
0	9.6
1	29.0
2	71.1
3	174.6
4	350.7
5	513.3
6	594.4
7	640.8
8	655.9
9	661.8

Figure 6 Actual yeast growth compared to exponential function.

be to put $\log(1 + r)$ for $\log 2.5$, and for simplicity we will call this s. Thus, the equations we are going to tinker with are

$$x(t + 1) - x(t) = rx(t) \qquad (10)$$

$$\frac{dx}{dt} = sx(t) \qquad (11)$$

Our replacements for these will be

$$x(t + 1) - x(t) = \left\{ r\left[1 - \frac{x(t)}{K} \right] \right\} x(t) \qquad (12)$$

$$\frac{dx}{dt} = \left\{ s\left[1 - \frac{x(t)}{K} \right] \right\} x(t) \qquad (13)$$

These are called *logistic equations*. The terms in the braces are density-dependent growth rates. K is the maximum population size for the given environment. The rates are specifically tailored so that, when $x(t)$ is small in relation to K (low population density), the growth rate is nearly equal to the constant in the original equation, Equation (10) or (11). But as the population increases and nears its maximum size K, the growth rate approaches 0.

Do these properties force the solutions to look like Figure 6? In the continuous case they do, but in the discrete case the story is complicated. We shall now give a complete analysis of differential Equation (13), saving the more problematical difference Equation (12) for Section 2F.

In analyzing Equation (13), we first observe that the constant functions $x(t) = 0$ and $x(t) = K$ are solutions. In seeking the remaining solutions, we can assume that $x(t) \neq 0$ and $x(t) \neq K$. We rewrite Equation (13) as

$$\frac{x'(t)}{x(t)[1 - x(t)/K]} = s$$

By the method of partial fractions, we transform this to

$$\frac{x'(t)}{x(t)} + \frac{x'(t)}{K - x(t)} = s$$

$$\int \frac{x'(t)}{x(t)} dt + \int \frac{x'(t)}{K - x(t)} dt = \int s \, dt + c$$

$$\log x(t) - \log[K - x(t)] = st + c$$

$$\frac{x(t)}{K - x(t)} = e^{st+c} = e^c e^{st} = c' e^{st}$$

A little more algebra yields the "logistic" function

$$x(t) = \frac{Kc'}{c' + e^{-st}} \tag{14}$$

To convince ourselves that an equation of this type, with suitable values for K, c', and s, will fit the data of Figure 6, we need to make a few observations:

1. The data of Figure 6 suggests a horizontal asymptote of about 665 for $x(t)$. The graph of Equation (14) also has a horizontal asymptote for, as $t \to \infty$, $e^{-st} \to 0$ (provided we have $s > 0$, as it will be; see below), and so $x(t) \to Kc'/c' = K$. Thus, by choosing K to be whatever value seems indicated by the data, say 665, we can produce the same asymptote in the solution of the differential equation as we seem to have in the data.
2. We can easily arrange for the graph of Equation (14) to pass through the first data point of Figure 6, $(0, 9.6)$, by insisting that $x(0) = 9.6$. Substituting $t = 0$ into Equation (14) and solving for c' produces $c' = x(0)/[K - x(0)] = 9.6/(665 - 9.6) = 0.01465$. Taking this value for c' ensures $x(0) = 9.6$.
3. The data of Figure 6 suggests a curve which is concave up to start with ($d^2x/dt^2 > 0$), then concave down ($d^2x/dt^2 < 0$), with the inflection point (the point where $d^2x/dt^2 = 0$) occurring at an x value (vertical coordinate) of about half the maximum population size. This is an easily verified characteristic of our model also. Differentiating Equation (14) twice and setting $d^2x/dt^2 = 0$ gives

$$-st^* = \log c' \tag{15}$$

where t^* is the t value for the inflection point. Substituting into Equation (14) gives an x value of $K/2$ at the inflection point. Therefore, our model always gives an inflection point with the x value equal to half the maximum population.

Furthermore, if we wish to specify a value for t^*, the time at which the inflection occurs, we can do so by substituting this t^* value into Equation (15) and solving for s. For example, in Figure 6 the inflection point appears to have $t^* = 4$. Since $c' = 0.01465$, Equation (15) yields $s = -\frac{1}{4}\log 0.01465 = 1.05583$.

The upshot of our three observations is that, by taking $K = 665$, $c' = 0.01465$, and $s = 1.05583$, we can make our model reproduce three of the main features of Figure 6: the asymptote, the initial value (where the curve crosses the vertical axis), and the concavities, including the location of the inflection point. Figure 7 shows the graph of the logistic equation with the specified parameters,

$$x(t) = \frac{9.74225}{0.01465 + e^{-1.05583t}} \tag{16}$$

as compared with the actual data points. The fit appears to be excellent.

The parameters 9.74225, 0.01465, and 1.05583 appearing in Equation (16) were tailored to fit our data. What would the graph of Equation (14) look like

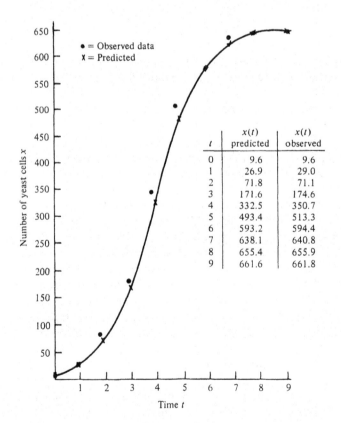

t	$x(t)$ predicted	$x(t)$ observed
0	9.6	9.6
1	26.9	29.0
2	71.8	71.1
3	171.6	174.6
4	332.5	350.7
5	493.4	513.3
6	593.2	594.4
7	638.1	640.8
8	655.4	655.9
9	661.6	661.8

Figure 7 Actual yeast growth compared to logistic function.

for other values of K, s, and c'? In all biologically reasonable cases, the population size approaches K as $t \to \infty$. Here are the mathematical details.

Theorem 4

If $x(t)$ is given by Equation (14) and $s > 0$, then $\lim_{t \to \infty} x(t) = K$.

Proof If $s > 0$, then $\lim_{t \to \infty} e^{-st} = 0$. Thus

$$\lim_{t \to \infty} \frac{Kc'}{c' + e^{-st}} = \frac{Kc'}{c' + \lim_{t \to \infty} e^{-st}}$$

$$= \frac{Kc'}{c'} = K$$

Sometimes, it is convenient to work with an alternate version of Equation (14), where c' is expressed in terms of an initial value. If $x(t)$ has value x_0 at time t_0, then from Equation (14)

$$x_0 = x(t_0) = \frac{c'K}{c' + e^{-st_0}}$$

Solving for c' gives (provided $x_0 \neq K$)

$$c' = \frac{x_0 e^{-st_0}}{K - x_0}$$

Substituting into Equation (14) gives

$$x(t) = \frac{x_0 K}{x_0 + (K - x_0)e^{-s(t - t_0)}} \tag{17}$$

a form we will use in the next section.

To conclude this section, let us turn to difference Equation (12) and ask whether we can solve it and find a formula for the solution, which displays the three main features of the observed data (the asymptote, the initial condition, and the nature of the concavities). As it happens, we are stumped before we start. There is no way presently known to find a simple formula for the solution of difference Equation (12).

Needless to say, this gives us a powerful incentive to prefer the continuous model for the growth of small organisms in a test tube. However, this is not the end of the story. The plot thickens in the next sections, where the discrete model reveals itself to have exciting possibilities not possessed by the more mathematically convenient continuous model.

EXERCISES

● **1** Find the equation of a logistic curve with asymptote $x = 200$, initial value $x(0) = 50$, and inflection point at $t = 7$.

2 Find a formula for the solution of differential Equation (13) with initial condition $x(t_0) = 80$, where t_0 is not necessarily 0.

3 Show that the solution to differential Equation (13) is an increasing function of t provided $x(0) < K$. Show also that, if $x(0) > K$, then the solution is a decreasing function of t. Do you think these results hold for difference Equation (12) also?

4 The steps taking us from Equation (13) to (14) are not valid if $x(t)$ ever assumes the value K.

    (a) Explain why this is so. [It turns out that, if $x(0) \neq K$, then it can be shown that $x(t) \neq K$ for all t.]

    (b) Can you state what the solution of Equation (13) is that corresponds to the initial condition $x(0) = K$?

5 In tinkering with Equation (11) to obtain Equation (13), we wanted to create a variable growth rate (depending on x), which is nearly s for small x and approaches 0 as x approaches K. We chose $s = 1 - x/K$ to fulfill these two conditions.

    (a) Show that $se^{-x/(K-x)}$ would also be a variable growth rate satisfying these conditions.

    (b) Can you think of any reason we should not use the growth rate of part (a) instead of the one we actually used?

6 Is there any indication in this section that the logistic equation obtained to fit the data of Figure 6 tells us anything of biological interest that we could not find out by studying the table and graph of observed data in Figure 6?

7 Find the equation of a logistic curve which gives a good rough fit to the data in the following table.

t	$x(t)$	t	$x(t)$
0	10	7	79
1	15	8	90
2	22	9	98
3	31	10	103
4	42	11	106
5	55	12	108
6	67	13	109

8 One way to measure the fit between a function $x(t)$ and a set of observed x values, say $x_0, x_1, \ldots, x_n$ corresponding to $t = 0, 1, \ldots, n$, is by the *sum of squared deviations formula*:

$$SS = \sum_{t=0}^{n} [x(t) - x_t]^2$$

Calculate this for $x(t)$ given by Equation (16). (Use the data in the table accompanying Figure 7.)

Computer exercises

9 We have no guarantee that the values s, K, and c', which we chose and which appear in Equation (16), give the lowest value of SS, as defined in Exercise 8. See whether small changes in one or more of these parameters might lower SS. First try a 1 percent increase and a 1 percent decrease in each of these in turn, leaving the other two unchanged. Then try all possible combinations of 1-percent changes and no changes.

10 The expression $s[1 - x(t)/K]$ in Equation (13) is called the growth rate. Using $s = 1.05583$ and $K = 665$, compute it for $t = 0, 1, \ldots, 10$ in each of the following two ways:

    (a) Substitute actual observed values (Figure 6) for $x(t)$ into the expression for the growth rate.

    (b) Replace $x(t)$ by formula (16). Then substitute $t = 0, 1, \ldots, 10$. Compare the results. Is the fit between them as impressive as the fit between theory and observation shown in Figure 7?

11 Theorem 4 doesn't say how quickly $x(t)$, as given by Equation (14), approaches K. It turns out that $x(t)$ will be within 1 percent of K as soon as t exceeds the critical value $(1/s) \log 0.99/(0.01c')$. Here are two ways you could try to justify this assertion. Can you carry out one or the other? Which one is best?

(*a*) Prove it algebraically from formula (14).

(*b*) Using various choices and combinations of values for s, c', and K, compute $x(t)$ for t values in the neighborhood of the critical value.

BIBLIOGRAPHY

Hamblin, Robert L., R. Brooke Jacobsen, and Jerry L. L. Miller: "A Mathematical Theory of Social Change," Wiley-Interscience, New York, 1973. Chapters 3 and 4 contain a lot of interesting theory and data involving logistic equations and their use in sociology.

2D EQUILIBRIUM, STABLE AND UNSTABLE

Abstract We define equilibrium and stability and show that the logistic differential equation always has a large neighborhood of stability, regardless of the growth parameter. By contrast, in the discrete logistic equation there can be instability.

Prerequisites The solution of the logistic differential equation (as in Section 2C).

Intuitively, a dynamic system is said to be in equilibrium if it does not change as time proceeds. Thus, a population is in equilibrium if it stays the same size. The mathematical way to put this would be: let $x(t)$ denote the population at time t; if $x(t)$ is a constant function, equivalently if $dx/dt = 0$ for all t, then the population is in equilibrium. Here is the formal definition used for the kinds of differential equations of interest to us.

Definition

For a differential equation of the form

$$\frac{dx(t)}{dt} = f(x(t)) \tag{18}$$

the value $\bar{x}$ is called an *equilibrium level* if $f(\bar{x}) = 0$. Observe that if $\bar{x}$ is an equilibrium level, then the constant function $x(t) = \bar{x}$ satisfies Equation (18), and so $x(t) = \bar{x}$ is called an *equilibrium solution* of differential Equation (18).

Example 6

1. $dx(t)/dt = \sin x(t)$ has 0, $\pm \pi$, $\pm 2\pi, \ldots$ as equilibrium levels. The corresponding equilibrium solutions are $x(t) = 0$, $x(t) = \pi$, $x(t) = -\pi$, etc.
2. What are the equilibrium levels and equilibrium solutions of $dx(t)/dt = s[1 - x(t)/K]x(t)$? The equilibrium levels are obtained by setting $s(1 - x/K)x = 0$. We obtain 0 and K as equilibrium levels and $x(t) = 0$ and $x(t) = K$ as the equilibrium solutions.

We also have a similar concept of equilibrium for difference equations.

Definition

A difference equation of the form

$$x(t + 1) - x(t) = f(x(t)) \tag{19}$$

has $\bar{x}$ as an *equilibrium level* if $f(\bar{x}) = 0$. In that case, the constant function $x(t) = \bar{x}$ is a solution of Equation (19) and is called an *equilibrium solution*.

Example 7

1. $x(t + 1) - x(t) = \sin x(t)$ has equilibrium levels 0, $\pm \pi$, $\pm 2\pi$, etc. The corresponding equilibrium solutions are $x(t) = 0$, $x(t) = \pi$, $x(t) = -\pi$, etc.
2. $x(t + 1) - x(t) = r[1 - x(t)/K]x(t)$ has equilibrium levels 0 and K. The corresponding equilibrium solutions to the difference equation are $x(t) = 0$ and $x(t) = K$.

If a population governed by logistic Equation (12) or (13) is at the equilibrium level K at some point in time, say $t = 0$ for convenience, then it will continue at the equilibrium level. The reason is that, for either Equation (12) or (13), the function $x(t) = K$ is a solution consistent with the initial condition $x(0) = K$. Furthermore, it is the unique solution consistent with that initial condition. (In the case of the difference equation, Theorem 1 proves this; in the case of the differential equation, the proof requires more advanced analysis.) In summary, we may say that the equilibrium level is "sticky" in our model: if the population hits equilibrium, it sticks there.

However, this property of "stickiness" is not enough to make the concept of equilibrium truly useful. In the first place, why should we expect a population to start at, or hit an equilibrium level? Even if we could control the population under laboratory conditions, it may be difficult to be sure how many organisms there are at the start. Imagine trying to count exactly 14,397 bacteria.

The other shortcoming of the equilibrium concept lies in the fact that our model is not a perfect description of the biological reality. For example, our model assumes the rate of increase [dx/dt or $x(t + 1) - x(t)$] to be totally dependent on the existing population size. If all other factors are equal, this might be a good assumption, but other factors are rarely equal. The weather may change for a period of months, make food either easier or harder to find, and therefore increase or decrease the growth rate. Pollution of the organism's environment, predation by other species, and a host of other fluctuating factors may influence the growth rate from time to time, in ways that are not reflected in the model. Factors like these may constantly bump the ecosystem away from its equilibrium level. There are three ways to cope with these perturbations:

1. Pretend they don't exist. This is the method of last resort.
2. Scrap the original model in exchange for a new one which takes the other factors into account. That is, in place of a model of the form $dx/dt = f(x)$, we try to develop a model of the form $dx/dt = f(x, E_1, E_2, \ldots, E_n)$, where $E_1, \ldots, E_n$ represent the extra factors. In principle this is a good idea. However, it is often difficult to accomplish and may yield a model which is mathematically very complex and hard to draw conclusions from.
3. Regard the additional factors as being of relatively minor significance, merely providing perturbations in the basic model.

Alternative 3 is, in many ways, the most attractive. However, it is only reasonable when the equilibrium has a property of *attraction*. Intuitively, an equilibrium has this characteristic if, whenever the population is not too far from equilibrium, it will tend to approach the equilibrium as time proceeds. In other words, the equilibrium acts as a magnet. In particular, if the population gets "bumped away" from equilibrium by some small temporary environmental fluctuation, for example a temporary food shortage, after conditions return to normal, the population will be drawn back toward the equilibrium. For this reason, the technical term for this property of attraction is *stability*.

Just as a magnet can only attract objects not too far away, an equilibrium is often only stable provided that the fluctuations away from it are not too large. This is one of the points to be illustrated in Example 8 which follows. Part 2 of Example 9 shows that an equilibrium may not be stable at all.

Example 8

Imagine a population governed by the logistic equation

$$\frac{dx}{dt} = 2\left(1 - \frac{x}{100}\right)x \qquad (20)$$

and initially ($t = 0$) at the equilibrium level $x = 100$. The population remains at this level till $t = 10$, when a severe storm perturbs the population down to 80.

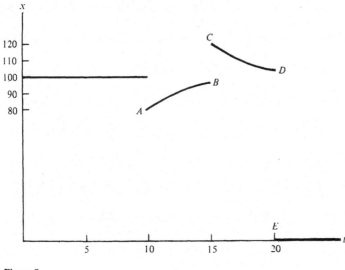

Figure 8

The population will now follow the solution to Equation (20), corresponding to the initial condition $x(10) = 80$. By Equation (17) of Section 2C, this is

$$x(t) = \frac{(80)(100)}{80 + 20e^{-2(t-10)}} \tag{21}$$

Theorem 4 implies that the population now increases toward 100. This is represented by the arc AB in Figure 8. Now suppose that at $t = 15$ there is another perturbation, this time a favorable one, which moves the population to the level of 120. The population continues along yet another solution to Equation (20), as indicated by arc CD in Figure 8. Once again, Theorem 4 tells us that the population will ultimately return toward 100. Finally, suppose that at $t = 20$ there is a devastating perturbation that wipes out the whole population. From here on the population is described by $x(t) = 0$. Now there will be no return to the equilibrium level of 100.

Summarizing this example: the equilibrium $x = 100$ is stable because solutions that start not too far away, such as at A or C, return toward 100 in the long run. But the stability property is *local*, not *global* because, if we go far enough, for example to $x = 0$, there will be no return to the original equilibrium. Thus, in this example, any perturbation which does not completely wipe out the population can be regarded as minor because, owing to the stability of $x = 100$, the perturbation will be corrected back toward equilibrium.

To define stability formally, we first need the concept of a neighborhood.

Definition

A *neighborhood* of a number x is any open interval (a, b) where $a < x < b$.

Definition

An equilibrium level $\bar{x}$ for the differential equation $dx/dt = g(x(t))$ [or a difference equation $x(t+1) - x(t) = g(x(t))$] is *stable* if there exists a neighborhood N of $\bar{x}$ with the property that, whenever $x_0 \in N$, then the solution $x(t)$, with the initial condition $x(t_0) = x_0$,

1. Is finite for all $t > t_0$
2. Has $\lim_{t \to \infty} x(t) = \bar{x}$

N is called the *neighborhood of stability*.

Figure 9 illustrates these notions. $\bar{x}$ is an equilibrium level, and (a, b) is a neighborhood of $\bar{x}$. If the population graph $x(t)$ ever finds itself between the dashed lines [i.e., $a < x(t) < b$], then $x(t) \to \bar{x}$ as $t \to \infty$.

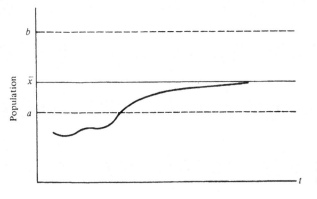

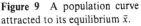

Figure 9 A population curve attracted to its equilibrium $\bar{x}$.

Example 9

1. Consider the differential equation $dx/dt = -2x(t)$ with the initial condition $x(0) = x_0$. The solution of this well-known equation (see Theorem 3 of Section 2B in this chapter) is $x(t) = x_0 e^{-2t}$. The only equilibrium level is $x = 0$. Regardless of what the initial condition is, $x(t)$ is finite for all t and approaches 0 as $t \to \infty$. Thus the neighborhood of stability consists of all the real numbers.
2. If we consider the analogous difference equation $x(t+1) - x(t) = -2x(t)$ with $x(0) = x_0$, the equilibrium level is still 0, but the solution is $x(t) = x_0(-1)^t$ (see Exercise 5 of Section 3A). Now, no matter what nonzero initial value x_0 we start with, $x(t)$ merely alternates between x_0 and $-x_0$ as t increases. (Remember, in our difference equations t takes on only the values $0, 1, 2, \ldots$.) Therefore there is no neighborhood of stability. Consequently, 0 is not a stable equilibrium.

We now carry out a complete stability analysis for logistic differential Equation (13).

Theorem 5

1. K is a stable equilibrium level of

$$\frac{dx}{dt} = s\left(1 - \frac{x}{K}\right)x$$

and its neighborhood of stability is $(0, \infty)$.
2. The only other equilibrium, 0, is not stable.

Proof Let x_0 be an initial value corresponding to t_0. We begin with part 2 and show that, if $x_0 < 0$, then the corresponding $x(t)$ violates condition 1 of the definition of stability.

In Equation (17) of Section 2C we showed that, when $x_0 \neq K$, the solution of the differential equation is

$$x(t) = \frac{x_0 K}{(K - x_0)e^{-s(t - t_0)} + x_0} \tag{22}$$

In this case, the denominator of Equation (22) has a *singularity*, i.e., a value of t for which the denominator is 0. We can solve for it:

$$(K - x_0)e^{-s(t - t_0)} + x_0 = 0$$

$$e^{s(t - t_0)} = \frac{x_0 - K}{x_0}$$

Since $x_0 < 0$, the right side is > 1 and therefore has a positive logarithm. That is,

$$s(t - t_0) = \log\frac{x_0 - K}{x_0} > 0 \tag{23}$$

Now solving for t is a simple matter, and the value we obtain, say t', is $> t_0$ (see Figure 10). The upshot is that as t increases from its initial value t_0 toward t', the denominator of Equation (22) is positive and approaches 0; so $x(t) \rightarrow -\infty$. In particular, the solution is undefined when $t = t'$. According to condition 1 of the definition of stability, this means that a negative x_0 cannot be part of any neighborhood of stability. Since any neighborhood of the equilibrium 0 must contain negative values, 0 is not stable and part 2 of our Theorem is proved.

To prove part 1, let $x_0 > 0$ but $x_0 \neq K$. Once again, Equation (22) is the form of the solution. Again we look for a singularity for the denominator by attempting to carry out the algebraic steps which led to Equation (23). However, if $x_0 < K$, then $(x_0 - K)/x_0 < 0$ and has no real logarithm, and so we can't solve Equation (23) for a real t. If $x_0 > K$, Equation (23) can be solved. However, since $(x_0 - K)/x_0 < 1$, $\log[(x_0 - K)/x_0] < 0$ and, since $s > 0$, the solution t' of Equation (22) will be such that $t' < t_0$. Since we are only concerned with time values after the initial condition, we again fail to find a singularity. Thus condition 1 of the definition of stability holds for any solution corresponding to an initial condition x_0, where $x_0 > 0$ and $x_0 \neq K$. That condition 2 of the definition holds in relation to the equilibrium K has already been stated and shown in Theorem 4. Therefore

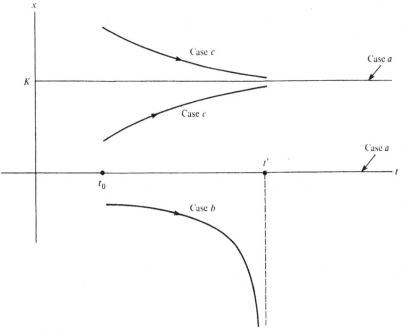

Figure 10

the set of all positive real numbers forms a neighborhood of stability for the equilibrium K. This proves part 1 of the theorem.

Figure 10 shows three types of solutions to the logistic differential equation and may illuminate the proof of Theorem 5:

Case a. The equilibrium solutions, $x(t) = 0$ and $x(t) = K$
Case b. Solutions with negative initial value ($x_0 < 0$)
Case c. Solutions with positive initial value ($x_0 > 0$)

Notice, by the way, that our discussion of negative initial conditions is a purely mathematical exercise. A negative number of organisms has no biological meaning.

The next logical question is whether the situation for the difference-equation analog of our logistic differential equation is the same: are K and 0 stable and unstable equilibria, respectively, for

$$x(t + 1) - x(t) = r\left[1 - \frac{x(t)}{K}\right]x(t)$$

The answer, surprisingly, is that the situation is quite different. For example, there are values of K and r for which K is not a stable equilibrium for the entire neighborhood of positive initial conditions. Here is an example.

Example 10

Consider the logistic difference equation

$$x(t + 1) - x(t) = 3\left[1 - \frac{x(t)}{6}\right]x(t)$$

which we may rewrite

$$x(t + 1) = [4 - 0.5x(t)]x(t)$$

If we start with the initial condition $x(0) = 5 + \sqrt{5}$, then

$$x(1) = [4 - 0.5(5 + \sqrt{5})](5 + \sqrt{5})$$
$$= 5 - \sqrt{5}$$
$$x(2) = [4 - 0.5(5 - \sqrt{5})](5 - \sqrt{5})$$
$$= 5 + \sqrt{5}$$

Clearly the sequence $x(1)$, $x(2)$, $x(3)$, ... alternates indefinitely between the values $5 \pm \sqrt{5}$. Therefore, if the equilibrium, which is $x(t) = 6$, has a neighborhood of stability, then it is not large enough to include $5 + \sqrt{5}$ or $5 - \sqrt{5}$. This is quite different from the behavior of the analogous differential equation. Actually the differences between these two equations are even more far-ranging than this example indicates. More details will be found in Section 2F.

Sidelight: History of Stability

One of the first people to work on the mathematical concept of stability was George Biddell Airy (1801–1892). He was faced with the problem of keeping an astronomical telescope moving uniformly, immune to small random shocks from the environment.

Perhaps the most spectacular early application of attractiveness and stability notions is due to James Clerk Maxwell (1831–1879), who used these ideas to study the composition of the rings which surround the planet Saturn. He showed that the rings could not be solid because in that event the motion would be an unstable equilibrium and thus some small perturbation would long ago have destroyed the rings. According to Maxwell,

> The only system of rings which can exist is one composed of an indefinite number of unconnected particles revolving around the planet with different velocities, according to their respective distances.

The earliest comprehensive mathematical theory of stability was created by Edward John Routh (1831–1907) in an essay entitled "Stability of Motion," which won the Adams prize competition of Cambridge University in 1877. In describing the subject of the competition, the organizers explained the concept of stability in the following picturesque way:

> To illustrate the meaning of the question, imagine a particle to slide down inside a smooth inclined cylinder along the lowest generating line, or to slide down the highest generating line. In the former

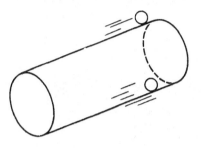

> case a slight derangement of the motion would merely cause the particle to oscillate about the generating line, while in the latter case the particle would depart from the generating line altogether. The motion in the former case would be, in the sense of the question stable, in the latter unstable.

EXERCISES

● **1** Find the equilibrium level or levels of each of the following. Then check for the existence of a neighborhood of stability.

   (a) $dx/dt = -1.5x(t)$.
   (b) $x(t + 1) - x(t) = -1.5x(t)$.

Compare to Example 9 of this section.

● **2** Consider the difference equation

$$x(t + 1) - x(t) = -rx(t) \qquad r \geq 0$$

Show that there is a certain critical value r_0 for r such that, when $r < r_0$, the equation has a neighborhood of stability, but, when $r \geq r_0$, it does not. What is the critical value of r_0?

● **3** Can you conceive of a way, in a population model, in which there can be a perturbation that disturbs the equilibrium level of 0 to some positive level? Could this only happen if a species came into being overnight? Bear in mind that we are dealing with "test-tube" models, describing a physically delimited space, such as a dish of nutrient in a laboratory, or, by extension, an island, an oasis in a desert, or a patch of greenery in an urban area.

4 (a) Find the solution of the differential equation

$$\frac{dx}{dt} = K(x - 1)(x - 2)$$

corresponding to initial condition $x(t_0) = x_0$. (*Hint*: Try partial fractions, as in Section 2C.)

   (b) What are the equilibrium values?

(c) Describe the behavior of the solution as $t \to \infty$ in each of the following four cases. Don't forget to look for singularities, as in the proof of Theorem 5.

(i) $x_0 < 0$.

(ii) $0 < x_0 < 1$.

(iii) $1 < x_0 < 2$.

(iv) $x_0 > 4$.

(d) Are there any initial conditions for which the solution has a singularity at some t value $t' < \infty$?

(e) State which equilibria are stable and what the neighborhood of stability is in each case.

Computer exercise

5 (a) Verify the calculations of Example 10 using a computer or pocket calculator. In entering the constants, have the machine compute $\sqrt{5}$. Do you get $x(0) = x(2)$ and $x(1) = x(3)$ when you start with $x(0) = 5 + \sqrt{5}$?

(b) Now repeat part (a) but on paper and following the rule that after each operation of arithmetic you round the result to two decimal places. In particular, take $5 + \sqrt{5}$ to be 7.24 (a slight overestimate). For a number whose last nonzero digit is 5, use the round-even rule: round up or down according to which direction makes the last nonzero digit even; e.g., 3.755 becomes 3.76. This exercise mimics the behavior of a fictitious computer with very little storage for a number.

BIBLIOGRAPHY

Eisen, Martin: Graphical Analysis of Some Difference Equations in Biology, UMAP Module 553, available from COMAP, 271 Lincoln St., Lexington, MA 02173. Section 5 deals with a pair of interacting species—a two-species version of logistic growth.

2E STABILITY OF LINEAR DIFFERENCE EQUATIONS— COMMODITY PRICES AND COBWEB ANALYSIS

Abstract A linear difference equation is about the simplest kind there is, but we show that even here the equilibrium may be unstable. We describe a classic economic model that uses this instability to explain the erratic nature of commodity prices. The technique of cobweb analysis is explained and used to analyze stability.

Prerequisites Sections 2A and 2D.

We continue our study of the stability of equilibria for difference equations with the simplest type of equation, one of the form

$$p(t + 1) = mp(t) + k \tag{24}$$

where m and k are constants. An equation of this form is called *linear*. Analyzing this kind of equation can be a useful warm-up for the more complicated logistic difference equation and other nonlinear difference equations.

Equation (24) arises when economists try to describe the price fluctuations of agricultural commodities. Economists often divide goods into two categories: commodities and manufactured items. Commodities are the primary products of the earth, such as oil, corn, tin, lumber, and so on. In both of these categories there are year-to-year fluctuations in price. But whereas the prices of manufactured goods usually follow fairly smooth trends, the prices of commodities often fluctuate up and down sharply. Figure 11a and b shows price fluctuations for hogs and milk cows.

Where do these fluctuations come from? Economists look for the answer in the concepts of supply and demand. The supply of a commodity in a given time period is simply the amount available for sale in that period. But how does the supply come to exist in exactly that amount? A fundamental fact about commodities is that one must plan far in advance for their production. To get more wheat in the fall, you must plant more in the spring. Therefore, in the period between planting and harvesting, there is little that can be done to affect the supply.

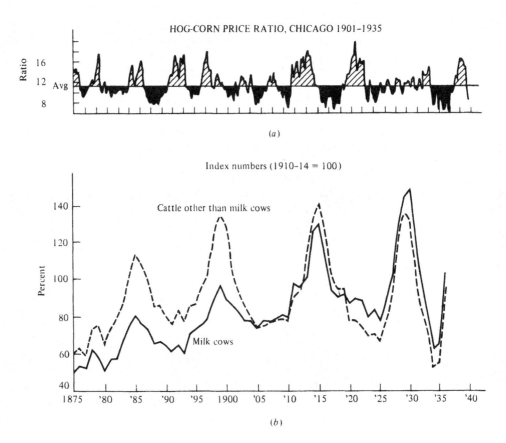

Figure 11 (a) How-corn price ratios. (b) Purchasing power per head of milk cows and cattle other than milk cows. [*Adapted from M. Ezekiel, The Cobweb Theorem, Quart. J. Eco., vol. 52, February, 1938.* Copyright © (1938 John Wiley & Sons, Inc.). Reprinted by permission of John Wiley & Sons, Inc.]

We call this the *production lag*. The production lag for hogs is 11 months since it takes that long for the gestation and growth of a hog. Nearly all commodities have substantial production lags. Some manufactured goods (e.g., oil tankers) have big lags too, but most do not. If a sudden fad for pumpernickel seizes the country, you can be sure that the amount of pumpernickel produced will shoot up overnight.

We will deal with a single unspecified commodity in this section, and for convenience we will scale our time measurement so that the production lag for that commodity is 1 time unit. Let $s(t)$ denote the supply at time t. $s(t + 1)$ is completely determined by decisions made at time t. To make such a decision, a farmer (or other commodity producer) would like to know what the price will be at time $t + 1$, but without a crystal ball the farmer cannot know this. We will assume that the farmer decides solely on the basis of the price at t, which we denote $p(t)$. Therefore, the supply at $t + 1$ depends on $p(t)$. We will assume the simplest possible relationship, a linear one:

$$s(t + 1) = ap(t) + b \tag{25}$$

Since higher prices will induce more people to produce the commodity, $a > 0$ (see Figure 12).

But this is only half the story about supply and price; the other half involves the concept of demand. The demand for a commodity is the amount that will be bought at a given price. When the price goes up, demand goes down, and vice versa. Furthermore, the response of demand to price changes is immediate. There is no lag, as in the relation between price and supply described in Equation (25). Therefore, if we denote demand at t by $d(t)$, then $d(t)$ is some function of $p(t)$. Likewise $d(t + 1)$ is that same function of $p(t + 1)$. Once again we pick the simplest possible type of function, a linear one:

$$d(t + 1) = cp(t + 1) + e \tag{26}$$

where $c < 0$.

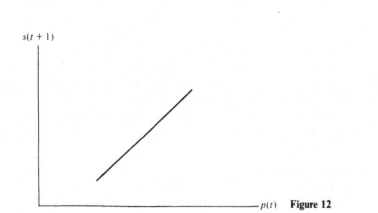

$s(t + 1)$

$p(t)$ **Figure 12**

Finally, we need the relation of supply to demand at a given time. Since you cannot buy something that does not exist, $d \leq s$ at all times. But we assume the reverse inequality, $s \leq d$, also because it is impractical to keep a commodity beyond the time when it is ready for sale. If necessary, sellers will reduce prices to achieve a balance in supply and demand. Therefore, we can set $d(t + 1) = s(t + 1)$ and obtain from Equations (25) and (26)

$$ap(t) + b = cp(t + 1) + e$$

which reduces to

$$p(t + 1) = mp(t) + k \tag{27}$$

where $m = a/c$ is a negative constant, since $a > 0$ and $c < 0$, and $k = (b - e)/c$. In practice, $k > 0$. (Can you explain this?)

Computing the equilibrium price can be done by setting $p(t + 1) = p(t)$ and solving simultaneously with Equation (27) to obtain $p(t) = k/(1 - m)$. This can be given a useful geometric interpretation. We graph the price equation, Equation (27), on axes, where the horizontal axis represents $p(t)$ and the vertical axis represents $p(t + 1)$. This gives a line of slope m and intercept k. We also graph the line of slope 1 and intercept 0. All points on this line have $p(t + 1) = p(t)$; so these points represent possible equilibria. Therefore, this line is called the *equilibrium line*. The actual equilibrium is the point where this equilibrium line crosses the price line of Equation (27) (see Figure 13).

A glance at Figure 11a and b shows that, in the real world, equilibrium is seldom maintained. One reason for this is that there are plenty of factors which are not taken into account in the model and which fluctuate and serve to bump our commodity model away from equilibrium. For example, since hogs are fed on corn, a sudden increase in corn prices may cause farmers to decrease hog production, even though this is not predicted by Equation (25). We now ask what our model predicts about the response to such a perturbation. Do the prices return to equilibrium or not? Here are some numerical examples that suggest that this return to equilibrium (stability) may or may not occur.

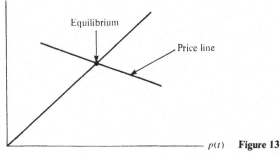

Figure 13

Example 11

Take $m = -2$ and $k = 900$ and start with $p(0) = 295$, 5 below the equilibrium value of 300. The last column shows a definite pattern of doubling the distance from equilibrium. If this is not just an accident, then perhaps 295 is not in any neighborhood of stability.

t	$p(t)$	$p(t)$ − equilibrium
0	295	−5
1	310	+10
2	280	−20
3	340	+40

Example 12

Take $m = -\frac{1}{2}$, $k = 600$, and $p(0) = 432$. The equilibrium is 400. This time the pattern in the last column suggests convergence to the equilibrium as $t \to \infty$. If this were true for a neighborhood of price values around the equilibrium, we would have stability.

t	$p(t)$	$p(t)$ − equilibrium
0	432	32
1	384	−16
2	408	8
3	396	−4

Because Equation (27) is linear, we can investigate the stability theoretically, using simple algebra.

Theorem 6

If $m \neq 1$, the difference equation

$$p(t + 1) = mp(t) + k$$

has the solution

$$p(t) = m^t p(0) + k\left(\frac{m^t - 1}{m - 1}\right)$$

Proof Successively substituting $t = 0, 1, 2$, etc., into Equation (27) gives:

$$p(1) = mp(0) + k$$

$$p(2) = mp(1) + k = m[mp(0) + k] + k = m^2 p(0) + mk + k$$

$$\vdots$$

It is evident that the general formula is

$$p(t) = m^t p(0) + k(m^{t-1} + m^{t-2} + \cdots + 1)$$

The last term is a geometric series and can be summed by a well-known formula so that

$$p(t) = m^t p(0) + k\left(\frac{m^t - 1}{m - 1}\right)$$

$$= m^t\left[p(0) + \frac{k}{m - 1}\right] - \frac{k}{m - 1} \tag{28}$$

If $m = 1$, the formula given by the theorem is incorrect. In that case, $p(t + 1) = p(t) + k$; so the solution is clearly $p(t) = p(0) + kt$. There is no equilibrium, hence no question of its stability. In our commodity model $m < 0$; so the special case $m = 1$ need not concern us.

The following corollary tells us that the size of the absolute value of m determines whether we have stability or not.

Corollary

1. If $|m| > 1$ and $p(0) \neq k/(1 - m)$, then $\lim_{t \to \infty} |p(t)| = \infty$.
2. If $|m| < 1$, then $\lim_{t \to \infty} p(t) = k/(1 - m)$ for any initial condition $p(0)$.

Proof
1. If $m > 1$, $m^t \to \infty$ as $t \to \infty$, and so $p(t) \to \infty$. If $m < -1$, m^t oscillates but $|m^t| \to \infty$, and the result follows.
2. If $|m| < 1$, then $m^t \to 0$ as $t \to \infty$; so $p(t) \to k/(1 - m)$.

We can understand this corollary and the reason it is true, using the graph of the price equation shown in Figure 14. Given any initial price $p(0)$, we find $p(1)$ as follows:

1. Lay off the length $p(0)$ on the horizontal axis.
2. From this point, extend a vertical segment to the price curve $p(t + 1) = mp(t) + k$. The height of this segment is $p(1)$, and its endpoint on the price curve is therefore $(p(0), p(1))$.

If we now repeat this sequence of steps with $p(1)$ in place of $p(0)$, we can get $p(2)$. To do this, we first have to take the vertical segment of height $p(1)$ found in step 2 and lay it off on the horizontal axis so that we can apply step 1. This is followed by moving up to the price curve, as in step 2. It turns out that we can use a shortcut procedure that combines the two steps of transferring a vertical to a

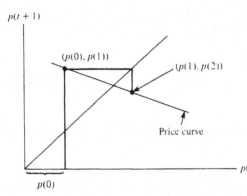
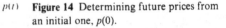

Figure 14 Determining future prices from an initial one, $p(0)$.

horizontal segment and then moving up to the price curve. We use the equilibrium line to do this as follows:

1. Starting from the point $(p(0), p(1))$ found in step 2, move horizontally to the equilibrium line. This brings you to the point $(p(1), p(1))$. This point is directly over the point you would have obtained by laying off $p(1)$ on the horizontal axis. Since laying off $p(1)$ is only a preliminary to moving up to the price curve, we now do this directly. In other words, the next step is:
2. From the point just found on the equilibrium line, move up or down to the price curve. The point where you meet it has height $p(2)$.

By repeating these steps, alternately moving horizontally to the equilibrium line and vertically to the price curve, we find $p(3), p(4), \ldots$, all as a series of heights of points on the price curve.

The procedure just described is called *graphical iteration*. It can be applied for any difference equation where $p(t + 1)$ is a function of $p(t)$. This function [$mp(t) + k$ in our commodity-price example] is called the *iteration function* and its graph is called the *iteration curve*. In this section, iteration curves are always straight lines; we take up more complicated cases in the next section.

Figure 15a seems to suggest that, when graphical iteration is carried out, the price sequence converges to the equilibrium price, represented by the intersection of the equilibrium line and the price curve. In this case, if we were to plot the price sequence against time, as in Figure 15b, we would get a series of prices which oscillate but converge to the equilibrium. This is what is described by part 2 of the corollary. If we make the price curve steeper ($m < -1$), as in Figure 16, our graphical procedure produces an oscillation that does not converge to equilibrium. This illustrates part 1 of the corollary.

Economists and mathematicians are fond of this graphical iteration procedure, because a little geometric intuition demonstrates that the steepness of the price curve [the constant m in Equation (27)] is the key to determining whether the equilibrium is stable (Figure 15) or unstable (Figure 16). Because of their appearance, figures such as Figures 15a and 16a are called *cobwebs*, and drawing and analyzing them is called *cobweb analysis*.

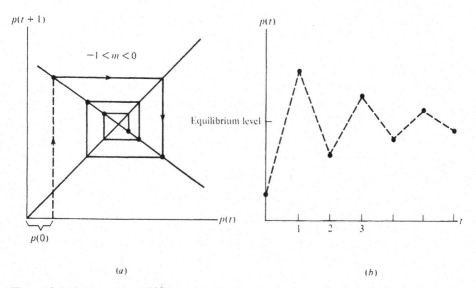

Figure 15 (*a*) A price cobweb. (*b*) Prices plotted against time, as implied by the cobweb.

The model we have presented here for commodity prices is quite simple. Few economists would rely upon it for numerical predictions. Its value is that it captures the essential reason for commodity-price fluctuation in a way that is easy to understand. Much of the more sophisticated work builds on the basis of this model. One common modification is to replace one or both of the linear

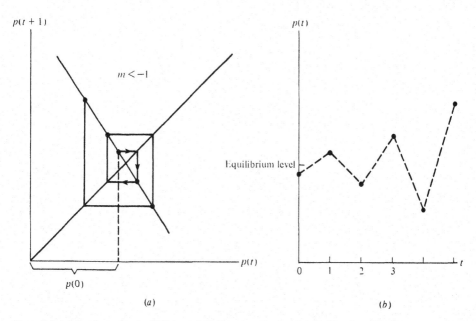

Figure 16 (*a*) A price cobweb. (*b*) Prices plotted against time, as implied by the cobweb.

functions in Equations (25) and (26) with quadratic functions of price. Then one arrives at a difference equation of the following sort in place of Equation (27):

$$p(t + 1) = ap(t)^2 + bp(t) + c \tag{29}$$

This difference equation is a good deal like the logistic equation of Section 2D, and so its stability analysis is also similar. We take up the stability analysis of such equations in the next section.

EXERCISES

● **1** (a) If $0 < m < 1$ and $p(0) < k/(1 - m)$, prove that $p(0) < p(1) < \cdots < k/(1 - m)$.
　(b) If $0 < m < 1$ and $p(0) > k/(1 - m)$, prove that $p(0) > p(1) > \cdots > k/(1 - m)$.
Try these problems two ways: geometrically by cobwebbing and using formula (28).

2 Another way to prove Theorem 6 is simply to substitute formula (28) into the difference equation and see if it satisfies it. Carry this out.

● **3** Why is the model of this section not so applicable to manufactured goods? In particular, how can pumpernickel production increase overnight, whereas wheat production cannot? Isn't this a paradox since pumpernickel is made of wheat?

4 (a) If $a = 400$ in Equation (25), what does a \$1 increase in price do to the supply at $t + 1$? (This change in supply is called the *elasticity of supply*.) What does a \$1 decrease do?
　(b) If $c = 600$ in Equation (26), what does a \$1 increase in price do to the demand? (This is called the *elasticity of demand*.) What does a \$1 decrease do?

● **5** Can you rephrase the corollary of Theorem 6, using the economic concepts of elasticity of supply and demand (see Exercise 4)?

6 If $m = 1$ in Equation (27), carry out a graphical iteration; i.e., draw the cobweb. You may use any convenient values for k and $p(0)$.

● **7** Suppose a farmer decides about raising hogs at time t on the basis of the current price and the price at $t - 1$. Formulate a model based on this assumption. Can you get an equation like Equation (27)?

8 Figure 11b differs from Figure 16b in that it seems to have an upward tilt. Can you modify Equation (27) so that it can produce a solution with an upward tilt?

● **9** In Equation (25) would b be positive, negative, or zero in a realistic example? What about e in Equation (26)?

10 The corollary to the theorem in this section makes no mention of what happens if $m = -1$. Can you carry out the analysis? Is the equilibrium stable?

Computer exercises

11 Program a computer to calculate $p(1)$, $p(2)$, ..., using $m = -\frac{1}{2}$, $k = 600$, and $p(0) = 432$ as in Example 12, but with this change in the model: $p(t + 1) = -\frac{1}{2}p(t) + 600 + R(t)$, where $R(t)$ has the following properties. Except when t is divisible by 3, $R(t) = 0$. When t is divisible by 3, $R(t)$ has a random value, determined by the following probability rule: $R(t)$ can be either 0, +20, or −20, each with probability $\frac{1}{3}$. This is meant to simulate random weather influences.

12 Suppose the price equation is

$$p(t + 1) = \begin{cases} m_1[p(t) - 50] + 70 & \text{if } p(t) \leq 50 \\ m_2[p(t) - 50] + 70 & \text{if } p(t) \geq 50 \end{cases}$$

so that the graph has a bend in it at (50, 70) but is straight on each side of the bend. If $m_1 = -\frac{1}{2}$ and $m_2 = -2$, what can you say about the sequence $p(0), p(1), \ldots$? Write a computer program to calculate 10 terms of this sequence for various initial values $p(0)$. Can you make any generalizations? Now suppose $m_1 = -2$ but $m_2 = -\frac{1}{2}$. Illustrate your calculations with graphical-iteration diagrams.

BIBLIOGRAPHY

Beach, E. F.: "Economic Models," John Wiley & Sons, Inc., New York, 1957.

Ezekiel, M.,: The Cobweb Theorem, *Quar. J. Eco.*, vol. 52, pp. 255–280, February 1938.

Meadows, Dennis L.: "Dynamics of Commodity Production Cycles," Wright-Allen Press, Inc., Cambridge, Mass., 1970.

Salert, Barbara: Public Support for Presidents, UMAP Module 299–300, available from COMAP, 271 Lincoln St., Lexington, MA 02173. The ebb and flow of public opinion is described with linear difference equations.

Sherbert, Donald R.: Difference Equations with Applications, UMAP Module 322, available from COMAP, 271 Lincoln St., Lexington, MA 02173. Repeats some of the material of this section, but with additional nice examples, and then goes on to second-order equations.

2F STABILITY OF SOME NONLINEAR DIFFERENCE EQUATIONS—LOGISTIC POPULATION GROWTH

Abstract We study the stability of a category of difference equations which includes the logistic equation introduced in Section 2C. The stability behavior is shown to depend on the shape (and not the exact equation) of the iteration curve. In the case of the logistic, this shape is governed by the size of r. The theory suggests that a high growth rate may make it hard for a species to stay near a constant (equilibrium) population size.

Prerequisites The concept of stability, as in Section 2D, and cobweb analysis, as in Section 2E. Section 2C is useful motivation, but not mathematically necessary. Differential calculus.

In this section, we study the stability of the logistic difference equation

$$y(t + 1) = \left[r + 1 - \frac{ry(t)}{K}\right]y(t) \tag{30}$$

and other related difference equations. We can simplify the notation in Equation (30) by setting

$$x(t) = \frac{ry(t)}{(r + 1)K}$$

$$b = r + 1$$

Then Equation (30) becomes

$$x(t + 1) = b[1 - x(t)]x(t) \tag{31}$$

where $b > 1$ since r is a positive growth rate.

Our approach is to analyze the iteration function, which is $f(x) = b(1 - x)x$. The graph of this function is an upside-down parabola. See Figure 17 for examples.

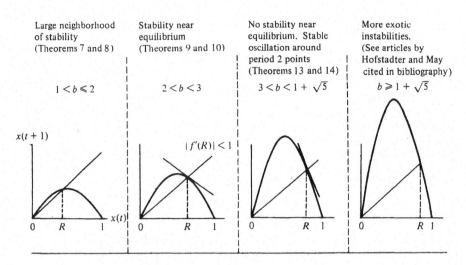

Figure 17 Stability behavior for $f(x) = b(1 - x)x$.

It is not hard to see that, no matter what the value of b is ($b > 1$), the parabola crosses the x axis at $x = 0$ and $x = 1$. As in Section 2E, we can find the equilibrium levels for Equation (31) by seeing where the iteration curve crosses the equilibrium line. If we call the equilibrium level R, we must have

$$f(R) = R$$

$$b(1 - R)R = R$$

$$b(1 - R) = 1$$

$$R = \frac{b - 1}{b}$$

(This is the equilibrium for x. When we undo the change of variables, we obtain K as the equilibrium for y.)

A routine calculus exercise shows that the parabola has its maximum at $x = \frac{1}{2}$ and the value of the maximum is

$$f\left(\frac{1}{2}\right) = b\left(1 - \frac{1}{2}\right)\left(\frac{1}{2}\right) = \frac{b}{4}$$

This makes clear the role b plays. It is a parameter that "tunes" the height of the iteration curve. As we shall see, it also determines the stability behavior of the difference equation. This is the main theme of this section, and the results are displayed in Figure 17. Roughly speaking, small values of b ("short" parabolas) guarantee stability for the equilibrium R, but large values ("tall" parabolas) do not.

The results of this section are not restricted to logistic Equation (31). They apply to any difference equation whose iteration curve has a shape which is not too different from an upside-down parabola. This is extremely important for the

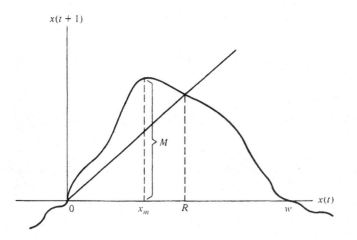

Figure 18 A "nearly logistic" iteration function.

relevance of the ideas in this section. There are many biological populations whose iteration curves are nearly parabolas, but probably few or none with exact parabolas.

Our results will apply to any difference equation of the form

$$x(t + 1) = f(x(t)) \tag{32}$$

where f is any continuous function with the following resemblances to the logistic function $b(1 - x)x$.

1. There is an interval of values $[0, w]$ where $f(x) \geq 0, f(w) = 0$, and $f(0) = 0$.
2. f achieves its maximum value M at a single positive value x_m. For $x \leq x_m$, f is monotone increasing. For $x_m \leq x, f$ is monotone decreasing.
3. There is a single positive equilibrium value R, where $R < w$ and $f(R) = R$.
4. For $0 < x < R, f(x) > x$; i.e., the graph lies above the equilibrium line for values below the equilibrium.

It is not hard to show that, when $b > 1$, logistic Equation (31) has these characteristics. Figure 18 shows a function which is not a logistic but which has these four characteristics.

We shall present our theorems in pairs: first the special case of the logistic, followed by the generalization to the wider class of "nearly logistic" iteration functions, defined by properties 1 through 4. In each case it is simplest to present the proof of the general theorem and derive the logistic case as a corollary.

Theorem 7 (Logistic Version)

If $1 < b \leq 2$ in logistic Equation (31), then the interval $(0, 1)$ is a neighborhood of stability for the equilibrium $R = (b - 1)/b$.

The significance of the condition $1 < b \leq 2$ in Theorem 7 is that this implies that $R = (b - 1)/b \leq \frac{1}{2} = x_m$. This means that the equilibrium is on the rising part of the iteration curve. This is what makes Theorem 7 follow from Theorem 8.

Theorem 8 (General Version)

Let f be the iteration function for a difference equation, as in Equation (32). Suppose conditions 1 through 4 hold. If $R \leq x_m$, then the interval $(0, w)$ is a neighborhood of stability for the equilibrium R.

Informal proof of Theorem 8 Beginning with an initial value x_0, we carry out a graphical iteration, as described in Section 2E. As we can see from Figure 19,

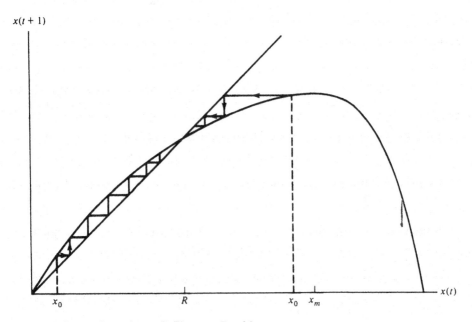

$x(t + 1)$

$x(t)$

$x_0 \qquad R \qquad x_0 \quad x_m$

Figure 19 Converging staircases in Theorems 7 and 8.

if $x_0 < R$, this iteration creates an upward-trending "staircase" that clearly converges (because f is continuous) to the equilibrium point (R, R). If $R < x_0 \leq x_m$, it is clear that graphical iteration produces a downward-trending staircase that converges to (R, R). Finally if $x_0 > x_m$, then after one iteration we find ourselves at a point on the iteration curve where an upward- or downward-trending staircase begins.

The graphical iteration used in Theorem 8 is best understood in comparison with some examples where the equilibrium of the iteration function is not on the rising part of the curve. Example 13 shows such a case.

Example 13

Figure 20 shows the graph of the iteration function for a logistic difference equation where $b = 2.9$ and $R = 0.66$. We start with an initial value of $x(0) = 0.09$. As we generate the points $P_1, P_2, \ldots$, we appear initially to get a staircase, as in Figure 19. But instead of continuing as a staircase, our staircase turns into something which looks like a cobweb that spirals in toward the equilibrium. It looks like we have stability.

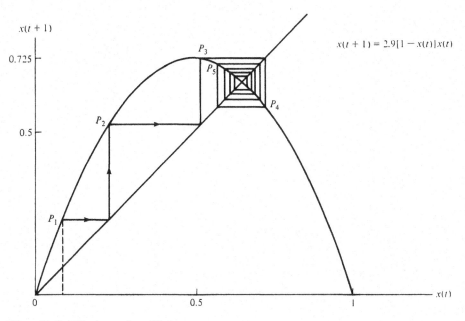

Figure 20 Spiraling toward equilibrium.

Our geometric intuition tells us that as long as the peak of the iteration curve lies above and to the left of the equilibrium, we will get a spiraling cobweb. This is true. But there is more that needs to be said: the spiraling may be in towards the equilibrium, as in Figure 20, or out away from it, as in Figure 21. In Figure 21 we have a logistic with $m = 4$ and $R = 0.75$. We start with $x(0) = 0.74$ and observe that the points $P_1, P_2, \ldots, P_6$ spiral away from equilibrium. As we continue with P_7, P_8, and P_9 a staircase temporarily appears. But if we were to continue beyond P_9, we would apparently spiral away from equilibrium. It does not appear as if the equilibrium is stable.

What accounts for the difference between Figure 20 and Figure 21? In the case of the logistic, it seems to be a matter of the size of b, as Theorem 9 will show. However, there is a deeper reason, involving the derivative of the iteration function, and this is displayed by Theorem 10.

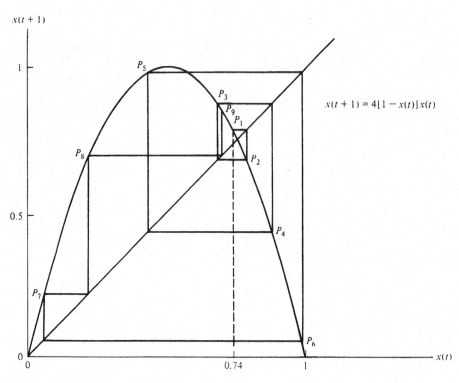

Figure 21 Spiraling away from equilibrium.

Theorem 9 (Logistic Version)

If $2 < b < 3$ in logistic Equation (31), then there is some neighborhood of stability for the equilibrium $R = (b - 1)/b$.

Theorem 10 (General Version)

Let f be the iteration function for a difference equation, as in Equation (32). Suppose conditions 1 through 4 hold. If, in addition, f and f' are continuous and $|f'(R)| < 1$, then there is a stable neighborhood of R.

Informal proof of Theorem 10 Instead of giving a formal proof involving epsilons and deltas, we shall appeal to intuition with the following argument. The tangent line to a curve at a point is a close approximation to the curve near the point of tangency. Therefore, when constructing a cobweb with an iteration function f, if we are close enough to the equilibrium R, then it is almost the same to use the tangent at R instead of the iteration function itself. In other words, Figure 22a is replaced by the approximately equivalent Figure 22b.

When our iteration function is a line, as in Figure 22b, the fate of a cobweb is easy to work out. In fact we did it in Section 2E of this chapter. We discovered that, if the slope of the line is < 1 in absolute value, every cobweb converges to the

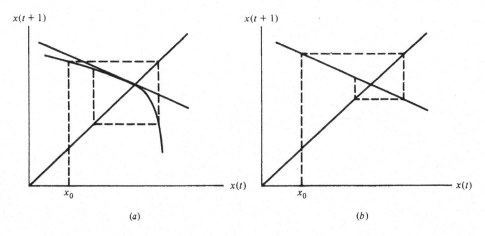

(a) (b)

Figure 22 Approximating an iteration curve by a tangent line.

equilibrium. In the present case, the slope of our tangent line is $f'(R)$, and so $|f'(R)| < 1$ implies convergence.

All this only holds when we are close enough to the equilibrium that the tangent line is a good enough approximation to the iteration function. Unfortunately, the theorem gives no information about how close we have to be (i.e., how big the stable neighborhood is).

Proof of Theorem 9 For the logistic, $f(x) = b(1 - x)x$; so $f'(x) = b - 2bx$. At equilibrium $f'((b - 1)/b) = b - 2(b - 1) = 2 - b$. Thus, if $2 < b < 3$, we have $-1 < 2 - b < 0$, and so $|f'((b - 1)/b)| = |2 - b| < 1$. Therefore, Theorem 10 applies.

For an illustration of Theorem 9, see Figure 20 in which the iteration function is the logistic with $b = 2.9$. Notice that Theorems 9 and 10 do not tell us anything about the size of the neighborhood of stability. Compare this with Theorems 7 and 8.

When $b > 3$, as in Figure 21, the previous theorems do not guarantee any neighborhood of stability around the equilibrium. However a different and fascinating phenomenon makes its appearance: a pair of points of period 2.

Definition

u and v are said to be a *pair of points* (or values) *of period 2* for the iteration function f if $f(u) = v$ and $f(v) = u$ (see Figure 23).

The algebraic significance of a pair of points of period 2 is that, if at time t we have $x(t) = v$, then the time series starting at t: $x(t)$, $x(t + 1)$, $x(t + 2), \ldots,$ is merely the alternating series $v, u, v, u, \ldots$.

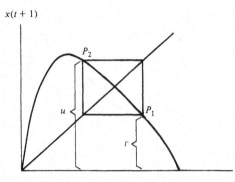

$x(t+1)$

$x(t)$ **Figure 23** Points of period two.

Geometrically this means that, when we start a graphical iteration whose first point on the iteration-function graph has height v (P_1 in Figure 23), then the cobweb comes back to this point after an intermediate stop on the iteration-function graph at a point P_2 whose height is u. If the cobweb is spun out further, it repeatedly traces out the rectangle shown in Figure 23.

Definition

f^2 is the function defined by $f^2(x) = f(f(x))$.

Example 14

If $f(x) = b(1 - x)x$, then

$$f^2(x) = b[1 - b(1 - x)x][b(1 - x)x]$$

f^2 plays an important role in the next theorems because, if u and v are a pair of points of period 2, then $f^2(v) = f(f(v)) = f(u) = v$; so v is an equilibrium level for the iteration function f^2. Likewise, u is an equilibrium level for f^2.

Theorem 11 (Logistic Version)

If $b > 3$ in logistic Equation (31), then there exists a pair of points u and v of period 2. Furthermore, the equilibrium $R = (b - 1)/b$ lies between them.

The proof of Theorem 11 is an easy deduction from the following more general theorem.

Theorem 12 (General Version)

Let f be the iteration function for a difference equation, as in Equation (32). Suppose conditions 1 through 4 hold. If, in addition, $|f'(R)| > 1$, then there are points u and v of period 2. Also, R lies between u and v.

Proof of Theorem 12 To find period-2 points for f, we need to find equilibria of f^2. Graphically, this means points where the graph of f^2 meets the equilibrium line.

Since $f(R) = R$, $f^2(R) = R$ also. Furthermore, if we differentiate f^2 using the chain rule,

$$\frac{df^2}{dx}(R) = f'(f(R))f'(R)$$

$$= [f'(R)]^2 > 1$$

Therefore, the graph of f^2 crosses the equilibrium line from below as in Figure 24a, not from above as in Figure 24b.

Since $f(w) = 0$ and $f(0) = 0$, $f^2(w) = 0$. This means $(w, 0)$ lies on the curve. In order to pass through this point, the curve must turn down for another crossing of the equilibrium line after the crossing at (R, R). Call this additional crossing point (u, u).

Because $u > R$ and f has only the single positive equilibrium R, $f(u) \neq u$. Also $f(f(u)) = f^2(u) = u$. Thus u and $f(u)$ are a pair of points of period 2 because they are distinct and f maps each into the other.

Since f is monotonically decreasing for $x > R$, $u > R$ implies that $f(u) < R$. Thus R is between u and $f(u)$.

Proof of Theorem 11 We have already observed that the iteration function f for the logistic satisfies conditions 1 through 4. By direct computation, $f'(R) = 2 - b$ (see the proof of Theorem 9). Therefore, when $b > 3$, $f'(R) < -1$ and the remaining condition of Theorem 12 is satisfied.

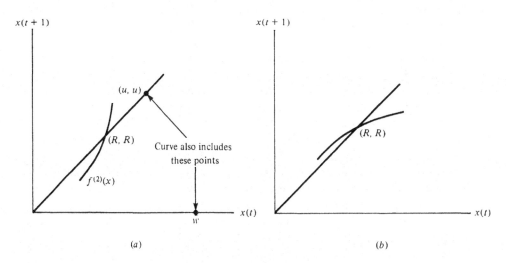

Figure 24 (a) Graph of f^2 crosses equilibrium line like this. (b) Graph of f^2 does not cross equilibrium line like this.

The existence of a pair of period-2 points does not, by itself, exclude a neighborhood of stability for the equilibrium R, although it does mean that such a neighborhood cannot include the period-2 points. But for some of the functions of the sort we are studying, the period-2 points have a kind of stability of their own, which rules out stability for the equilibrium R, as we shall see.

Theorem 13 (Logistic Version)

If $3 < b < 1 + \sqrt{5}$ in logistic Equation (31), then there is a neighborhood (v, u) of the equilibrium R with these properties:

1. If $x(0) \in (v, R)$, then $x(0), x(2), x(4), \ldots \to v$.
2. If $x(0) \in (R, u)$, then $x(0), x(2), x(4), \ldots \to u$.
3. v and u are period-2 points.

Theorem 14 (General Version)

If f satisfies the conditions of Theorem 12 and, in addition, $x_m < f(M)$, then there is a neighborhood (v, u) of R with these properties:

1. If $x(0) \in (y, R)$, then $x(0), x(2), x(4), \ldots \to v$.
2. If $x(0) \in (R, u)$, then $x(0), x(2), x(4), \ldots \to u$.
3. v and u are period-2 points.

Here is an informal outline of a proof.

Proof outline
1. The points u and v we are looking for would be equilibria of f^2. Thus, we must study the intersections of $y = f^2(x)$ and $y = x$. We will show that there are two intersections (v, v) and (u, u), separated by (R, R). See Figure 26.
2. The graph of $f^2(x)$ has a camel-backed shape, with two maxima and a minimum at $(x_m, f(M))$. [We deduce this by studying the derivative of $f^2(x)$, which is $f'(f(x))f'(x)$.] Since $x_m < f(M)$, this minimum lies above the line $y = x$.
3. Using the monotonicity properties of f, we can show that $x_m < f(M)$ also implies that the second maximum at (x_2, M) lies below the line $y = x$.
4. The condition $|f'(R)| > 1$ implies that the crossing at (R, R) is as shown in Figure 26 and not as in Figure 25. In order for the curve to connect up properly, additional crossings must exist, say at (u, u) and (v, v). Figure 26 shows this in the case of the logistic. In the more general case there could be numerous crossings of $y = f^2(x)$ and $y = x$ between x_m and R and numerous crossings between R and x_2. In that case (u, u) is chosen as the closest crossing to the left of (R, R) and (v, v) is the closest crossing to the right of (R, R).
5. Clearly any cobweb starting between (v, v) and (R, R) converges to (v, v). This proves condition 1. Any cobweb starting between (u, u) and (R, R) converges to (u, u), and this proves condition 2. Condition 3 holds because (v, v)

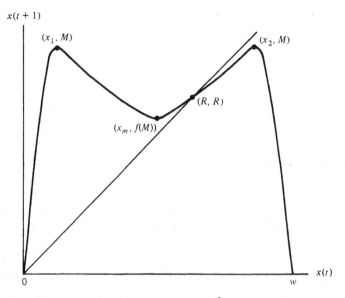

Figure 25 A plausible but incorrect picture of f^2 in Theorem 14.

and (u, u) are on the graph of $y = x$ and $y = f^2(x)$. This makes them equilibria for f^2 and hence period-2 points for f.

Proof of Theorem 13 In the proof of Theorem 11 we showed that $b > 3$ implies that all the conditions of Theorem 14 hold, except $x_m < f(M)$. Thus we need

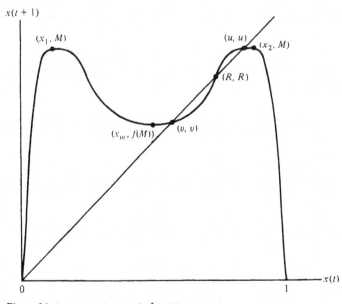

Figure 26 A correct picture of f^2 in Theorem 14.

show only that $b < 1 + \sqrt{5}$ implies $x_m < f(M)$. We have calculated earlier that for logistic Equation (31) $M = b/4$. Therefore, $f(M) = b(1 - b/4)(b/4) = b^2(4 - b)/16$. The condition $x_m < f(M)$ becomes

$$\frac{1}{2} < \frac{b^2(4 - b)}{16}$$

or
$$8 < 4b^2 - b^3$$

or
$$b^3 - 4b^2 + 8 < 0$$

Simple algebra shows this to be equivalent to

$$(b - 1)^2(b - 2) - 5(b - 2) < 0$$

Since $b > 3$, we have $b - 2 > 0$, and so we may divide by $b - 2$ without changing the sense of the inequality. Doing this shows that $x_m < f(M)$ is equivalent to

$$(b - 1)^2 < 5$$

The hypothesis $b < 1 + \sqrt{5}$ implies this inequality and thus implies $x_m < f(M)$. Thus Theorem 14 applies and proves that conditions 1, 2, and 3 hold.

These results have a somewhat surprising consequence for populations that grow according to a logistic difference equation, Equation (30), or the equivalent version, Equation (31). Recall that in these equations r and b are growth parameters. If we had two populations with the same equilibrium R and both are bumped away from the equilibrium by the same amount, the one with the higher b (equivalently, higher r) will bounce back more vigorously (perhaps even overshooting the equilibrium in the other direction). Theorem 7 shows that, if there is not too much of this vigor ($b \leq 2$), the equilibrium will be stable. But if b is larger, as in Theorem 13, this greater vigor will make the equilibrium unstable. In such a case the population would only rarely be found near its equilibrium level. Therefore, the equilibrium would be just a mathematical curiosity with little biological relevance.

EXERCISES

1 Show that the graph of the iteration function for $x(t + 1) = b[1 - x(t)]x(t)$ satisfies condition 4 when $b > 1$.

2 Consider the difference equation

$$x(t + 1) = f(x(t)) = ax(t)^2 + bx(t)$$

where $a < 0$ and $b > 0$. Show that:
    (a) $f(0) = 0$, and there is a $w > 0$ such that $f(w) = 0$.
    (b) f has a relative maximum at $x_m = -b/a$ and $x_m > 0$. Furthermore, between 0 and x_m, f is monotonically increasing, while between x_m and w, f is monotonically decreasing.
    (c) There is a unique positive equilibrium value R, such that $f(R) = R$.
    (d) Between 0 and R the graph of f lies above the equilibrium line [i.e., $f(x) > x$].

● 3 (a) Explain why Theorem 10 gives no useful information about whether 0 is a stable equilibrium for $x(t + 1) = \sin x(t)$.

(b) Choose a value at random for $x(0)$. Enter it on a pocket calculator and then compute $x(1)$, $x(2), \ldots$, by repeatedly punching the "sin" button. Do the values approach 0? Can you explain this by graphical iteration? (*Hint*: In drawing the graph, pay special attention to the relation between the iteration function and the equilibrium line.)

● 4 (a) Show that $x(t + 1) = 2 + [x(t) - 2]^3$ has an equilibrium at 2, which has a neighborhood of stability.

(b) By considering the initial condition $x(0) = 3$, show that the stable neighborhood cannot include the value 3.

(c) Determine all equilibrium values for this difference equation.

5 Are there any b values (possibly negative) for which Theorem 10 shows that the equilibrium 0 has a neighborhood of stability in the logistic difference equation?

6 (a) If f is the iteration function for $x(t + 1) = b[1 - x(t)]x(t)$, show that the sufficient condition for x to have period 2, $f^2(x) = f(f(x)) = x$, implies that

$$b^3x^4 - 2b^3x^3 + b^2(b + 1)x^2 + (1 - b^2)x = 0$$

(b) Since $f(0) = 0$ and $f(R) = R$, it follows that $x = 0$ and $x = (b - 1)/b$ should be roots of the equation in part (a). Verify this.

(c) Since 0 and $(b - 1)/b$ are roots, x and $x - (b - 1)/b$ should be factors of the polynomial in part (a). Factor them out. The roots of the remaining quadratic are period-2 points u and v. Find formulas for them.

(d) Show that $f(v) = u$ and $f(u) = v$ by actual calculation with the polynomial formula for f.

● 7 (a) If u and v are points of period 2 for the equation $x(t + 1) = f(x(t))$, they are said to be stable provided they are stable equilibria for the equation $u(t + 1) = f^2(u(t))$. Use Theorem 10 to show that $|f'(u)f'(v)| < 1$ implies that u and v are stable.

(b) Apply the result of part (a) to Equation (30) to show that, if $3 < b < 1 + \sqrt{5}$, then the period-2 points have a neighborhood of stability. {*Hint*: The period-2 points are $[b + 1 \pm \sqrt{(b + 1)(b - 3)}]/2b$.}

8 Let $x(t + 1) = c/x(t)$, where c is a constant.

(a) What is the equilibrium?

(b) Show that, except for the equilibrium, all values have period 2.

(c) Is the equilibrium stable?

(d) Are any of the period-2 points stable? (See Exercise 7a.)

9 In the logistic difference equation describing population growth, Equation (30), if we start with an initial condition $x(0) > (r + 1)K/r$, will $x(1)$ be positive or negative? What about $x(2)$ and $x(3)$? What biological principle does this suggest about large perturbations away from an equilibrium?

● 10 If our logistic equation, Equation (31), models population growth, then negative values of x are unrealistic. Show that, if $b \leq 4$, then, whenever $0 \leq x(t) \leq 1$, we also have $0 \leq x(t + 1) \leq 1$.

11 Show by example that the result in Exercise 10 is not true for $b > 4$.

Computer exercises

12 Try to estimate the extent of the stable neighborhood for the equilibrium in the example of Figure 20. Try different initial conditions $x(0)$, and calculate $x(1)$, $x(2)$, $\ldots$, $x(10)$ in each case.

13 In 1960 Ulam showed that, when $b = 4$ in Equation (31), the iterates of almost any initial condition $x(0)$ have a random character and are distributed on the real number line according to the probability density.

$$g(y) = \frac{1}{\pi \sqrt{y(1 - y)}}$$

This means that, if one computes a large number of iterates, $x(1), x(2), \ldots, x(n)$, then the fraction of them between y and $y + \Delta y$ is approximately $g(y)\Delta y$, when Δy is small. For large Δy the fraction is

$$\int_{y}^{y+\Delta y} g(y)\, dy$$

Verify this by computational experiments. Pick $x(0)$ at random between 0 and 1. Divide the interval $[0, 1]$ into N equal categories of size Δy (you choose N) and tabulate how many $x(t)$ fall into each category. Compare with the integral formula. Alternatively, if Δy is small enough, compare with $g(y_i)\Delta y$, where the y_i are the midpoints of the categories. Try this for various initial conditions and values of N.

BIBLIOGRAPHY

Frauenthal, James: Introduction to Population Modeling, UMAP Monograph, available from COMAP, 271 Lincoln St., Lexington, MA 02173. An outline of a large variety of population models.

Haberman, Richard: "Mathematical Models," Prentice-Hall, Englewood Cliffs, N.J., 1977. About a third of this book is a detailed exposition of population models, with examples and exercises.

Harmon, Kathryn Newcomer: The Diffusion of Innovation in Family Planning, UMAP Module 303, available from COMAP, 271 Lincoln St., Lexington, MA 02173. The fundamental equation in this module is of the type discussed in our Section 2F. The analysis is algebraic and avoids calculus.

Hofstadter, Douglas: Metamagical Themas, Sci. Amer., vol. 245, no. 5 p. 22, November, 1981. Excellent expository account of how the stability of the logistic difference equation depends on the growth parameter. Goes beyond our Section 2F into really exotic territory, but with no proofs.

May, Robert: Simple Mathematical Models with Very Complicated Dynamics, Nature, vol. 261, June 10, 1976. Similar to Hofstadter's article but with more mathematics and more biology.

Witten, Matthew: Fitness and Survival in Logistic Models, J. Theor. Bio., vol. 74, pp. 23–32, 1978.

3 DETERMINISTIC VERSUS STOCHASTIC MODELS— FORECASTING EPIDEMICS

Abstract Often scientific predictions or estimates have a misleading impression of certainty about them; for example, predictions that 27,432 people will be hospitalized with influenza in the United States next year. The reason is that manageable models are very often deterministic; that is, they leave out chance factors. In this section we compare deterministic models with stochastic ones, which are based on probabilities rather than definite rates. The advantages and disadvantages of each type of model are pointed out.

Prerequisites Calculus, a little about differential equations. The required probability is introduced in Section 3A.

3A AN OUTLINE OF PROBABILITY

The world around us is a rather uncertain place. Even our attempts to predict the weather often misfire, and in the social sciences it is a rare thing to be able to make definite predictions with confidence.

Scientists disagree about why this is so. One point of view is called *determinism.* This view holds that, if we had enough information and a complete understanding of the laws of nature, we could predict perfectly. The alternative view is that the laws of nature have a built-in uncertainty. This point of view has the advantage of having experimental results to support it. These results indicate that the behavior of an electron is, in principle, unpredictable. Its present position and velocity do not determine its position and velocity an instant from now.

Whether or not nature is inherently uncertain, it certainly appears uncertain, and so we need mathematical models to deal with this apparent uncertainty. Such models are based on the theory of probability.

The basic concept in probability theory is the *experiment.* Any process whose outcome is (in a practical sense) uncertain is called an experiment. Flipping a coin is an experiment, as is determining whether or not it will rain tomorrow. The set of all possible outcomes of an experiment is called the *sample space.*

Example 1

1. If we flip a coin once, the sample space is $\{H, T\}$, where $H =$ heads and $T =$ tails.
2. If we flip the coin twice, the sample space is $\{HH, HT, TH, TT\}$.
3. If we test seven people to determine how many have high blood pressure, the sample space is $\{0, 1, 2, \ldots, 7\}$.

It is possible for a sample space to be infinite, but we will not deal with such examples in this chapter.

The key idea of probability theory is that each outcome in the sample space has a number attached to it that measures its likelihood of occurring. If x denotes an outcome, $p(x)$ denotes the *probability* of x occurring. The assignment of probability must satisfy the following requirements. If the sample space is $\{x_1, x_2, \ldots, x_n\}$, then:

1. $0 \leq p(x_i) \leq 1$, for each i.
2. $p(x_1) + p(x_2) + \cdots + p(x_n) = 1$.

Intuitively, we interpret a probability of 0 as representing an impossibility, while 1 represents a certainty.

Assigning probabilities to outcomes is very important when we apply probability theory. Here is a brief outline of three methods which are used:

1. *The relative-frequency method.* If the experiment can be repeated many times, we tabulate the fraction of times each outcome comes up and use this as the probability. For example, if you shoot at a bull's-eye 100 times and hit it 17 times, then the probability of a hit is 0.17.
2. *The equal-probability method.* In certain cases it seems reasonable that all outcomes have the same probability. If there are n outcomes, then the fact that all the probabilities add to 1 implies that each outcome should be assigned the probability $1/n$. For example, the probability of rolling a 3 with a fair die is $\frac{1}{6}$ since the sample space is $\{1, 2, \ldots, 6\}$. This method is not reasonable for the bull's-eye example because the chances of hitting and missing are unlikely to be equal.
3. *Subjective-guess method.* If you want to know the probability that one of your friends will come to your party, neither of the preceding methods is applicable. But you might assign a probability as a guess. If you know the person well and understand the factors that might prevent your friend attending, your guess might be useful.

Any subset of a sample space is called an *event*. For example, $\{HH, TT\}$ is an event in the experiment of tossing a coin twice. This event could be described in words as "the result of the first flip was the same as the result of the second."

Definition

If E is an event, the *probability of that event* $p(E)$ is defined as the sum of the probabilities of the outcomes making up that event.

Example 2

What is the probability of getting a 5, 6, or 7 when drawing a card from a standard deck?

The sample space has 52 elements since a standard deck has 52 cards. The equiprobable assignment of probabilities seems reasonable; so each card has probability $\frac{1}{52}$ of being drawn. The event we are interested in has 12 members: four 5s, four 6s, and four 7s. Therefore, the probability of this event is $12(\frac{1}{52}) = 0.23$.

The applications of probability in this book deal mostly with multistage experiments, in which each stage is some simpler experiment.

Example 3

1. A coin is tossed 3 times, and we determine how many heads occur. In this example, the simple experiment is tossing a coin once. It has sample space $\{H, T\}$. The multistage experiment has sample space $\{HHH, HHT, HTH, THH, HTT, THT, TTH, TTT\}$.
2. Ten cards are chosen from a deck and the number of spades is recorded. The simple experiment is choosing one card and seeing if it is a spade.
3. John takes a blood-pressure reading each day for a week and records how many times it was above normal. Here the simple experiment is to take the blood pressure once.
4. Indiana and San Francisco State play a series of three soccer games to determine the national championship. Each game is a simple experiment.

A multistage experiment is often described by a kind of diagram called a *tree*. Figure 1a shows a tree for the experiment of flipping two fair coins. Point R, called the *root* of the tree, represents the state of affairs before either experiment is performed. The branching from R represents the first experiment, whose outcomes are H or T. The next level of branching represents the next experiment.

Of special interest to us are multistage experiments where the stages are said to be *independent*.*

---

* Conventionally one speaks of events being independent. Our definition could be put into that framework, but for our restricted purposes the more specialized definition seems easier.

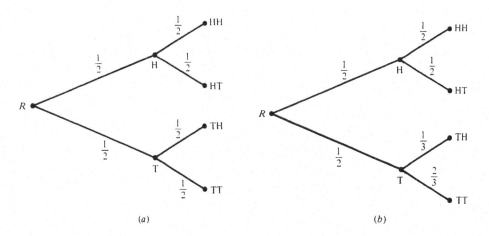

Figure 1 (*a*) Two stage experiment of tossing a fair coin twice. (*b*) Two stages of coin tossing where second coin may or may not be fair.

Definition

Stages of a multistage experiment are said to be independent if the probabilities for one stage of the experiment do not depend on the outcome of the previous stage.

Example 4

1. We flip two different coins, once each. We decide in advance that the probabilities of heads and tails are each $\frac{1}{2}$ for both coins. Suppose we get a head in the first toss. We still believe that $p(H) = p(T) = \frac{1}{2}$ for the second. In this case the coin tosses are independent. (How would you analyze the situation if the same coin were being tossed twice?)

2. Indiana and San Francisco State have never played; so just before a two-game series you decide that the probability of Indiana winning the first game is $\frac{1}{2}$. If it wins the first game, you think that then Indiana will have a probability of $\frac{3}{4}$ of winning the second game. But you also decide that, if Indiana loses the first game, it only has a $\frac{1}{3}$ probability of winning the second. In this example, the stages are not independent (see Figure 3).

3. Hereditary characteristics of a human being are determined by the type of sperm and the type of egg that combine to create the individual. In the simplest case, there is only one "word" of the genetic code involved and it comes in two varieties (biologists call them genes or alleles) A and B. The sperm and egg each contribute either an A or a B gene. The characteristic of the organism is determined by which combination it gets: AA, AB, BA, or BB. Determining the type of an organism can be thought of as a two-stage experiment: first a sperm is chosen with either gene A or gene B; then an egg is chosen with either gene A or gene B. These two stages appear to be independent. However, the probabilities that apply to the sperm selection may be different from those for the egg selection.

Independence can be visualized in terms of the tree diagram. In the case of flipping two coins (Figure 1a), it means that the two second-level branchings

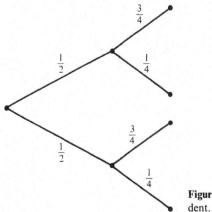

Figure 2 Two-stage experiment where stages are independent.

(*H* branching to *HH* or *HT*, and *T* branching to *TH* or *TT*) are labeled the same way with probabilities. If the labels had been as in Figure 1*b*, we would not have independence because the $(\frac{1}{2}, \frac{1}{2})$ distribution on the upper branching differs from the $(\frac{1}{3}, \frac{2}{3})$ distribution on the branching below it. Figure 1*b* might be appropriate for an experiment where we use a fair coin for the first toss, and, if (and only if) it comes up tails, we substitute an unfair coin for the second toss. In this case, we call the probabilities for *H* or *T* on the second toss *conditional probabilities* because they depend on what happened on the first toss. Example 7 gives another illustration of conditional probability; Figure 2 gives another illustration of independence.

Multiplication Rule

Let E_1 and E_2 be experiments that are performed separately as stages of an experiment *E*. Suppose the stages E_1 and E_2 are independent. If *A* is an event in the sample space of E_1, while *B* is an event in the sample space of E_2, then

$$p(A \text{ occurs in } E_1 \text{ and } B \text{ occurs in } E_2) = p(A \text{ occurs in } E_1)p(B \text{ occurs in } E_2)$$

Here is an example to justify the multiplication rule. Suppose we perform the two-stage experiment consisting of E_1 followed by E_2 1000 times (e.g., if $E_1 = E_2 = $ a coin flip, we do a total of 2000 flips). Relying on the relative-frequency interpretation of probability, we can assert that the number of times we get *A* in experiment E_1 (e.g., heads on first flip) is $1000\, p(A)$. How many *of these* are followed by a *B* in E_2 (e.g., tails on second flip)? The independence condition says that our answer should be unaffected by knowing that the first-stage outcome was *A*. In other words, we may just as well ask: What fraction of times do we get *B* when we do E_2? Once again, the relative-frequency interpretation suggests the answer is $p(B)$. Thus the total number of occurrences of *A* followed by *B* is $1000\, p(A)p(B)$. The fraction of all 1000 trials resulting in *A* followed by *B* is $p(A)p(B)$. According to the relative-frequency interpretation of probability, this is a good estimate of $p(A \text{ in } E_1 \text{ and } B \text{ in } E_2)$.

Example 5

Two cards are drawn from a deck in an ESP experiment, and a student is asked to guess the suits of the cards. On each experiment the probability of success *S* is $\frac{1}{4}$ and the probability of failure *F* is $\frac{3}{4}$, provided there is no ESP at work. The outcomes of the two-stage experiment are *SS*, *SF*, *FS*, and *FF*. Assuming independence, the probabilities are calculated by the multiplication rule as follows:

$$p(SS) = (\tfrac{1}{4})(\tfrac{1}{4}) = \tfrac{1}{16}$$

$$p(SF) = (\tfrac{1}{4})(\tfrac{3}{4}) = \tfrac{3}{16}$$

$$p(FS) = (\tfrac{3}{4})(\tfrac{1}{4}) = \tfrac{3}{16}$$

$$p(FF) = (\tfrac{3}{4})(\tfrac{3}{4}) = \tfrac{9}{16}$$

The multiplication rule can be extended to a series of more than two experiments. If we have experiments $E_1, E_2, \ldots, E_n$ and the probability of event A_i in experiment E_i is $p(A_i)$, then the probability A_1 occurs in E_1 *and* A_2 occurs in E_2 *and* $\cdots$ *and* A_n occurs in E_n is $p(A_1)p(A_2) \cdots p(A_n)$.

Example 6

Three voters selected at random are asked whether they voted for Carter C or Reagan R in the 1980 Presidential election. Since Reagan won 60 percent of the popular vote in the state where these voters live, it seems reasonable that $p(R) = 0.6$ in each of the three experiments. The probabilities of various combinations are computed by the multiplication rule like this:

$$p(RRR) = (0.6)^3 = 0.216 \qquad p(RCC) = (0.6)(0.4)^2 = 0.096$$
$$p(RRC) = (0.6)^2(0.4) = 0.144 \qquad p(CRC) = (0.4)(0.6)(0.4) = 0.096$$
$$p(RCR) = (0.6)(0.4)(0.6) = 0.144 \qquad p(CCR) = (0.4)^2(0.6) = 0.096$$
$$p(CRR) = (0.4)(0.6)(0.6) = 0.144 \qquad p(CCC) = (0.4)^3 = 0.064$$

The event "one of the three voters voted for Carter" is $\{RRC, RCR, CRR\}$ and has probability $0.144 + 0.144 + 0.144 = 0.432$.

There is also a multiplication rule, called Bayes' law, for multistage experiments where the stages are not independent. For example, in Figure 1b the probability of obtaining the outcome TT is $(\frac{1}{2})(\frac{2}{3}) = \frac{1}{3}$. Here is another example.

Example 7

What is the probability that Indiana wins both soccer games in the situation described in part 2 of Example 4? Figure 3 shows the tree diagram with the conditional probabilities. The desired probability, as well as the probabilities of all the other outcomes, are calculated at the right.

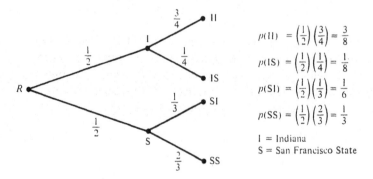

$$p(\text{II}) = \left(\frac{1}{2}\right)\left(\frac{3}{4}\right) = \frac{3}{8}$$

$$p(\text{IS}) = \left(\frac{1}{2}\right)\left(\frac{1}{4}\right) = \frac{1}{8}$$

$$p(\text{SI}) = \left(\frac{1}{2}\right)\left(\frac{1}{3}\right) = \frac{1}{6}$$

$$p(\text{SS}) = \left(\frac{1}{2}\right)\left(\frac{2}{3}\right) = \frac{1}{3}$$

I = Indiana
S = San Francisco State

Figure 3 Multiplication rule for conditional probabilities.

Example 8

Carbon 14 is a variety (isotope) of carbon which, over time, undergoes radioactive decay and becomes transformed into carbon 12. The decay is probabilistic: if we monitor a given carbon-14 atom over any interval of time, there is no way we can tell in advance whether or not it will decay. We consider a two-stage experiment: the first stage is to see whether a decay occurs in time interval $[0, t]$, and the second is to see whether the atom decays in the interval $[t, t + \Delta t]$. Figure 4 shows the tree diagram. (Note that, if a decay occurs in the first interval, then there can't be another in the next because carbon 12 doesn't decay.)

The fundamental assumption of the theory is that a carbon-14 atom has no memory: the probability of no decay in some time interval depends only on the length of that interval and not on the length of the previous interval of the carbon-14 atom's existence. (There is no such thing as being "overdue" for decay.) Thus we can speak of the probability of no decay in a time period of length x as being a function of x, say $F(x)$. Since we assume our atom is not decayed to start with, $F(0) = 1$. The multiplication rule applied to the top branch (N followed by N) of the tree gives

$$F(t + \Delta t) = F(t)F(\Delta t)$$

Therefore

$$\frac{F(t + \Delta t) - F(t)}{\Delta t} = - F(t)\frac{1 - F(\Delta t)}{\Delta t}$$

$$= -F(t)\frac{F(0) - F(0 + \Delta t)}{\Delta t}$$

Assuming that F is differentiable and letting $\Delta t \to 0$ gives

$$\frac{dF}{dt} = - \lambda F(t)$$

where $\lambda = F'(0)$, a constant that needs to be found through experiment. The solution of this differential equation is

$$F(t) = F(0)e^{- \lambda t} = e^{- \lambda t}$$

This is the fundamental law of radioactive decay.

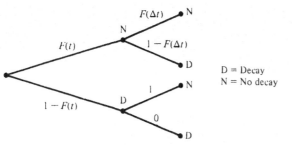

D = Decay
N = No decay

Figure 4 Radioactive decay.

If we have a sample of N carbon-14 atoms, then, unless N and t are quite small, we can deduce that the number of carbon-14 atoms remaining is $Ne^{-\lambda t}$. Exercise 11 of Section 3B of this chapter has more about this. See also Example 4 of Section 2 in Chapter 3.

When faced with uncertainties we often try to cut through the uncertainty by saying: "Yes, but what happens on the average?" The mathematical concept that lies behind this approach is the *mean* (also called the *expectation*). Here's the definition.

Definition

Suppose $S = \{x_1, x_2, \ldots, x_n\}$ is a sample space, with probabilities $p(x_i)$ assigned to the outcomes. Suppose also there are other numbers, representing some important quantity, denoted $r(x_i)$, associated with the outcomes. Then the *mean* (or expected value) of the quantity represented by the $r(x_i)$ is

$$\mu = p(x_1)r(x_1) + p(x_2)r(x_2) + \cdots + p(x_n)r(x_n)$$

The function r is called a *random variable*.

The intuitive interpretation of the mean is that, if the experiment was repeated many times, the long-term average of the random variable should equal the mean.

Example 9

What is the mean number of heads if a fair coin is flipped three times?

There are eight outcomes in the sample space, and they are listed below. The probabilities are calculated by the multiplication rule. The value of the random variable "number of heads" is in the third column.

$$\mu = 3(\tfrac{1}{8}) + 2(\tfrac{1}{8}) + 2(\tfrac{1}{8}) + 2(\tfrac{1}{8}) + 1(\tfrac{1}{8}) + 1(\tfrac{1}{8}) + 1(\tfrac{1}{8}) + 0(\tfrac{1}{8}) = 1.5$$

Outcome	Probability	Number of heads	Length of longest streak
HHH	$\tfrac{1}{8}$	3	3
HHT	$\tfrac{1}{8}$	2	2
HTH	$\tfrac{1}{8}$	2	1
THH	$\tfrac{1}{8}$	2	2
HTT	$\tfrac{1}{8}$	1	2
THT	$\tfrac{1}{8}$	1	1
TTH	$\tfrac{1}{8}$	1	2
TTT	$\tfrac{1}{8}$	0	3

A single experiment may have a number of random variables of interest to us. For example, when tossing three coins, we may be interested in the random variable "longest consecutive streak of identical outcomes." This is tabulated in column 4. The mean of this random variable is

$$\mu = 3(\tfrac{1}{8}) + 2(\tfrac{1}{8}) + 1(\tfrac{1}{8}) + 2(\tfrac{1}{8}) + 2(\tfrac{1}{8}) + 1(\tfrac{1}{8}) + 2(\tfrac{1}{8}) + 3(\tfrac{1}{8}) = 2$$

Example 10

Certain birth defects can be diagnosed before birth. This allows women the option of abortion. For example, neural-tube defects (anencephaly and spina bifida), which occur in one to two births per thousand, can be reliably diagnosed. However, there may be as many as five stages in the testing (as of 1980). Of all women, 95 percent require only one test, at a cost of $20; 2 percent require an additional test, for a total cost of $40; etc. The table below gives full details.

Number of tests required	Probability	Total cost, $
1	0.95	20
2	0.02	40
3	0.015	120
4	0.013	200
5	0.002	400

In this example, the random variable that interests us is the cost of testing. Here is its mean:

$$\mu = (0.95)(20) + (0.02)(40) + (0.015)(120) + (0.013)(200) + (0.002)(400)$$

$$= 19.0 + 0.8 + 1.8 + 2.6 + 0.8$$

$$= 25.0$$

As we have observed, the mean is an average, and, in computing an average, extreme values can roughly cancel each other out if they are extreme in opposite directions. For example, if we wager $100 on a coin toss, the mean is $(\tfrac{1}{2})(100) + (\tfrac{1}{2})(-100) = 0$. But if we wager only $1, we get the same mean: $(\tfrac{1}{2})(1) + (\tfrac{1}{2})(-1) = 0$. From the point of view of the mean, these wagers are identical, but one is much riskier than the other. To try to capture this numerically, we calculate the standard deviation σ.

Definition

Let a random variable take on values $x_1, x_2, \ldots, x_n$ with probabilities $p_1, p_2, \ldots, p_n$. Let μ be the mean. Then the standard deviation is

$$\sigma = \sqrt{\sum_{1}^{n} p_i(x_i - \mu)^2}$$

Example 11

For the \$100 wager discussed above, the values of the random variable are 100 and -100, with probabilities $\frac{1}{2}$ and $\frac{1}{2}$. $\mu = 0$; so

$$\sigma = \sqrt{\tfrac{1}{2}(100 - 0)^2 + \tfrac{1}{2}(-100 - 0)^2}$$
$$= 100$$

For the \$1 wager,

$$\sigma = \sqrt{\tfrac{1}{2}(1 - 0)^2 + \tfrac{1}{2}(-1 - 0)^2}$$
$$= 1$$

The previous example shows that the riskiness of the wager is reflected in a high value of σ. In general, a low value of σ indicates that the values of the random variable are clustered near the mean. The larger σ is, the more deviation from the mean there is among the random-variable values.

Example 12

We toss a fair coin twice. What are μ and σ for the random variable "number of heads"?

The event "0 heads" is $\{TT\}$ and has probability $\frac{1}{4}$, the event "1 head" is $\{TH, HT\}$ and has probability $\frac{1}{2}$, etc. Therefore,

$$\mu = \tfrac{1}{4}(0) + \tfrac{1}{2}(1) + \tfrac{1}{4}(2)$$
$$= 1$$
$$\sigma = \sqrt{\tfrac{1}{4}(0 - 1)^2 + \tfrac{1}{2}(1 - 1)^2 + \tfrac{1}{4}(2 - 1)^2}$$
$$= \sqrt{\tfrac{1}{2}}$$

A multistage experiment is called a *binomial* experiment if:

1. It consists of the same experiment repeated a number of times.
2. Each stage has two outcomes, arbitrarily called success S and failure F. The probability of success in each stage is denoted p, and the probability of failure is then $1 - p$.
3. The stages are all independent.

Example 13

Tossing a coin 10 times is a binomial experiment. Tossing a coin and then rolling a die is not.

In a binomial experiment one is usually interested in the random variable "number of successes out of N trials." The main questions of interest are:

1. What is the probability that there are exactly r successes in N trials?
2. What is the mean number of successes in N trials?
3. What is the standard deviation of the number of successes in N trials?

Example 14

Three babies are born on a certain day, and we check each one for hair color. Assume that light L has probability $\frac{1}{3}$ and dark D has probability $\frac{2}{3}$.

1. What is the probability that two of the babies are light-haired?
2. What is the mean number of light-haired babies?
3. What is the standard deviation of the number of light-haired babies?

First we list the outcomes of this three-stage experiment, along with their probabilities (computed by the multiplication rule).

LLL	$(\frac{1}{3})^3 = \frac{1}{27}$	LDD	$(\frac{1}{3})(\frac{2}{3})^2 = \frac{4}{27}$
LLD	$(\frac{1}{3})^2(\frac{2}{3}) = \frac{2}{27}$	DLD	$(\frac{2}{3})(\frac{1}{3})(\frac{2}{3}) = \frac{4}{27}$
LDL	$(\frac{1}{3})(\frac{2}{3})(\frac{1}{3}) = \frac{2}{27}$	DDL	$(\frac{2}{3})^2(\frac{1}{3}) = \frac{4}{27}$
DLL	$(\frac{2}{3})(\frac{1}{3})^2 = \frac{2}{27}$	DDD	$(\frac{2}{3})^3 = \frac{8}{27}$

To answer question 1, we list the outcomes that make up the event "2 out of 3 are light." They are: LLD, LDL, and DLL. The probability of this event is therefore $\frac{2}{27} + \frac{2}{27} + \frac{2}{27} = \frac{2}{9}$. In a similar way we could determine the probabilities of the other events listed below.

Event	Probability
0 light: {DDD}	$\frac{8}{27}$
1 light: {LDD, DLD, DDL}	$\frac{12}{27}$
2 light: {LLD, LDL, DLL}	$\frac{6}{27}$
3 light: {LLL}	$\frac{1}{27}$

To answer question 2, we compute

$$\mu = 0(\tfrac{8}{27}) + 1(\tfrac{12}{27}) + 2(\tfrac{6}{27}) + 3(\tfrac{1}{27})$$
$$= 1$$

For question 3,

$$\sigma = \sqrt{\tfrac{8}{27}(0-1)^2 + \tfrac{12}{27}(1-1)^2 + \tfrac{6}{27}(2-1)^2 + \tfrac{1}{27}(3-1)^2}$$
$$= \sqrt{\tfrac{18}{27}}$$

The arithmetic in the previous example is a little tedious, but it is very little compared to the work involved when the number of trials is much more than three. Fortunately, we need not follow the steps of that example, because there are formulas that give the answers directly.

Theorem 1

In a binomial experiment with N trials, where the probability of success on a single trial is p,

1. The probability of exactly r successes is $\binom{N}{r} p^r (1 - p)^{N-r}$.

2. The mean number of successes is $\mu = Np$.
3. The standard deviation of the number of successes is $\sigma = \sqrt{Np(1 - p)}$.

All of the computations of the preceding example could have been carried out with the formulas of this theorem. Try it.

Example 15

A drug is used on 1000 patients. If the probability of improvement is 0.85, the mean number of improved patients is $1000(0.85) = 850$. The standard deviation is $\sigma = \sqrt{1000(0.85)(0.15)} = 11.29$.

Example 16

During presidential election years, attention is often directed to "bellweather" or "weather-vane" counties. These are counties which have a long history of always voting for the winner in past presidential elections. For example, between 1896 and 1968, Laramie County, Wyoming; Palo Alto County, Iowa; and Crook County, Oregon all voted for the winner in each presidential election. What makes them such good indicators? Are these counties ideal cross sections of America? An alternative and perhaps more likely answer is that it is just luck. As an analogy, suppose 100 people toss a coin four times, "trying" to get heads. None of them have any more skill at getting heads than the others. But out of 100 people, some will appear to have skill because they get four heads just by luck. Now here is an analysis for the voting situation.

In the 19 elections between 1896 and 1968, the number of times any given county voted for the eventual winner was about 70 percent of all possible cases. So let's suppose that each county has the same 0.7 chance of voting for the winner in each election. (In other words, no county is better than any other at picking winners.) Assuming that each election is independent of the others, then the probability of one county being right 19 times is $(0.7)^{19} \approx 0.001$. Here we are applying Theorem 1 to the 19-stage experiment of checking one county's record in 19 different elections. Now we consider a new experiment in which we check each of the 3000 counties in the United States to see if they did or did not vote for the winner all 19 times. Success in a single trial of this experiment has probability 0.001; so the mean number of counties who were right 19 times is $3000(0.001) = 3$. This was exactly what was observed. Therefore it is entirely possible that the observed results were due to chance.

A histogram for a random variable is a graphical representation of the probabilities of the various values the random variable can take on. When there are finitely many values (as in all the examples in this chapter), it is constructed as follows:

1. We plot the values of the random variable on a horizontal axis.
2. Above each value we erect a rectangle whose area equals the probability of that value.

Figure 5 shows a histogram for the random variable "number of light-haired babies," described in Example 14.

For binomial experiments with large values of N, the histogram will have many bars, and we sometimes indicate its shape with a smooth curve that approximates it. For binomial experiments these histograms always have the characteristic mound shape of Figure 6: a single peak in the middle and a trailing off toward 0 to the right and left of the peak. The mean always falls nearly (sometimes exactly) under the peak. When $Np > 5$ and $N(1 - p) > 5$, the standard deviation has this interpretation: 68 percent of the area under the histogram lies within σ units of μ. In probability terms this means that, if you carried this experiment out many times, then 68 percent of the time the number of successes would be within σ of μ.

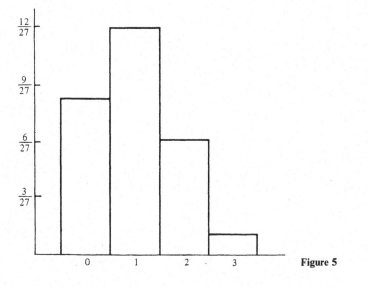

Figure 5

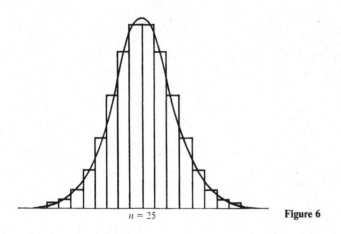

$n = 25$

Figure 6

Example 17

A coin is tossed 100 times. Suppose the probability of heads on a single toss is 0.6. Sketch the histogram for the random variable "number of heads." Interpret μ and σ in probability terms.

$$\mu = 100(0.6) = 60$$

$$\sigma = \sqrt{100(0.6)(0.4)} = 4.9$$

A smooth approximation to the histogram is shown in Figure 7. In carrying out this experiment of 100 tosses, the probability is about 0.68 that the number of heads will be between 55.1 and 64.9.

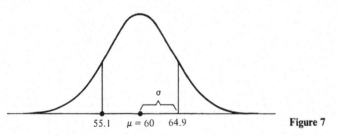

55.1 $\mu = 60$ 64.9

Figure 7

EXERCISES

● 1 Which method of assigning probabilities is most reasonable in the following experiments?
  (a) What will be the sex of an unborn child?
  (b) Will it snow on Christmas this year?
  (c) Will this coronary bypass operation be successful or not?
  (d) What number a person, asked to pick a number from 1 to 10, will choose?
  (e) What will your grade be in this course?
  (f) What number will come up if we spin a roulette wheel?

2 A coin has been tested and found to come up twice as often heads as tails. What would be a reasonable assignment of probabilities for the experiment of tossing this coin?

● 3 The odds of an event are said to be r to s if the probability is $r/(r + s)$.
  (a) Show that 3-to-1 odds means the same as 6-to-2 or, for any x, 3x-to-x.

(b) This means we can always quote odds where $s = 1$. Given an event with probability p, find a set of odds with $s = 1$; i.e., find a formula for r in terms of p.

● **4** A single card is drawn from a well-shuffled deck. If we assume the probabilities of all cards are equal, what is:

(a) The probability a spade or heart is drawn?

(b) The probability a 4 or an 8 is drawn?

(c) The probability a 5 or a heart is drawn?

5 A fair coin is flipped four times:

(a) What is the sample space of outcomes? (There should be 16 all together.)

(b) What is the probability of getting two heads?

(c) What is the probability of getting at least one head?

(d) What is the probability of getting more heads than tails?

6 A die is rolled once.

(a) What outcomes make up the event "getting an even number"? Let's call this set A. What is $p(A)$?

(b) What outcomes make up the event "getting a number larger than 3"? Let's call this set B. What is $p(B)$?

(c) Show that $p(A \cup B) = p(A) + p(B) - p(A \cap B)$.

● **7** (a) A die is rolled once, and you win a payoff equal to the number rolled. What is the mean and standard deviation of your payoff?

(b) With the same die we increase the payoff to be k dollars more than the number rolled. Show that the mean is k higher than before but the standard deviation is unchanged.

8 In producing a computer memory chip, there is a 0.0001 chance of making a defective one. If 10,000 are produced, what is the mean and standard deviation of the number of defective ones? What is the mean and standard deviation of the number of good ones?

9 In the game of craps, a player rolls two dice. If a 7 or 11 comes up, the player wins immediately, while if a 2, 3 or 12 comes up, the player loses immediately. For any other outcome the player continues rolling until either the number repeats ("makes the point"), in which case it is a win, or a 7 comes up, in which case it is a loss.

(a) What is the probability of winning immediately?

(b) What is the probability of losing immediately?

(c) What is the probability of having to continue rolling to try to make the point?

(d) If the player is trying to make the point, which is 5, what is the probability of doing so on any one roll?

● **10** If you bet on horse number 1 you have a 0.2 chance of winning $10, a 0.3 chance of winning $5, and a 0.5 chance of losing your $2 bet. If you bet on horse number 2 you have a 0.3 chance of winning $8, a 0.3 chance of winning $4, and a 0.4 chance of losing your $2 bet. Calculate the mean payoff of each bet and see which is higher.

11 Calculate the standard deviation of the amount of money required for testing in Example 10.

12 For N tosses of a fair coin, find a formula for σ/μ (in terms of N) for the random variable "number of heads." Can you place any interpretation on this ratio? What happens to σ/μ as $N \to \infty$?

13 A political candidate believes that the electorate is evenly divided pro and con on the merits of a certain legislative proposal. The candidate plans to interview 300 voters. Sketch a histogram for the random variable "number of voters polled who favored the proposal."

Computer exercises

14 Write a computer program that provides some evidence for the 68-percent rule that, in a binomial experiment where $Np > 5$ and $N(1 - p) > 5$, about 68 percent of the area under the histogram falls between $\mu - \sigma$ and $\mu + \sigma$. Here's an outline to follow:

1. Choose a fixed value of p and an N so that $Np > 5$ and $N(1 - p) < 5$.
2. Use part 1 of Theorem 1 to compute the heights of the various bars of the histogram. (If your calculation of factorials takes too long, look up Stirling's formula in Section 4 of Chapter 2.)

3. Compute μ and σ by Theorem 1.
4. Decide which bars or portions of bars fall between $\mu - \sigma$ and $\mu + \sigma$ and are to be added.
5. Carry out the addition and compare with 0.68. Keeping p fixed, do the calculation for various N values. What happens as N increases? If you have a graphics plotter, plot the histograms.

15 Write a computer program to simulate 100 sets of tosses of 5 fair coins (use a random-number generator). Tally the number of sets in which there were 5 heads, the number with 4 heads, etc. Compute the mean number of heads over all 100 sets. Compute the standard deviation. Compare these experimental values to those predicted by the formulas of Theorem 1.

BIBLIOGRAPHY

Cohen, Joel: Mathematics and Myth, *SIAM News*, vol. 6, no. 3, June, 1973. This is the origin of our problem about bellwether counties in Example 16 of Section 3A.

Hoel, Paul G., Sidney C. Port, and Charles J. Stone: "Introduction to Probability Theory," Houghton Mifflin, Boston, 1971. An advanced and complete account.

Malkevitch, Joseph, and Walter Meyer: "Graphs, Models and Finite Mathematics," Prentice-Hall, Inc., Englewood Cliffs, N.J., 1974. Chapter 8 has an elementary account of probability.

Mullenix, Paul: Randomized Response Technique: Getting in Touch with Touchy Questions, UMAP Module 576. Available from COMAP, 271 Lincoln St., Lexington, MA 02173. A nice application of conditional probability to find out from strangers what you wouldn't dare ask your friends.

3B A DAY IN THE LIFE OF AN EPIDEMIC—STOCHASTIC AND DETERMINISTIC VIEWS

Abstract The role of chance in epidemics is discussed and related to the size of the population involved. Our results illustrate what is sometimes loosely called the *law of averages*, the fact that chance factors become less significant as the number of cases becomes larger.

This section is a jumping-off point for the following two. These following sections demonstrate two different ways in which the probabilistic rate derived here can be used to predict the "mean behavior" of the epidemic, as it develops over time.

Prerequisites The binomial probability model.

It is difficult to understand why statisticians commonly limit their enquiries to averages and do not revel in more comprehensive views. Their souls seem as dull to the charm of variety as that of the native of one of our flat English counties, whose retrospect of Switzerland was that, if its mountains could be thrown into its lakes, two nuisances could be got rid of at once.

Sir Francis Galton

Epidemics of infectious disease have been part of man's history for many years. The Old Testament makes frequent reference to infectious diseases, and evidences

Table 1 Crimean war casualties (1853–1856)

	Wounded	Died of wounds	Sick	Died of disease
French	39,869	20,356	196,430	49,815
English	18,283	4,947	144,390	17,225
Russians	92,381	37,958	322,097	37,454

of various infections have been found on some of the mummies of Egyptian pharaohs.

Not only have epidemics been commonplace, but their effects have often been profound. Bubonic plague is thought to have destroyed one-quarter of the population of Europe during the Middle Ages and must, therefore, be counted as one of the major calamities of history. Armies have been a particularly good breeding ground for disease, with the result that, in many wars, disease has killed more men than bullets have (see Table 1). Some authors have gone so far as to claim that "Justinian's Plague" was the final push that toppled the Roman Empire.

A glance at Figure 8 will show how dramatically many of the great killer diseases have been tamed in New York City. Most developed countries show similar patterns. However, there is little justification for complacency. The occurrence of venereal diseases is currently increasing by leaps and bounds in the

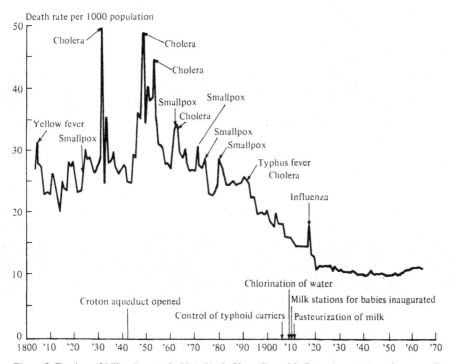

Figure 8 Taming of killer diseases in New York City. (*From M. Spiegelman, "Introduction to Demography," Harvard University Press, Cambridge, Mass., 1968. Reprinted by permission.*)

United States, and even such minor diseases as influenza contribute to mortality, especially among the elderly and those weakened by other illnesses. In the under-developed countries infectious disease is still a major public health problem. Finally, new diseases and new variants of old ones can be expected to appear on the scene. It is possible, for example, that bubonic plague, whose disappearance is largely a mystery, may return.

The mathematical theory of epidemics attempts to predict the course of an epidemic, including such factors as the total number who will be infected, time when the epidemic will peak, etc. We will present some simple models of an epidemic, which spreads from persons with the disease, called *infectives*, to persons who do not yet have it, called *susceptibles*. In particular, we will try to work out how many susceptibles will be infected in one day by the group of people who are infective at the beginning of that day.

In most diseases there is a third group of persons who are immune, and a fourth group of persons who have recovered but in the interests of simplicity we will suppose that no one is immune and no one recovers. Another assumption we make to simplify matters is that there is no latency period between when a person contracts the disease and when the person is able to spread it.

These simplifying assumptions make our model too unrealistic to be used as a tool in improving public health. However, our goal is pedagogical and fairly limited: to study the stochastic nature of an epidemic. In this respect, the story our simple model tells is not too different from the story one would obtain from more complicated models.

Diseases differ greatly in the speed with which they spread. Respiratory diseases (flu, the common cold) are often spread by droplet infection and a strong uncovered sneeze in a crowd could infect quite a few people. By contrast, diseases that spread by body contact, such as venereal diseases, generally spread more slowly. To pin things down a bit better, we make the following definition and assumptions about our disease.

Definition

Potentially infectious contact is any contact, between an infective and another person, which is normally capable of transmitting the disease.

For example, for a venereal disease, potentially infectious contact means sexual contact. For a disease spread by sneezing and coughing, potentially infectious contact may mean that the person contacted is standing within 3 feet of the sneezer and in the direction of the sneeze.

Potentially infectious contact need not result in the spread of the disease to the person contacted. For example, the person contacted may already have the disease. Or any number of chance factors may intervene to prevent infection. For example, a person sneezed upon may just happen not to breathe in any of the germs launched in that direction.

If potentially infectious contact doesn't always spread disease, why do we care about it? Because we can't observe the actual movement of germs, whereas we can observe potentially infectious contact and thereby get data for the model.

Definition

Infectious contact is a potentially infectious contact which actually spreads the disease from an infective to a susceptible.

Assumption 1 (Homogeneous Mixing)

Let E_{ij} denote the event that an infective individual X_i has potentially infectious contact during a certain day with another person X_j (who may be either infective or susceptible). Then all the events E_{ij} are independent and have the same probability p in any given day.

Assumption 2

The probability p is such that it causes a given infective to have an average of 20 potentially infective contacts in one day.

Assumption 3

The probability that a potentially infectious contact becomes an infectious contact is 0.1, provided the person contacted is susceptible (and, of course, 0 otherwise).

We now work out some consequences of these assumptions for a population in which the set of infectives numbers I and the set of susceptibles numbers S at the start of that day. Our goal is to find out how many susceptibles will have an infectious contact with an infective (and therefore become infective) during one day. We work up to this with a series of simpler questions.

1. What is the value of p? Consider a particular original infective X_1. There are $I + S - 1$ others in the population with whom that infective may have potentially infectious contact. For any single one of them the contact may either occur or not occur, and the probability of occurrence in each case is p, according to Assumption 1. Since these contact–no-contact experiments (the E_{ij}) are independent, the binomial probability model applies. Therefore, the mean number of potentially infective contacts involving X_1 is $(I + S - 1)p$. By Assumption 2 we must have

$$(I + S - 1)p = 20$$
$$p = \frac{20}{I + S - 1}$$

The total population size $I + S$ will be sufficiently larger than 1 that we will make only an insignificant error by writing instead

$$p = \frac{20}{I + S} \tag{1}$$

2. If X_1 is infective and X_2 is susceptible, what is the probability of infectious contact between them? Using the multiplication rule and Assumptions 1 and 3, we see that this probability is $0.1p$.

3. What is the probability that a given susceptible will avoid any infectious contact with infectives in a day? For each of the I infectives, the probability of having infectious contact with that infective is $0.1p$. Therefore the probability of not being infected by that infective is $1 - 0.1p$. By the multiplication rule, the probability of avoiding infectious contact with all of the I infectives is $(1 - 0.1p)^I$.

4. What is the infection probability C for a given susceptible, i.e., the probability that this individual will have at least one (maybe more) infectious contact with an infective in one day and therefore be converted to an infective? Since this event is the complement of "having no infectious contacts," for which we just worked out the probability to be $(1 - 0.1p)^I$,

$$C = 1 - (1 - 0.1p)^I$$

5. How many susceptibles will be converted to infectives by infectious contact with infectives during one day? Naturally, we cannot give a definite answer. We can, however, use the binomial probability model to deduce that:

   a The probability of k conversions is $\binom{S}{k} C^k (1 - C)^{S-k}$.

   b The formula for μ, the mean number of conversions, is

   $$\mu = SC \tag{2}$$

   c The formula for σ, the standard deviation of the number of conversions, is

   $$\sigma = \sqrt{SC(1 - C)} \tag{3}$$

It is helpful to approximate C by expanding its formula by the binomial theorem and then using only the first two terms. Thus

$$C = 1 - \left[1 - 0.1Ip + \binom{I}{2}(0.1)p - \cdots \right] \tag{4}$$

$$\approx 1 - (1 - 0.1Ip)$$

$$= 0.1Ip$$

$$= 0.1I \frac{20}{I + S}$$

$$= \frac{2I}{I + S}$$

Using this approximation, we obtain that the mean number of conversions in that day is

$$\mu = \frac{2IS}{I + S} \tag{2a}$$

and the standard deviation is

$$\sigma = \sqrt{\frac{2IS}{I + S}\left(1 - \frac{2I}{I + S}\right)} \tag{3a}$$

The number 2 in these formulas arises from the 20 in Assumption 2 and the 0.1 in Assumption 3. These values are, of course, arbitrary. A more general model would replace the foregoing formulas with

$$\mu = \frac{rIS}{I + S} \tag{2b}$$

$$\sigma = \sqrt{\frac{rIS}{I + S}\left(1 - \frac{rI}{I + S}\right)} \tag{3b}$$

However, for the sake of definiteness we will proceed with $r = 2$.

The size of the standard deviation, in relation to that of the mean, is often used as a measure of how significant the deviations from the mean are likely to be. In particular, if σ/μ is very small, one may be justified in ignoring the probabilistic character of the epidemic and asserting that μ will definitely be the number infected in a day. To investigate this, we now calculate σ/μ, called the *coefficient of variation*, from the formulas we just found:

$$\frac{\sigma}{\mu} = \frac{\sqrt{SC(1 - C)}}{SC}$$

$$= \sqrt{\frac{1 - C}{SC}}$$

$$= \sqrt{\frac{1 - 2I/(I + S)}{2IS/(I + S)}}$$

$$= \sqrt{\frac{I + S - 2I}{2IS}}$$

$$= \sqrt{\frac{S - I}{2IS}} \tag{5}$$

Now let's specialize further and suppose we are dealing with an epidemic at the stage where the I infectives constitute 10 percent of the population. If $I + S = N$, then $I = 0.1N$ and $S = 0.9N$. From Equations (2a) and (3a) we get $\mu = 0.18N$ and $\sigma = \sqrt{(0.18)(0.8)}N$. Finally, Equation (5) gives

$$\frac{\sigma}{\mu} = \frac{2.108}{\sqrt{N}} \tag{6}$$

Table 2

Context	N	$\mu = 0.18N$	$\sigma = \sqrt{(0.18)(0.8)N}$	$\sigma/\mu = 2.108/\sqrt{N}$	σ/μ as percent (rounded to nearest integer)
First grade of an elementary school	100	18	3.79	0.211	21
A high school	1000	180	12	0.067	7
A university	10,000	1800	37.95	0.021	2
A small city	100,000	18,000	120	0.007	1
A large city	1,000,000	180,000	379.47	0.002	0

Table 2 shows typical values of this function of N for various settings in which we might be studying epidemics. The lesson of Table 2 is that, if N is large enough, one can ignore the probabilistic character of the epidemic because σ/μ is very small. Of course the question of how small σ/μ should be before we consider that μ will definitely be the number infected in a day is not really a mathematical one. The user of the model must determine what the tolerance for error will be.

Sidelight: Real-World Epidemic Modeling

The epidemic models in this book are very simple, but more complicated versions of them are in use and function as early-warning systems to detect and project epidemics while they are just getting under way. One of the most ambitious of these efforts is a computerized influenza detection and prediction scheme deployed in the Soviet Union and intended to cover 99 percent of the population of that country. This system requires daily input of data about the number of cases of influenza seen by doctors or admitted to hospitals. The basic equations of the model are not based on homogeneous mixing (Assumption 1). Instead, the entire Soviet Union is divided into 128 geographical areas and the amount of mixing between them has been painstakingly determined (in a 3-year research project) from various kinds of information, including the number of bus and train tickets from one area to another that have been sold.

How accurate is this model? Eighty percent of the time the model achieves the following level of accuracy:

1. The peak day of the epidemic occurs within 5 days of when it is predicted to occur.
2. The ratio of the highest predicted number of cases in a region to the highest actual number of cases in a region is between 0.7 and 1.5.

EXERCISES

● **1** Suppose we modify Assumption 2 so that the average number of potentially infectious contacts in a day is 10.

(*a*) How does this change our estimate of p?

(*b*) How do formulas (1), (2*a*), (3*a*), (5), and (6) change for this new p value?

(*c*) Work out the new values of σ/μ in Table 2 for this new p.

● **2** Suppose we start with a group of infectives which is 20 percent of the entire population.

(*a*) What formula replaces Equation (6) under this circumstance?

(*b*) Work out the new values of σ/μ in Table 2 for this new assumption.

3 Using Equation (6), plot σ/μ as a function of N.

4 Work out the formulas for μ, σ, and σ/μ which arise by approximating $(1 - 0.1p)^I$ by the first three terms of the binomial expansion instead of the first two terms as in Equation (4).

5 Rank the following in terms of how well you think homogeneous mixing (Assumption 1) applies. In each case, indicate what factors are involved.

(*a*) A class in a grade school

(*b*) An entire school

(*c*) A crowd at a soccer stadium

(*d*) A family

(*e*) A city

6 Which do you think is easier to observe occurring, infectious contact or potentially infectious contact? Explain your answer.

7 Can you think of a way in which one could experimentally determine the probability that a given potentially infectious contact would turn into an infectious contact?

8 In everyday life, we are faced with many uncertainties. But often they are minor and, for practical purposes, we feel certain despite the uncertainty. For example, if our watch says that it takes 36 minutes to drive to school, we are not bothered by the fact that our watch may be a few seconds off. For example, 10-seconds error out of 36 minutes is less than 0.5-percent error. How big would the percentage uncertainty have to be in this and the other examples below before (in your opinion) the average person would be bothered by the uncertainty?

(*a*) Estimating the cost of a car repair

(*b*) Estimating how much money is in your checking account

(*c*) Estimating how much paint is needed to paint a room

(*d*) Estimating how long it will take you to do your homework

(*e*) Estimating how many hours a date will last

(*f*) Estimating how long it will take to drive to the stadium for the ball game.

● **9** Criticize the following alternative argument. Since there are I infectives and S susceptibles, the maximum number of infectious contacts that may occur is IS. Each of these has a probability of $0.1p$ of occurring. Therefore, by the binomial model, we have a mean number of infectious contacts equal to $0.1pIS$. Compare this with the formula we obtained in the text, which was $SC = S[1 - (1 - 0.1p)^I]$, which is not the same. However, after introducing an approximation, we did arrive at $0.1pIS$ in formula (2*a*).

10 Suppose we are dealing with a time period Δt, which is not necessarily 1 day, but possibly much shorter. It seems reasonable to revise Assumption 2 so that the average number of potentially infectious contacts is $20(\Delta t)$. Assuming no other changes, except that in Assumption 1 the probability will now be denoted $p_{\Delta t}$ and will refer to the time period of Δt days:

(*a*) Find a formula for $p_{\Delta t}$ to replace Equation (1).

(*b*) If $C_{\Delta t}$ is the probability a given susceptible will be converted to an infective in a time period of Δt, find an approximate formula for $C_{\Delta t}$ to replace Equation (4).

(*c*) Find formulas to replace Equations (2*a*), (3*a*), and (6).

11 If there are N atoms of carbon 14 in a sample at $t = 0$ and the probability of any particular one of them not being decayed by time t is $e^{-\lambda t}$, then:

(a) Compute μ and σ for the number of atoms not decayed at time t.

(b) Show that

$$\frac{\sigma}{\mu} = \sqrt{\frac{e^{\lambda t} - 1}{N}}$$

(c) Taking into account that λ is negative (about -0.1210), what can you conclude about the only conditions on N and t under which σ/μ is not completely negligible?

Computer exercise

12 Study the accuracy of the approximation $C = 0.1Ip$ obtained by using the first two terms of the binomial expansion of $(1 - 0.1p)^I$. In particular:

(a) Compute the value of $(1 - 0.1p)^I$ for $N = 100$, $N = 1000$, and $N = 10,000$ and compare with $1 - 0.1pI$. In each computation, assume $I = 0.1N$ and $p = 2/(I + S) = 2/N$.

(b) Work out the values for μ, σ, and σ/μ for $N = 100$, $N = 1000$, and $N = 10,000$, using the correct values for $(1 - 0.1p)^I$, which you computed in part (a). Compare with the corresponding values found in Table 2.

BIBLIOGRAPHY

Zinsser, Hans: "Rats, Lice and History," Bantam Books, New York, 1971. A classic, nontechnical account of disease in human affairs. Charmingly written for such a gruesome subject.

3C THE LIFETIME OF A DETERMINISTIC EPIDEMIC

Abstract We build on the framework of the previous section to get a model of how the epidemic develops over time. We ignore the probabilistic aspect of the epidemic at the very outset: we take the mean rate determined in the last section and use it to develop a differential equation.

Prerequisites Calculus, some very elementary differential equations.

In the previous section we derived the formula $\mu = rIS/(I + S)$ for the mean number of conversions of susceptibles to infectives per day. Here I = number of infectives, S = number of susceptibles, and r is a constant which reflects the rate of potentially infectious contact and the chance that a potentially infectious contact becomes an infectious contact. Since μ is only the mean of a random variable, other outcomes are possible in any given day. However, the larger the population size N $(= I + S)$, the smaller σ/μ will be and the less significant deviations from μ will be. We base our work in this section on:

Assumption 4

N is so large that σ/μ is small enough that we are willing to ignore the probabilistic aspect of the epidemic. In other words, we take it to be definite that $rIS/(I + S)$ gives the exact number of susceptibles who will be turned into infectives in a day.

Our aim now is to consider how the number of infectives and the number of susceptibles change over time periods other than a single day. Therefore, let us use the function $I(t)$ to denote the number of infectives at time t and $S(t)$ to denote the number of susceptibles at time t. Then the daily conversion rate is, according to formula (2b) of the previous section, $rI(t)S(t)/N$, where $N = I(t) + S(t) =$ total population size. We assume that N does not change during the course of the epidemic; so we can set $r/N = \beta$, a constant. The final step is to assume that one day is a short enough time period that one can take this as the instantaneous rate of change in $I(t)$. Thus,

$$\frac{dI(t)}{dt} = \beta I(t)S(t) = \beta I(t)[N - I(t)] \tag{7}$$

Notice that this equation no longer expresses any probabilistic aspects of the epidemic.

We can find out a lot about the shape of the graph of $I(t)$, even without solving the differential equation.

1. In the real world, the number of infectives can never be negative. It could be 0, but then there is no epidemic, and so we rule this out too. If we assume our model is a good one, then it must reflect this, and so $I(t) > 0$ for all t. Likewise, the number of infectives can never exceed N. If it equals N, then everyone is infected and the epidemic is over. Therefore, during the course of the epidemic $I(t) < N$. As a result of these inequalities, the right side of Equation (7) is never zero or negative. Therefore dI/dt is always positive, which means that I is monotone increasing.
2. To obtain information about whether the graph is concave up or concave down, we need information about the second derivative. Differentiating Equation (7), we obtain

$$\frac{d^2I}{dt^2} = \beta \frac{dI}{dt}(N - I) - \beta I \frac{dI}{dt}$$

$$= \beta \frac{dI}{dt}(N - 2I) \tag{8}$$

Since $\beta \, dI/dt > 0$, the sign of d^2I/dt^2 depends entirely on $N - 2I$. In particular:

1. When $I < N/2$, for example in the early stages of the epidemic, the graph is concave up.
2. When $I = N/2$, the graph has an inflection point.
3. When $I > N/2$, for example in the late stages of the epidemic, the graph is concave down.

What all this suggests is a graph like Figure 9.

Unfortunately, the exact shape of the graph, including such details as the height of the asymptote, position of the inflection point, etc., cannot be determined without knowing the formula for $I(t)$. To this end we now solve differential Equation (7). (This equation is essentially the same as the logistic equation for population growth, discussed in Section 2C of this chapter. We use the same method of solution.)

Separating the variables and integrating, we get

$$\int_0^t \frac{dI}{I(N - I)} = \int_0^t \beta \, dt$$

The right side equals βt and the left can be handled by partial fractions. Thus,

$$\frac{1}{N}\int_0^t \frac{dI}{I} + \frac{1}{N}\int_0^t \frac{dI}{N - I} = \beta t$$

$$\frac{1}{N}\left\{\log I(t)\Big|_0^t - \log[N - I(t)]\Big|_0^t\right\} = \beta t$$

$$\log I(t) - \log I(0) - \log[N - I(t)] + \log[N - I(0)] = \beta N t$$

Finally, combining logs, we get

$$\log \frac{I(t)[N - I(0)]}{I(0)[N - I(t)]} = \beta N t$$

$$\frac{I(t)[N - I(0)]}{I(0)[N - I(t)]} = e^{\beta N t}$$

$$I(t)[N - I(0)] = I(0)e^{\beta N t}[N - I(t)]$$

$$I(t)[N - I(0) + I(0)e^{\beta N t}] = NI(0)e^{\beta N t}$$

$$I(t) = \frac{NI(0)e^{\beta N t}}{N - I(0) + I(0)e^{\beta N t}} \tag{9}$$

In practice one is often less interested in the graph of $I(t)$ than in what is called the *epidemic curve* $w(t)$, the curve that shows the rate at which new cases of infection

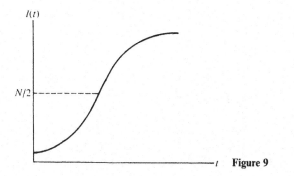

Figure 9

are appearing. $w(t)$ is considered a good measure of the strain placed on health-care resources at time t: it measures new cases that show up at the doctor's office or at the hospital. $w(t)$ is just another notation for dI/dt, the formula for which can be obtained by differentiating Equation (9) or by substituting Equation (9) into the right side of Equation (7). We obtain

$$w(t) = \frac{dI}{dt} = \frac{\beta N^2 I(0) e^{\beta Nt}[N - I(0)]}{[N - I(0) + I(0)e^{\beta Nt}]^2} \qquad (10)$$

A graph of this function, for $\beta = 0.0001$, $N = 11{,}000$, and $I(0) = 1000$, is shown in Figure 10.

The gruesome-looking expression on the right side of Equation (10) can be analyzed rather straightforwardly by means of standard calculus techniques. First, set $a = \beta N^2/I(0)$ and $b = [N - I(0)]/I(0)$; so Equation (10) becomes

$$w(t) = \frac{dI}{dt} = \frac{abe^{\beta Nt}}{(b + e^{\beta Nt})^2} \qquad (11)$$

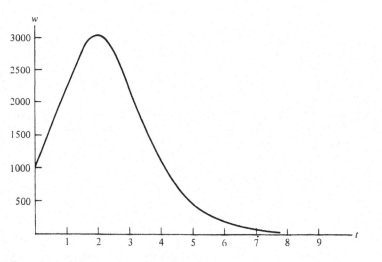

Figure 10 Epidemic curve for $\beta = 0.0001$, $N = 11{,}000$, and $I(0) = 1000$.

Applying L'Hospital's rule shows that, as $t \to \infty$, $w(t) \to 0$. This makes good sense: epidemics always peter out eventually.

Differentiating $w(t)$ and setting the derivative equal to 0 shows a single relative extremum at

$$t_{max} = \frac{1}{\beta N} \log b$$

and, by direct substitution, $w(t_{max}) = \beta N^2/4$. A little more work (calculate the second derivative) shows that this is a relative maximum. If $b > 1$, then $\log b > 0$ and t_{max} is positive, which means that the epidemic curve rises to a peak at t_{max}, before declining toward 0. On the other hand, if $b \leq 1$, then $t_{max} \leq 0$ and the epidemic curve is monotonically decreasing.

EXERCISES

● **1** Can you infer that $I(t)$ should be a monotonically increasing function even without using differential Equation (7)? (*Hint*: Take into account that we have ruled out the possibility of being cured of this disease.)

2 Verify that Equation (9) is a solution to Equation (7) by direct substitution into Equation (7).

● **3** If $S(t)$ denotes the number of susceptibles, explain why $dS/dt = -dI/dt$.

4 Verify the derivation of Equation (10).

● **5** Show that $S(t)$ drops below 1 when $t - t_{max} > [1/(\beta N)] \log (N - 1)$. [Since $S(t)$ counts people, $S(t) < 1$ can be taken to mean the epidemic is over.]

● **6** Why would the condition of Exercise 5 be a better criterion for when the epidemic is over than waiting for $S(t) = 0$?

7 Show that the epidemic curve $w(t)$ is symmetric around the point t_{max}; i.e., $w(t_{max} + T) = w(t_{max} - T)$ for any T.

● **8** (*a*) Show that the epidemic curve is monotone decreasing if $I(0) > S(0)$.

(*b*) Show that, if $I(0) < S(0)$, the epidemic curve rises to a peak before declining.

9 Carry out the calculations summarized in the last two paragraphs of this section.

10 Can you conclude anything about the nature of the graph of $w(t)$ by studying Equation (8) in conjunction with what you know about dI/dt? In particular, can you do Exercise 8 this way without looking at formulas (10) and (11)?

11 If you double the rate β, will this cut in half the time required for dI/dt to reach its maximum? What effect will doubling β have on the height of the epidemic curve at its maximum?

12 Suppose our epidemic is a long-term one and we wish to take into account that new persons are being born. Specifically, suppose the population at t is $N_0 e^{rt}$. Assuming that newborns are never infective at birth, how would you modify Equation (7)? Can you solve your equation?

Computer exercise

13 For $N = 11,000$ and $I(0) = 1000$, compute a table of values for $I(t)$ (for $t = 0, 1, \ldots, 10$) in each of the following cases:

    (*a*) $\beta = 0.00005$
    (*b*) $\beta = 0.0002$
    (*c*) $\beta = 0.0004$

Plot these on the same axes for comparison. By examining these curves, can you describe in words the influence β has on the nature of the epidemic?

BIBLIOGRAPHY

Horelick, Brindell, and Sinan Koont: Epidemics, UMAP Module 73. Available from COMAP, 271 Lincoln St., Lexington, MA, 02173. Very much like our Section 3C, except it includes a category of people recovered from the disease. A deterministic model: differential equations but no probability.

3D THE LIFETIME OF A SIMPLE STOCHASTIC EPIDEMIC

Abstract Again we use the results of Section 3B to study how an epidemic unfolds over time. But this time we preserve the chance aspects a little further into the modeling before taking means and arriving at a definite prediction about how the epidemic develops over time. The results differ from those obtained from the deterministic version of this model in Section 3C of this chapter.

Prerequisites The binomial model, elementary differential equations.

Since the daily infection rate in Section 3B is a random variable, then it makes sense that the number of infectives 2, 3, or more days from now should also be a random variable. We should, for example, be able to predict the probability that three-quarters of the population will be infective at, say $t = 8$. The model of the previous section does not enable us to do this: it only makes a single definite prediction for how many infectives there will be at any value of t. This is because the probabilistic character of the epidemic was suppressed rather early in the modeling. In this section we proceed a little differently, retaining the stochastic nature of the infection rate a little longer before discarding chance aspects by taking means. In particular, we do not make Assumption 4 of Section 3C.

Figure 11 shows the difference between the two approaches. In each case, we start at the upper left corner, which corresponds to the situation at the end of Section 3B. The work of the last section is represented by the vertical arrow at the left, followed by the horizontal one at the bottom. The work we shall do in this section is represented by the top horizontal arrow, followed by the vertical one at the right. The exact epidemic curves reached by the two different methods are not the same. This is an example of the fact that the choice of a modeling technique affects the answers obtained.

To begin with, we need to generalize some of the ideas of Section 3B. There we were dealing with a time period of one day, but, to formulate the differential equations of a stochastic model, we need information about arbitrarily small times Δt. Let $C_{\Delta t}$ denote the probability a given one of the S susceptibles will be converted to an infective by infectious contact with one of the I infectives in a time period of Δt days. We assume

$$C_{\Delta t} = \beta I(\Delta t) \tag{12}$$

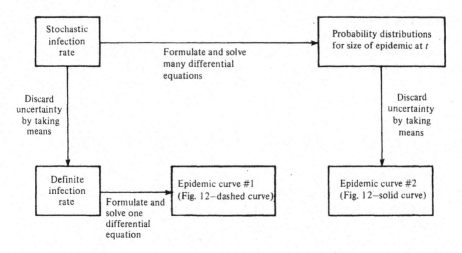

Figure 11 Two approaches to epidemic modeling.

This says that the probability is proportional to the number of infectives available to pass the infection on and the time period during which the susceptible individual is exposed to them. β is a constant of proportionality. This formula is a direct generalization of formula (4) of Section 3B, which is just the special case where $\Delta t = 1$. [It is possible to provide a more elaborate justification of formula (12). See Exercise 10 of Section 3B.]

The probability $C_{\Delta t}$ applies to each susceptible individual. The situation is very much like repeated coin tossings: each susceptible has probability $C_{\Delta t}$ of being converted to infective and probability $1 - C_{\Delta t}$ of remaining susceptible, just as each coin has a probability of $\frac{1}{2}$ of coming up heads and $1 - \frac{1}{2}$ of coming up tails. Therefore, we cannot predict how many conversions there will be. The binomial probability model could be used to determine the probabilities of various numbers of conversions occurring in the Δt time period, as in Example 18 on page 407. This example helps motivate the following assumption:

Assumption 5

If there are i infectives at time t and $N - i$ susceptibles, then the probabilities of various numbers of infectives at $t + \Delta t$ are as follows:

Number of infectives	Probability
$i + 1$ at $t + \Delta t$	$\beta i(N - i)\Delta t$
i at $t + \Delta t$	$1 - \beta i(N - i)\Delta t$
Any other number	0

Example 18

In a population of 100 people with 10 infectives at time t, what are the respective probabilities of 10, 11, or 12 infectives at $t + \Delta t$? In other words, what are the probabilities of 0, 1, or 2 conversions?

Since $N = 100$ and $i = 10$, $C_{\Delta t} = 10\beta\Delta t$. The formulas of Theorem 1, Section 3A, give the following for the probabilities:

$$\text{Probability of 0 conversions} = (1 - 10\beta\Delta t)^{90}$$

$$\text{Probability of 1 conversion} = \binom{90}{1}(10\beta\Delta t)(1 - 10\beta\Delta t)^{89} \tag{13}$$

$$\text{Probability of 2 conversions} = \binom{90}{2}(10\beta\Delta t)^2(1 - 10\beta\Delta t)^{88}$$

If we expand these formulas with the binomial theorem, we will discover that most terms have Δt to powers of 2 or higher. For example,

$$\text{Probability of 0 conversions} = 1 - \binom{90}{1}(10\beta\Delta t) + \binom{90}{2}(10\beta\Delta t)^2 - \cdots$$

We are interested in small values of Δt; in fact we will let $\Delta t \to 0$. When Δt is suitably small, $(\Delta t)^2$, $(\Delta t)^3$, ... are substantially smaller than Δt or any other terms in the formulas. If we discard all terms with second and higher powers of Δt, we get

$$\text{Probability of 0 conversions} = 1 - (90)(10)\beta\Delta t$$

$$\text{Probability of 1 conversion} = (90)(10)\beta\Delta t \tag{14}$$

$$\text{Probability of 2 or more conversions} = 0$$

(If you are nervous about the discarded terms, see Exercises 12 through 18.)

We turn now to the functions of primary interest to us.

Definition

$P_i(t)$ is the probability that the number of infectives at time t is i.

The functions $P_i(t)$ for $i = 1, 2, 3, \ldots$ are presently unknown to us, and we will formulate differential equations concerning them. But if we assume an initial condition of one infective at $t = 0$, this tells us something about the P_i at 0. This initial condition is equivalent to:

Assumption 6

$$P_i(0) = \begin{cases} 1 & \text{for } i = 1 \\ 0 & \text{for all } i > 1 \end{cases}$$

In order for the number of infectives at time $t + \Delta t$ to be 1, it is necessary for two conditions to hold:

1. The number is 1 at time t. (The number of infectives never decreases; so, if it is greater than 1 at t, it cannot be 1 at $t + \Delta t$.)
2. There must have been no new infectives between t and $t + \Delta t$.

The probability of event 1 is $P_1(t)$, and the probability of event 2, given that event 1 has occurred, is $1 - \beta(N - 1)\Delta t$, according to Assumption 5. Therefore, by the multiplication rule,

$$P_1(t + \Delta t) = P_1(t)[1 - \beta(N - 1)\Delta t]$$

Therefore,
$$\frac{P_1(t + \Delta t) - P_1(t)}{\Delta t} = - \beta(N - 1)P_1(t)$$

Letting $\Delta t \to 0$ yields

$$\frac{dP_1(t)}{dt} = - \beta(N - 1)P_1(t) \tag{15}$$

This is a standard differential equation whose solution is

$$P_1(t) = P_1(0)e^{-\beta(N - 1)t}$$
$$= e^{-\beta(N - 1)t}$$

We now pursue a similar type of reasoning to ascertain $P_2(t)$. Here are the ways the number of infectives can be 2 at $t + \Delta t$:

1. The number was 1 at t and increased by 1 in the next Δt time units.
2. The number was 2 at t and did not increase in the next Δt time units.

The probability of possibility 1 is $P_1(t)\beta(N - 1)\Delta t$, and the probability of possibility 2 is $P_2(t)[1 - 2\beta(N - 2)\Delta t]$. Thus,

$$P_2(t + \Delta t) = \beta(N - 1)\Delta t P_1(t) + [1 - 2\beta(N - 2)\Delta t]P_2(t)$$

$$\frac{P_2(t + \Delta t) - P_2(t)}{\Delta t} = \beta(N - 1)P_1(t) - 2\beta(N - 2)P_2(t)$$

Letting $\Delta t \to 0$ gives

$$\frac{dP_2(t)}{dt} = \beta(N - 1)P_1(t) - 2\beta(N - 2)P_2(t) \tag{16}$$

This is a differential equation involving two functions. Luckily, $P_1(t)$ is known because we have solved Equation (15). Making the substitution for $P_1(t)$ would allow us to use standard differential-equations methods and the initial condition $P_2(0) = 0$ to solve Equation (16). However, we shall not do so because, as we shall soon see, this approach is very clumsy and does not suit our purposes.

Our derivation of differential equations for $P_1(t)$ and $P_2(t)$ are indicative of a general pattern: for any integer $1 < i \le N$, we can find a differential equation involving dP_i/dt, $P_i(t)$, and $P_{i-1}(t)$. To carry this out, we first list the ways in which the number of infectives can be i at $t + \Delta t$:

1. It can be $i - 1$ at t and increase by 1 by $t + \Delta t$. The probability of this is $\beta(i - 1) \times (N - i + 1)\Delta t$.
2. It can be i at t and not increase by $t + \Delta t$. The probability of this is $1 - \beta i(N - i)\Delta t$.

Consequently,

$$P_i(t + \Delta t) = P_{i-1}(t)[\beta(i - 1)(N - i + 1)\Delta t] + P_i(t)[1 - \beta i(N - i)\Delta t]$$

From which we obtain

$$\frac{P_i(t + \Delta t) - P_i(t)}{\Delta t} = \beta(i - 1)(N - i + 1)P_{i-1}(t) - \beta i(N - i)P_i(t)$$

Letting $\Delta t \to 0$, we obtain

$$\frac{dP_i(t)}{dt} = \beta(i - 1)(N - i + 1)P_{i-1}(t) - \beta i(N - i)P_i(t) \tag{17}$$

Although we derived them separately, Equations (15) and (16) are special cases of Equation (17). Thus Equation (17) holds for $1 \le i \le N$ and represents a set of N differential equations. In two cases the two terms on the right reduce to one term because the other is 0. We have seen this for $i = 1$ [Equation (15)], but it is also true for $i = N$, where we get

$$\frac{dP_N(t)}{dt} = \beta(N - 1)P_{N-1}(t) \tag{18}$$

This set of N differential equations is rather like a set of N dominos standing in a row, so that, if one falls, it topples the next, which in turn topples the next, and so on. By solving the first equation, we are able, in principle, to solve the second. This enables us to solve the third, and so on. Unfortunately, for realistic values of N(in the thousands, say) this is impractical, even with the aid of computers.

No fully satisfactory elementary method has been found to obtain solutions to this system. However, it is a little more manageable to deal with the average number of infectives at time t. This function, called $\mu(t)$, is given by

$$\mu(t) = \sum_1^N iP_i(t) \tag{19}$$

$\mu(t)$ can be thought of as the stochastic model analog of the function $I(t)$, discussed in Section 3C. The stochastic analog of $w(t) = dI/dt$, which we called the epidemic curve, is naturally $d\mu/dt$. The good news about $\mu(t)$ is that there is a single equation that can be used to study it [instead of the system in Equations (17)].

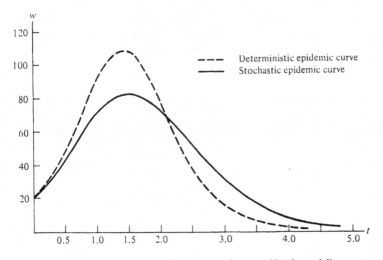

Figure 12 Comparing the results of two approaches to epidemic modeling.

The bad news is that this single equation is a partial differential equation. Exercises 7 through 11 outline some of the details. We'll skip over them and merely display the graph of $d\mu/dt$ for a particular selection of numerical parameters in Figure 12. The solid curve shows $d\mu/dt$ for the case where $N = 21$, one initial infective, and $\beta = 0.1$. For comparison, we plot the epidemic curve $w(t)$ for a deterministic epidemic with the same N, β, and initial condition of one infective.

It is striking that the two curves differ so much because, on the surface, it appears that the models differ only in the mathematical methods used and not in the underlying assumptions. Apparently, the point in the mathematics where one averages out the uncertainty can make an important difference to the result.

EXERCISES

1 Write out the complete system of differential equations of Equation (17) in the case $N = 4$.

● **2** (a) Show that $\dfrac{dP_N}{dt} + \dfrac{dP_{N-1}}{dt} = 2\beta(N-2)P_{N-2}(t)$.

(b) Find a formula for $\dfrac{dP_N}{dt} + \dfrac{dP_{N-1}}{dt} + \dfrac{dP_{N-2}}{dt}$.

3 Show that $\displaystyle\sum_{i=1}^{N} \dfrac{dP_i}{dt} = 0$. (*Hint*: One way is to extend the sort of calculations done in Exercise 2.)

● **4** A fundamental law about probabilities implies that $P_1(t) + P_2(t) + \cdots + P_N(t) = 1$. Can you use this in Exercise 3?

● **5** If we define $M_k(t) = \sum_k^N P_i(t)$, can you give a verbal interpretation of $M_k(t)$? Is it true that $M_k(t') \geq M_k(t)$ whenever $t' \geq t$?

● **6** How would the system of differential equations change if the initial number of infectives is two instead of one?

The following exercises concern the so-called *generating function* and a means of studying $\mu(t) = \sum_{i=1}^{N} iP_i(t)$, the mean number of infectives at t, through a partial differential equation involving the generating function.

Definition

The generating function is $F(X, t) = \sum_{i=0}^{\infty} X^i P_i(t)$, where X is a dummy variable with no previous significance for our problem.

7 Show that $\mu(t) = \partial F/\partial X\,|_{X=1}$ and $F(1, t) = 1$.

8 Show, by substitution from Equation (17), that

$$\frac{\partial F}{\partial t} = \sum X^i \frac{dP_i}{dt} = \beta \sum_{i=1}^{N} (X - 1)X^i i(N - i)P_i(t)$$

9 Find formulas (in terms of summations) for $\partial F/\partial t$ and $\partial^2 F/\partial X^2$.

10 By algebraic manipulation, rearrange the formula of Exercise 9 to:

$$\frac{\partial F}{\partial t} = \beta X(X - 1)\left[(N - 1)\sum_{i=1}^{N} iX^{i-1}P_i(t) - X \sum_{i=1}^{N} i(i - 1)X^{i-2}P_i(t)\right]$$

11 Using the results of Exercises 9 and 10, show:

$$\frac{\partial F}{\partial t} = \beta X(X - 1)\left[(N - 1)\frac{\partial F}{\partial X} - X \frac{\partial^2 F}{\partial X^2}\right]$$

(If one can solve this differential equation for F—not the easiest chore in the world—one can reach the goal of finding μ by differentiation via the result of Exercise 7.)

The following exercises show how to arrive at Equation (17) without making Assumption 5. In these exercises, $J(i, k, \Delta t)$ means the probability that the number of infectives jumps from i to $i + k$ in a period of length Δt.

12 Show that $J(i, k, t) = \binom{N - i}{i} (C_{\Delta t})^k (1 - C_{\Delta t})^{N-i-k}$.

13 Use the binomial theorem to show that, if $k \geq 2$, then $J(i, k, t)/\Delta t \to 0$ as $t \to 0$.

14 Use the binomial theorem to show that $J(i, 1, \Delta t) = \beta i(N - i)\Delta t + U_1$, where $U_1/\Delta t \to 0$ as $t \to 0$.

15 Show that $J(i, 0, t) = 1 - \beta i(N - i)\Delta t + U_2$, where $U_2/\Delta t \to 0$ as $\Delta t \to 0$.

16 (a) Use Exercise 15 to show

$$\frac{P_1(t + \Delta t) - P_1(t)}{\Delta t} = [-\beta(N - 1) + U_2/\Delta t]P_1(t)$$

(b) Now derive Equation (15) from part (a) of this exercise.

17 (a) Use Exercises 14 and 15 to derive an equation for

$$\frac{P_2(t + \Delta t) - P_2(t)}{\Delta t}$$

(b) Derive Equation (16) from part (a) of this exercise.

18 Use Exercises 13, 14, and 15 to derive Equation (17).

Computer exercises

19 Use a computer to calculate the probabilities in Equation (13) in Example 18 for $\beta = 0.1$ and a series of Δt values: $t = 1, 0.1, 0.01, 0.001$, etc. For these same values, calculate the estimates in Equation (14) in Example 18 and compare.

20 Write a computer program to simulate a discrete form of a stochastic epidemic with a random-number generator. Start with $N = 21$ and one initial infective. Loop through 500 time steps: at each time step t the probability of another susceptible being converted to infective should be $0.001I(t)$, where $I(t)$ denotes the number of infectives at that time t. Simulate this experiment by drawing a random number between 0 and 1. If the number drawn falls between 0 and $0.001I(t)$, this signifies a new infective and I must be updated $[I(t + 1) = I(t) + 1]$. Tabulate and print I for $t = 0, 10, 20, \ldots, 500$.

21 How would you compute the epidemic curve at $t = 0, 10, 20, \ldots, 500$ for the simulation of Exercise 20?

BIBLIOGRAPHY

Bailey, Norman T. J.: "The Mathematical Theory of Infectious Diseases," Hafner Press, New York, 1975. An extensive and often advanced account.

APPENDIX

DIFFERENTIAL EQUATIONS

A differential equation is an equation involving an unknown function and some of its derivatives. Some examples are:

$$\frac{dx}{dt} = x \tag{1}$$

$$\frac{d^2x}{dt^2} = x + \frac{dx}{dt} \tag{2}$$

In these examples, $x = x(t)$ is the unknown function of t.

Equation (1) is said to be a *first-order equation* because the highest derivative which occurs is the first derivative. Equation (2) is a second-order equation.

Just as an algebraic equation, like $x^2 + x = 4$ can be thought of as a puzzle in which the answer is one or more numbers, a differential equation can be regarded as a puzzle in which the answer is a function. For example, Equation (1) can be put into words as: "What function is its own derivative?" There are an infinite number of such functions: any function of the form

$$x(t) = Ce^t$$

where C is a constant. This can be checked by direct substitution of Ce^t into Equation (1).

The method of substitution can always be used to verify that a given function solves a differential equation. But finding the solution in the first place is harder. There are many specialized methods. To some extent, the subject of differential equations consists of diagnosing the given equation according to its special features and then applying a method that works for that type of equation. For the examples in this text, it is only necessary to understand the method of *separation of variables* for first-order equations. The equations to which this method applies are the ones of the form:

$$\frac{dx}{dt} = f(x)g(t)$$

Separating the variables means writing the equation in the forms

$$\frac{dx}{f(x)} = g(t)\, dt$$

$$\int \frac{dx}{f(x)} = \int g(t)\, dt$$

Carrying out these integrations is the only hard work (sometimes) in this method.

Example 1

1. From Equation (1) we proceed to

$$\frac{dx}{x} = dt$$

$$\int \frac{dx}{x} = \int dt$$

$$\log x = t + C$$

where C is the constant of integration. (It's important not to forget it; see below.) Thus

$$e^{\log x} = e^{t+C}$$

$$x = C'e^{t}$$

where $C' = e^{C}$, also an arbitrary constant.
2. For the equation

$$\frac{dv}{dt} = 1 - v^2$$

we obtain

$$\int \frac{dv}{1 - v^2} = \int dt$$

$$t = \arcsin v + C$$

3. For the equation

$$\frac{dx}{dt} = r(1 - x)xt^2$$

we obtain

$$\int \frac{dx}{(1 - x)x} = \int rt^2\, dt$$

We leave the integration to the reader.

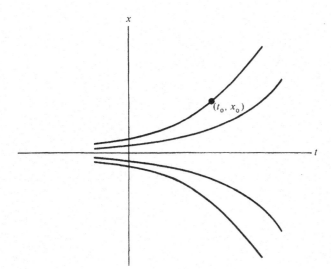

Figure 1 Solutions to a differential equation.

There are equations, such as $dx/dt = x + t$, where the variables can't be separated.

The constants of integration which appear in solving a differential equation may take on any value, and so, usually, there are infinitely many solutions to a differential equation. However, often there is an additional condition, called the *initial condition*, that allows us to pick out one particular solution from the solution set. The nature and role of the initial condition is best understood by reference to Figure 1, where we have plotted some of the solutions to Equation (1). One characteristic the entire set of solutions has is that, if one specifies any point (t_0, x_0) of the tx plane, then there is one and only one solution passing through it. This is called the *existence and uniqueness property* at (t_0, x_0).

Definition

An *initial condition* is a point (t_0, x_0) through which a solution of a differential equation must pass.

Normally, when we specify an initial condition, this allows us to pick out one solution from the infinite family of solutions to a differential equation. There are cases where there may not be a solution corresponding to a given initial condition or where there may be more than one (i.e., the existence or uniqueness property doesn't hold). However, this complication can be disregarded in this text.

The way an initial condition allows us to pick out one solution is that it enables us to find the value of the constant of integration.

Example 2

Find the solution of $dx/dt = x$, passing through the initial condition $(0, 2)$, in other words, satisfying $x(0) = 2$. As before, we find the general solution $x(t) = Ce^t$, but now

$$2 = x(0) = Ce^0 = C$$

and we have found the value of C. The particular solution we want is $x(t) = 2e^t$.

From the point of view of applications, the role of the initial condition is that most of the time the application forces a particular initial condition to accompany the differential equation, as in the following example.

Example 3

If a ball is dropped from a height, how fast will it be descending when 2 seconds have elapsed?

Let $v(t)$ denote the velocity of the ball t seconds after being dropped. Galileo's model of gravity (see Section 3 of Chapter 1 for more about it) asserts that, when distance is measured in feet,

$$\frac{dv}{dt} = 32 \tag{3}$$

Since the ball is "dropped" and not "thrown," it is at rest to start with and has zero velocity. In other words, $v(0) = 0$. This is an initial condition.

The solution to Equation (3) is

$$v(t) = 32t + C$$

Substituting $t = 0$ and $v = 0$, we find that $C = 0$. Therefore, the solution we want is

$$v(t) = 32t$$

By substituting $t = 2$, we find that after 2 seconds the velocity is 64 feet/second.

If the ball had been thrown with an initial velocity of 20 feet/second, then the initial condition would have been $v(0) = 20$, we would obtain $C = 20$, and after 2 seconds the ball would have a velocity of 84 feet/second.

BIBLIOGRAPHY

Brauer, Fred, and John A. Nohel: "Qualitative Theory of Ordinary Differential Equations," W. A. Benjamin, Inc., New York, 1969.

Simmons, George F.: "Differential Equations with Applications and Historical Notes," McGraw-Hill Book Company, New York, 1972.

ANSWERS TO SELECTED EXERCISES

Chapter 1, Section 2

1 (a) 144 feet; (b) 96 feet/second

3 (a) $t = \sqrt{x/16}$; (b) $t = \sqrt{64/16} = 2$ seconds; (c) $v = 32t$. From part (b) we find $t = 2$. Thus $v = (32)(2) = 64$ feet/second

5 $x = 16t^2$, so $t = \sqrt{x/16} = (\frac{1}{4})\sqrt{x}$. Since $v = 32t$, $v = 8\sqrt{x}$

8 If we keep the positive distance axis pointing down, then the initial velocity is -32, and so $v = 32t + v_0 = 32t - 32$. Therefore $x = 16t^2 - 32t$. If we reverse the axis, $v = 32 - 32t$ and $x = 32 - 16t^2$.

11 $T(32) = 150,000(0.8)^5 = 49,152$ hours

14 $T(x) = 32,000x^{-0.322}$

19 $y = mz + b$, so $\log T = m \log x + b$. Thus $T = e^{\log T} = e^{(m \log x + b)} = (e^{\log x})^m(e^b)$. Thus $T = cx^m$, where $c = e^b$.

Chapter 1, Section 3

1 (a) 400 feet; (b) 2 seconds; (c) 5 seconds

3 (a) Use $v_{\text{term}} = (32.2/0.329 \times 10^{-5})D^2$ to obtain the velocities: 0.025, 0.098, 0.220, and 0.391. They lie on a parabola.

(b) Use $v_{\text{term}} = 264\sqrt{D}$ to obtain the velocities: 18.7, 20.4, 22.1, 23.6, 25.0, and 26.4. They lie on a "sideways" parabola.

6 $x = (32.2 \times 10^5/0.329)D^2t$

13 (a) $\lim_{t \to \infty}[(e^{bt} - 1)/(e^{bt} + 1)] = 1$. Thus $v_{\text{term}} = \sqrt{32.2/r}$ and $v/v_{\text{term}} = (e^{bt} - 1)/(e^{bt} + 1)$.

(b) When $v = 0.99v_{\text{term}}$ the equation of part (a) gives $0.99 = (e^{bt} - 1)/(e^{bt} + 1)$. Solving for t gives $t = (1/b)\log[1.99/0.01]$

16 At the earth's surface we are 2.09×10^7 feet from the center. At this distance the acceleration is 55.0×10^{-20} feet/second^2. At a distance of 10^7 feet above the surface, it is 25.1×10^{-20} feet/second^2.

18 Weight on earth/weight on moon $= (mGm_e/R_e^2)/(mGm_m/R_m^2)$.
where $m_e = \rho_e(\frac{4}{3})\pi R_e^3$
$\quad\quad m_m = \rho_m(\frac{4}{3})\pi R_m^3$

Substituting these into the first formula gives the ratio of weights to be $\rho_e R_e/\rho_m R_m$.

Chapter 1, Section 4

1 $P(5) = (1 + 0.03 - 0.01)^5(1,500,000)$
$\quad\quad = 1,656,121.2$

3 If $b = d$, $P(t) = 1^t P(0) = P(0)$. The population size doesn't change.

5 Percent increase in k years is

$$\frac{P(k) - P(0)}{P(0)} \times 100 = \frac{(1 + b - d)^k P(0) - P(0)}{P(0)} \times 100$$

$$= [(1 + b - d)^k - 1] \times 100$$

7 $b^* = 1000 B(t)/P(t + \frac{1}{2})$
$\quad\quad = 1000 b(1 + b - d)^t P(0)/(1 + b - d)^{t + 1/2} P(0)$
$\quad\quad = 1000 b/(1 + b - d)^{1/2}$

Chapter 1, Section 5

1 (a)
$$M = \begin{bmatrix} \frac{1}{2} & 1 & 0 \\ \frac{3}{4} & 0 & 0 \\ 0 & \frac{1}{2} & 0 \end{bmatrix}; (b)\ F(\Delta) = \begin{bmatrix} 1400 \\ 600 \\ 500 \end{bmatrix}; (c)\ F(2\Delta) = \begin{bmatrix} 1300 \\ 1050 \\ 300 \end{bmatrix}$$

3 Each component of $F(\Delta)$ and $F(2\Delta)$ is twice as large as in Exercise 1.

5 (a) This is so because the last column of M is all zero. (b) M may not have an inverse.

Chapter 1, Section 6

1 Douche   31.5 mos.   Condom 71.5
  Rhythm   38.5   IUD   125.0
  Foam   41.5   Pill   250.0
  Diaphragm 55.5

3 (a) For $p = 0.1$, $\bar{w} = 10$. By Equation (4), $S_{10} = 1 - (1 - 0.1)^{10} = 0.65$. Thus 65 percent wait less than the mean.

(b) $\bar{w} = 1/p$, so we want $S_{1/p}$: $S_{1/p} = 1 - (1 - p)^{1/p}$.

5 (a) $S_M = 1 - (1 - p)^M$. Thus $0.5 = 1 - (1 - 0.1)^M$, from which $(0.9)^M = 0.5$ and $M = \log(0.5)/\log(0.9) = 6.58$.

(b) $0.5 = 1 - (1 - p)^M$, so $(1 - p)^M = 0.5$ and $M = \log(\frac{1}{2})/\log(1 - p) = -\log 2/\log(1 - p)$.

7 (a) People don't live infinitely long.

(b) $P\{\text{no conception in 300 mos.}\} = \{1 - p\}^{300}$

(c) Douche   0.00006   Condom 0.01456
    Rhythm   0.00037   IUD   0.08985
    Foam   0.00068   Pill   0.30047
    Diaphragm 0.00430

Chapter 1, Section 7

1 (a) $= x = \sqrt{\dfrac{(2)(40)(100)}{0.10}} = 282.84$

(b) The doublings occur in numerator and denominator of $2rk/s$, and so the effects cancel each other.

3 Compare $y(300)$ with $y(275)$ using Equation (1).

6 We could add the per ball cost for x balls, $0.06x$, to the total cost for each cycle. Since there are $365r/x$ cycles in a year, this adds $(0.06)(365r)$ to the formula for $y(x)$, giving

$$y(x) = 365\left(rkx^{-1} + \frac{sr}{2} + \frac{sx}{2}\right) + \frac{365r}{x}$$

9 If the deliveries occur on the mornings of days numbered $1, n + 1, 2n + 1, \ldots$, then in the n days between deliveries, rn balls are sold. Thus the delivery size should be $x = rn$. Substitute rn for x in Equation (1) and obtain the yearly cost to be

$$365\left(kn^{-1} + \frac{sr}{2} + \frac{srn}{2}\right)$$

Chapter 2, Section 1

1 (a) Soil and sand or sandstone; (b) $d = 0.6$ km; (c) $d = 0.56$ km

4 $T = 2d/(v \sin \theta) + (D - 2d \cot \theta)/v'$

7 (a) $d = \sqrt{(x - a)^2 + b^2} + \sqrt{(c - x)^2 + b^2}$
(b) $x = (a + c)/2$

12 No. The derivatives of $\sqrt{x}$ don't exist at $x = 0$, so the Taylor series doesn't exist.

15 (a) $10^{-(k-1)}$

(b) There is no way to tell. Consider 1.000000 and 0.999999.

(c) When $f(x) = \sin x$, $f^{n+1}(\bar{x})$ is either $\pm \sin \bar{x}$ or $\pm \cos \bar{x}$, depending on the value of n. In any case $|f^{n+1}(\bar{x})| \le 1$.

(d) If we have $x^{n+1}/(n + 1)! \le 10^{-2}$, by part (c), $|E| < 10^{-2}$. Thus we seek n so that $(\pi/4)^{n+1}/(n + 1)! \le 10^{-2}$. If $n + 2 \ge \pi/4$, the left side is a decreasing function of n (increasing n by 1 multiples the numerator by $\pi/4$ but the denominator by $n + 2$.) Therefore, compute the left side for $n = 1, 2, 3, \ldots$ and determine the first n for which the inequality holds, namely, $n = 4$.

17 No. For example, $10 \times 10 = 100$, $0.25 \times 0.04 = 0.01$. However, there will never be more than $2n$ digits.

Chapter 2, Section 2

1 (a) $\frac{1}{12}, \frac{1}{6}, 0, \frac{1}{6}$; (b) $(\frac{1}{12})^2$

2 (a) $\displaystyle\int_{1.4}^{1.5} \frac{1}{\sqrt{2\pi}} \exp[-\tfrac{1}{2}x^2]dx \approx \frac{1}{\sqrt{2\pi}} \exp[-\tfrac{1}{2}(1.45)^2] \approx 0.014$

$\displaystyle\int_{-0.2}^{0.2} \frac{1}{\sqrt{2\pi}} \exp[(-\tfrac{1}{2}x^2)dx] \approx 0.160$

(b) $(0.014)(0.160) \approx 0.002$

7 (a) $L(\mu) \approx (2\Delta/\mu)^n \exp(-(1/\mu) \sum_1^n x_i)$

(b) $L(\mu) = (\int_{x_1-\Delta}^{x_1+\Delta}(1(1/\mu)e^{-x/\mu} dx) \cdots (\int_{x_n-\Delta}^{x_n+\Delta}(1/\mu)e^{-x/\mu} dx)$

(c) $\mu = \sum_1^n x_i/n$

(d) No. The curves $(1/\mu)e^{-x/\mu}$ for various μ values are not translates of one another, so Gauss' theorem doesn't apply.

9 Differentiate the formula for $\log L(\mu, \sigma)$ with respect to σ, obtaining

$$-\frac{n}{\sigma} + \frac{1}{\sigma^3}\sum_1^n (x_i - \mu)^2$$

Setting this to 0 gives $\sigma = \sqrt{(1/n)\sum_1^n (x_i - \mu)^2}$.

Chapter 2, Section 3

1 (a) $\begin{bmatrix} \frac{2}{3} & \frac{1}{3} & 0 & 0 & 0 \\ \frac{1}{3} & \frac{1}{3} & \frac{1}{3} & 0 & 0 \\ 0 & \frac{1}{3} & \frac{1}{3} & \frac{1}{3} & 0 \\ 0 & 0 & \frac{1}{3} & \frac{1}{3} & \frac{1}{3} \\ 0 & 0 & 0 & \frac{1}{3} & \frac{2}{3} \end{bmatrix}$

(b) For $N = 3$ and $n = 6$, the equation is

$$\begin{bmatrix} x_1 \\ x_2 \\ x_3 \\ x_4 \\ x_5 \\ x_6 \end{bmatrix} = \begin{bmatrix} 1 & 0 & 0 & 0 & 0 & 0 \\ 0 & 1 & 0 & 0 & 0 & 0 \\ \frac{1}{3} & \frac{1}{3} & \frac{1}{3} & 0 & 0 & 0 \\ 0 & \frac{1}{3} & \frac{1}{3} & \frac{1}{3} & 0 & 0 \\ 0 & 0 & \frac{1}{3} & \frac{1}{3} & \frac{1}{3} & 0 \\ 0 & 0 & 0 & \frac{1}{3} & \frac{1}{3} & \frac{1}{3} \end{bmatrix} \begin{bmatrix} x_1 \\ x_2 \\ x_3 \\ x_4 \\ x_5 \\ x_6 \end{bmatrix}$$

3 In the case $n = 4$, we are looking for a vector that satisfies

$$\begin{bmatrix} x_1 \\ x_2 \\ x_3 \\ x_4 \end{bmatrix} = \begin{bmatrix} \frac{2}{3} & \frac{1}{3} & 0 & 0 \\ \frac{1}{3} & \frac{1}{3} & \frac{1}{3} & 0 \\ 0 & \frac{1}{3} & \frac{1}{3} & \frac{1}{3} \\ 0 & 0 & \frac{1}{3} & \frac{2}{3} \end{bmatrix} \begin{bmatrix} x_1 \\ x_2 \\ x_3 \\ x_4 \end{bmatrix}$$

Such a vector is called an eigenvector for the given matrix corresponding to an eigenvalue of 1. The matrix equation is equivalent to the system:

$$x_1 = (\tfrac{2}{3})x_1 + (\tfrac{1}{3})x_2$$
$$x_2 = (\tfrac{1}{3})x_1 + (\tfrac{1}{3})x_2 + (\tfrac{1}{3})x_3$$
$$x_3 = \qquad (\tfrac{1}{3})x_2 + (\tfrac{1}{3})x_3 + (\tfrac{1}{3})x_4$$
$$x_4 = \qquad\qquad (\tfrac{1}{3})x_3 + (\tfrac{2}{3})x_4$$

There are infinitely many solutions: x_1 may be chosen arbitrarily; then take x_2, x_3, x_4 all $= x_1$.

7 If we set the price at p, we sell $d = 10{,}516/p$ items for a total profit of $t(p) = (\tfrac{1}{2})(p - 6)d = (\tfrac{1}{2})(p - 6)(10{,}516/p)$. This function is monotone increasing to an asymptote of $\tfrac{1}{2}(10{,}516)$ as $p \to \infty$. This means that the higher the price, the higher the total profit. This is unreasonable. Where is the flaw in the model?

9 (a) $c = 102{,}496$; (b) $c = \sum_1^n (d_i/p_i)/\sum_1^n (1/p_i^2)$

12 We would have three equations involving m and b which would almost certainly be inconsistent.

Chapter 2, Section 4

1 S_1 and S_2 together contribute 18 votes to their coalition, which is enough to win whether or not their other partners stay in the coalition.

3 (a) If S_1, S_2, S_4 vote yes, S_4 is crucial.

(b) If S_1, S_3, S_4, S_5 vote yes, S_5 is crucial. If S_1, S_3, S_4, S_6 vote yes, S_6 is crucial.

(c) Yes coalitions where S_5 is crucial are $\{S_1, S_2, S_5, S_6\}$, $\{S_1, S_3, S_4, S_5\}$, $\{S_2, S_3, S_4, S_5\}$. No coalitions where S_5 is crucial are $\{S_3, S_4, S_5\}$, $\{S_2, S_5, S_6\}$, and $\{S_1, S_5, S_6\}$. Thus there are six crucial line-ups for S_5, so $\beta_5 = \frac{6}{64}$.

5 S_1 has 3 votes, S_2 and S_3 have 1, and 3 votes are needed to win.

7 (a) If a voter is on the losing side, that side will have even fewer votes—and so still lose—if he goes over to the winner.

(b) If B is on A's side in the line-up where A is crucial, then B is also able to change the outcome since he has even more votes than A. If B is not on A's side in the line-up where A is crucial, create a new line-up by interchanging A and B. Since A was originally crucial in a winning coalition in the old line-up, B will now be crucial in a winning coalition in the new line-up.

11 $\beta_i = c(A_i)/2^n$. $\beta_1 = \frac{6}{16}$, $\beta_2 = \frac{10}{16}$, $\beta_3 = \frac{2}{16}$, $\beta_4 = \frac{6}{16}$. The revised Banzhaf indices are the same as in Table 5.

Chapter 2, Section 5

1 (a) All the trigonometric functions: $\cos x$, $\sec 5x$, $\tan (x + 3)$, etc. $x - [x]$, where $[x]$ denotes the greatest integer in x.

5 We know that the values in Figure 1 are quite accurate, so there is no need to correct them. In addition, the periodic table and modern chemical theory explain why the figure looks the way it does.

Chapter 3, Section 1

1 No. A5 contradicts part (a) of A4.

3 Replace A5 by something like this: The fraction of the 18–22 age group enrolled is $0.46 + 0.008t$, where t is the number of years after the year represented by the first line of the table.

8 They are approximately similar in generality.

13 Let the actual error be Δs, so the percentage error is $(\Delta s/s)100$, where s is the measured value. The actual area is $(s + \Delta s)^2$, whereas the computed area is s^2. The difference, ΔA, is $s^2 + 2s\,\Delta s + (\Delta s)^2 - s^2 = 2s\,\Delta s + (\Delta s)^2$. The percentage difference in areas is $(\Delta A/A)100 = [s(2s + \Delta s)100]/s^2$, which is approximately $2[(\Delta s/s)100]$, which is twice the percentage error in s. We have computed the percentages based on measured values, s and A. Had we used true values, $s + \Delta s$ and $A + \Delta A$, the results would be similar.

Chapter 3, Section 2

1 Height $= 1.68$ (months) $- 1.05$

3 Temp. $= 0.0404$ (meters) $- 1.49$

5 Set $Y = \log H$ and $X = \log W$ so that $Y = \log H = 1.138 \log W + \log 0.0419 = 1.138X - 3.172$.

7 Set $Y = \sqrt{\text{distance}}$. The Y values corresponding to the five stopping distances are 3.81, 5.74, 8.47, 11.08, and 12.93. The regression line is $Y = 0.236$ (speed) $- 1.03$.

9 $X = -1.65/0.707 = -2.33$

11 Yes. If a person lives 0 distance from work it should take him 0 time to drive to work. In general, we should expect time = (1/rate)(distance), which is a straight line in the distance-time coordinate system.

15 (a) Slope = 1; (b) Slope < 1

(c) With the alternative definition, the distance from a set of points to a horizontal line would be undefined. But a horizontal line is a reasonable candidate for a regression line, so this is undesirable.

Chapter 3, Section 3

1 (a) When additional cars enter a traffic stream (increasing the density), drivers already in the stream slow down to let them enter. Also, the additional cars cause the average distance between cars to decrease. This causes drivers to reduce their speed for safety (to reduce their stopping distance to a safe level). Thus density seems to determine speed.

3 Both fouls and points are determined by a third variable: minutes played.

5 When people die during demonstrations, this can inflame the passions of the group to which the victim belongs.

Chapter 3, Section 4

1 If the plane (or line) goes through each point, then for each i, $z_i = \hat{z}_i$. Equation (11) shows that this implies $R^2 = 1$. Conversely, if $R^2 = 1$ then $\sum (\hat{z}_i - z_i)^2 = \sum (z_i - \bar{z}_i)^2$. Applying this to Equation (12) shows $\sum (z_i - \hat{z}_i)^2 = 0$. Therefore, each $(z_i - \hat{z}_i)^2 = 0$ (a sum of nonnegative quantities can be 0 only if each one is 0). Thus each $z_i = \hat{z}_i$.

3 When you have n explanatory variables and $n + 1$ data points, $R^2 = 1$.

5 Four

8 $R^2 = 0.77$ (approx.)

12 Set $z = \log A$, $x = \log l$, $y = \log w$. From $A = al^b w^c$ we get $\log A = \log a + b \log l + c \log w$. Thus $z = d + bx + cy$. The regression equation is $z = 0.995x + 1.06y - 0.103$.

Chapter 3, Section 5

1 (a) $r^t = e^{st}$, where $e^s = r$. Thus $d(r^t)/dt = se^{st} = (\log r)r^t$.

(b) When $r > 1$, $\log r > 0$ and $r^t > 0$. Therefore r^t is monotone increasing.

3 $0 < r < 1$ implies $1 < 1/r$, so $(1/r)^t \to \infty$ as $t \to \infty$.

5 To have $r^{t+T} = 2r^t$ we need $r^T = 2$, which means $T \log r = \log 2$. Thus take $T = \log 2/\log r$.

7 (a) $\{[P(1) - P(0)]/P(0)\}100 = \{(P_0 r - P_0)/P_0\}100 = (r - 1)100$

(b) $(r - 1)100$; (c) $(r - 1)100$

13 $P(0) = 4$, $P(1) = 12$, $P(2) = 24$, $P(3) = 48$, $P(4) = 96$

15 (a) The size of the tth generation (those just born at time t) will be $3 \cdot 3^t$

(b) $1 \cdot 3^t/(3 \cdot 3^t) = \frac{1}{3}$ for all t.

Chapter 3, Section 6

1 Figure 4a arises from both circles turning in the same direction. In Figure 4c the directions are opposite.

3 Figure 4b has the larger value of v_1/v_2.

5 $\angle BAC < 120°$. At $120°$ we have two equilateral triangles, so all lengths are equal.

7 Any orbit.

Chapter 3, Section 7

1 $S(C, d) = \dfrac{dC}{dd} \cdot \dfrac{d_0}{C_0} = \dfrac{360 \cdot d_0}{\theta_0 \cdot C_0}$

$= \dfrac{360}{\theta_0} \dfrac{d_0}{360(d_0/\theta_0)} = 1$

4 Let r be the radius of the earth. $m = 2\pi(r + h) = 2\pi r + 2\pi h = C + 2\pi h$. Thus $C = m - 2\pi h$.

5 (a) $S(C, h) = \dfrac{dC}{dh} \dfrac{h_0}{C_0} = \dfrac{-2\pi h_0}{m_0 - 2\pi h_0} = \dfrac{-6.2832}{24,866 - 6.2832}$

$= 0.00025$

$S(C, m) = \dfrac{dC}{dm} \dfrac{m_0}{C_0} = \dfrac{m_0}{m_0 - 2\pi h_0} = \dfrac{24,866}{24,866 - 6.2832}$

$= 1.00025$

(b) Eratosthenes' method is marginally better since $|-1| < |1.00025|$.

7 The star overhead Lysimachia plays the role of the sun in Eratosthenes' model. The star overhead Syene gives the vertical direction and so plays the role of the well. Thus we can use the same sort of figure and geometric reasoning as Eratosthenes used to obtain:

$$C = 20,000 \frac{360}{24} = 300,000 \text{ stadia}$$

14 Between E_2 and E_3.

15 Let C be the center of the moon. The angle sum of the quadrilateral $AEBC$ is $360°$, so $\angle AEB + \angle ACB = 180°$. But we also have $2\alpha + \angle ACB = 180°$, so $\alpha = (\frac{1}{2}) \angle AEB$.

19 Use the chain rule

$$S(z, x) = \frac{dz}{dx} \cdot \frac{x}{z} = \frac{dz}{dy} \frac{dy}{dx} \cdot \frac{x}{z}$$

$$= \left(\frac{dz}{dy} \cdot \frac{y}{z}\right)\left(\frac{dy}{dx} \cdot \frac{x}{y}\right) = S(z, y) \cdot S(y, x).$$

20 From $C_0 + \Delta C = 10,000\pi/(\theta_0 + \Delta\theta)$ obtain $\Delta C = 10,000\pi(-\Delta\theta)/\theta_0(\theta_0 + \Delta\theta)$. Use these two equations to substitute for $C_0 + \Delta C$ and ΔC in formula (3).

Chapter 4, Section 1

1 $dP/dx = [(r + x)^2 E^2 - 2xE^2(r + x)]/(r + x)^4$. Setting this to 0 gives $x = r$.

4 (a) $R = 0.01x(100 - x)$, where $0 \le x \le 99$. The critical value is $x = 50$, a relative maximum. At $x = 50, R = 25$. At the endpoints, 0 and 99, R is 0 and 0.99. The absolute maximum is $R = 25$, at $x = 50$ and $y = 50$.

(b) No.

6 (a) A 600-mile trip takes $600/s$ hours and costs $C = 600[0.01(20 + s/10)] + 6(600/s)$ dollars, where s is restricted to $(0, \infty)$. The maximum is $s = \sqrt{6000} = 77.5$ mph.

(b) Replace 600 by L and do the same calculation as in part (a). The answer won't involve L.

(c) Yes. 55 mph will be best.

(d) Replace 6 by w and compute the discrepancy between the cost at 55 mph and the cost at the optimum speed (ignoring the speed limit). Is this an increasing function?

8 $P = (x - 40)(3200 - 50x + 25y) + (y - 50)(25x - 25y)$. The critical point is at $x = 74, y = 79$.

12 No matter what route we take there will be 5 takeoffs and landings, and so the cost will be 250 dollars plus the mileage costs. Thus it is only in the mileage costs that savings can be made and this brings us back to the original problem of minimizing total distance.

Chapter 4, Section 2

1 The minimum is 8 and it occurs at $(8, 0)$.

3 Figure 5a is correct. Corners are $(0, 0)$, $(\frac{5}{2}, 0)$, $(0, \frac{22}{5})$, $(\frac{21}{13}, \frac{32}{13})$.

7 Let $x = $ amount of I, let $y = $ amount of II, and let $z = $ amount of III. Then minimize $4x + 7y + 5z$ over the feasible region described by

$$4x + \ \ y + 10z \geq 10$$

$$3x + 2y + \ \ \ z \geq 12$$

$$4y + \ \ 5z \geq 20$$

$$x, y, z \geq 0$$

8 Let x_{ij} be the number of tons to be shipped from source i to destination j. Minimize $5x_{11} + 7x_{12} + 9x_{13} + 6x_{21} + 7x_{22} + 10x_{23}$ over the feasible region described by

$$x_{11} + x_{12} + x_{13} \leq 120$$

$$x_{21} + x_{22} + x_{23} \leq 140$$

$$x_{11} + x_{21} = 100$$

$$x_{12} + x_{22} = \ \ 60$$

$$x_{13} + x_{23} = \ \ 80$$

$$\text{All } x_{ij} \geq \ \ 0$$

10 Let $x = $ number of type I experiments and let $y = $ number of type II experiments. Value $= x + 2y$. Maximize value subject to

$$5x + 2y \leq 100$$

$$x + 3y \leq \ \ 60$$

$$2x + 2y \leq \ \ 50$$

$$x, y \geq \ \ 0$$

The maximum is $42\frac{1}{2}$ and occurs at $(\frac{15}{2}, \frac{35}{2})$. But our answer must have integer coordinates to be acceptable. Rounding down to $(7, 17)$ might be reasonable.

13 Let $x = $ kg of pizza chips and let $y = $ kg of chili chips. Maximize $p = 0.12x + 0.10y$ subject to

$$3x + 3y \leq 240$$

$$5x + 4y \leq 480$$

$$2x + 3y \leq 360$$

$$x, y \geq 0$$

The corners are $(80, 0)$, $(0, 80)$, $(0, 0)$. The maximum is \$9.60 and occurs at $(80, 0)$.

17 This is a transportation problem where the sources are the four available blood types and the destinations are the patients. Not every route is available; e.g., we can't send type A to patient #3, so $x_{13} = 0$. Yes, it is possible to satisfy all needs.

Chapter 4, Section 3

1 $4x + y + u \quad = 10$

$18x + 5y \quad + v = 66$

$x, y, u, v \geq 0$

u is the amount of unused phosphate and v is the amount of unused nitrate.

3

	x_1	s_1	
p	-2	$\frac{3}{2}$	36
x_2	$\frac{1}{2}$	$\frac{1}{4}$	6
s_2	$\frac{9}{2}$	$-\frac{3}{4}$	12

6 There is only a finite number of ways to choose $n - m$ variables to set 0. Each of these ways gives either 1 or 0 BFS's, depending on whether a unique solution can be obtained for the other variables. The upper bound is

$$\binom{n}{n-m} = \frac{n!}{(n-m)!m!}$$

Chapter 4, Section 4

1 Exercises 6, 9, 10, 12, 15, 16, 18, 19, and 22 require integer solutions.

3 (a) One case would be where all inequalities are " $\geq$ " and all coefficients are nonnegative.

5 Maximize $p = \sum v_i x_i$.

7 Yes.

12 They are the same as in Example 4, except each x_{14} is replaced by the constant 1.

Chapter 4, Section 5

1

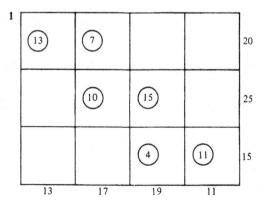

2 Here is an alteration circuit involving the (2, 4) square:

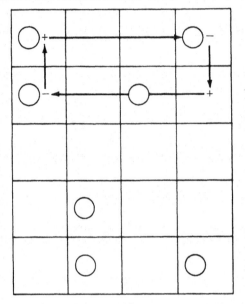

We can denote this circuit this way:

$$+(2, 4) \to -(2, 1) \to +(1, 1) \to -(1, 4) \to +(2, 4)$$

The alteration circuit involving the (5, 3) square is

$$+(5, 3) \to -(5, 4) \to +(1, 4) \to -(1, 1) \to +(2, 1) \to -(2, 3) \to +(5, 3)$$

3 Improvement indices:

$$(5, 3) \text{ square}: 4 - 3 + 3 - 5 + 1 - 3 = -3$$
$$(1, 2) \text{ square}: 2 - 5 + 3 - 3 \qquad\qquad = -3$$
$$(4, 2) \text{ square}: 1 - 3 + 3 - 5 + 1 - 5 = -8$$
$$(2, 1) \text{ square}: 2 - 3 + 5 - 1 \qquad\qquad = 3$$

etc.

4 The (1, 2) square has improvement index -3. 10 units can be diverted, producing the following BFS:

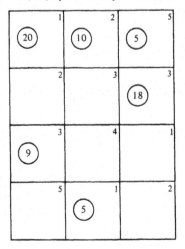

The (2, 2) square has improvement index $= -3$ and can have 9 units diverted to it. The (3, 3) square has improvement index $= -6$ and can have 9 units diverted to it.

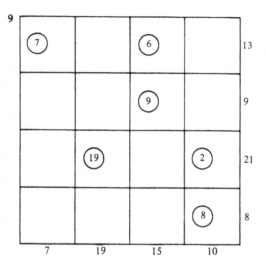

Chapter 4, Section 6

1

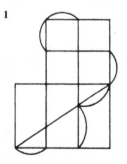

Added travel time = 6.6

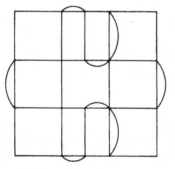

Added travel time = 6

3

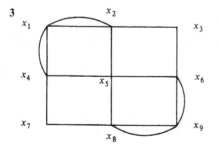

$x_1 x_2 x_1 x_4 x_7 x_8 x_9 x_8 x_5 x_2 x_3 x_6 x_9 x_6 x_5 x_4 x_1$ is a minimum duplication circuit.

4 There is no such circuit. The middle corner of the topmost street has only one way in. Once you have used that edge to get in, you can never come back again. Thus one of the outgoing edges can't be covered.

9 Minimize $\sum_1^{12} l_i x_i$ subject to:

$$x_1 + x_2 + x_9 \quad \text{is odd}$$
$$x_7 + x_3 + x_8 \quad \text{is odd}$$
$$x_5 + x_6 + x_{10} \quad \text{is odd}$$
$$x_4 + x_6 + x_{12} \quad \text{is odd}$$
$$x_1 + x_7 \quad \text{is even}$$
$$x_8 + x_5 \quad \text{is even}$$
$$x_6 + x_{12} \quad \text{is even}$$
$$x_2 + x_{11} \quad \text{is even}$$
$$x_3 + x_4 + x_9 + x_{10} \quad \text{is even}$$

Chapter 5, Section 1

3 Replace 1000 by 500 in the box labeled "Have 1000 pairs of random nos. been generated?"

4 Replace 75 by 750 in the box labeled "Are both random nos. between 0 and 75?"

5 The box entitled "Doctor takes next patient" should include: "IALARM = 15." Each time another minute ticks by we reduce this by 1; i.e., in the box I = I + 1 we also put "IALARM = IALARM − 1." Just before "Doctor busy" we insert a test to see if the doctor is done with the previous patient: If IALARM = 0 we branch to BUSY = 0 and after this go to the "Doctor busy" box. If IALARM ≠ 0 we go directly to the "Doctor busy" box.

9 Draw 10 random numbers between 000 and 999. To make this interval of size 1000 correspond to a length of 20, we have to scale down by a factor of 50. Thus, divide each random number by 50. Compute the maximum of these 10 scaled-down random numbers and record the value of the maximum. This is one trial and gives one outcome for the maximum duration of life. Perform as many trials as desired (about 100 would be good) and average the results.

10 (a) $(0.075)(1 - 0.075) = 0.069$; (b) $(1 - 0.075)^2(0.075) = 0.064$
(c) $(1 - 0.075)^{180} = 0.0000008$

12 Suppose a patient is in treatment and due to finish at, say, $t = 20$. At $t = 20$, if we check whether we can move a new patient from the waiting room to the treatment room before the previous patient has been cleared out (and before BUSY is set back to 0), we will get the answer "no." The new patient will not get to see the doctor till the next minute, $t = 21$.

13 (a) $(1 - p)p$; (b) $(1 - p)^{k-1}p$; (c) $(1 - p)^{180}$

Chapter 5, Section 2A

1 (a) $x(t + 1) = x(t)^2 + x(t) + t$, $x(0) = 1$, $x(1) = 2$, $x(2) = 7$, $x(3) = 58$
(c) $x(0) = 1$, $x(1) = 1$, $x(2) = 2$, $x(3) = 3$

2 Substitute $at + b$ for $x(t)$ in the difference equation: $a(t + 1) + b - (at + b) = a$. This is an identity (holds for all t), so the only task left is to check the initial condition. $x(0) = a(0) + b = b$ as required.

4 Try substituting t^2 for $x(t)$: $x(t + 1) - x(t)$ becomes $(t + 1)^2 - t^2$, which is $2t + 1$. This is not equal to $2t$.

Chapter 5, Section 2B

1 $x(t + \frac{1}{8}) = x(t) + 0.20x(t) = 1.2x(t)$

$x(t + 1) = (1.2)^8 x(t) = 4.3x(t)$

$x(t + 1) - x(t) = 3.3x(t)$

3 $x(t + 1) = 2x(t)$. Thus $x(t) = x(0)2^t = (100)(2^t)$

4 $x(t + 1) = 1.1x(t) - 0.2x(t) + 2[0.15y(t)]$

$y(t + 1) = 2[0.2x(t)] + 1.05y(t) - 0.15y(t)$

6 (a) The percent splitting is $[(1.2)^2 - 1]100 = 44$.

    (b) Let the correct fraction be s. Then $[(1 + s)^2 - 1]100 = 40$. Solving this gives $s = 0.183$.

9 (a) $M(t + 1/n) = (1 + 5/100n)M(t)$, $M(t + 2/n) = (1 + 5/100n)^2 M(t)$

    (b) $M(t + 1) = (1 + 5/100n)^n M(t)$

    (c) $M(t) = (1 + 5/100n)^{tn} M(0) = (1 + 5/100n)^{tn}(1000)$

    (d) Set $m = 100n/5 = 20n$ so $M(t) = 1000(1 + 1/m)^{tm/20}$:

$$\lim_{m \to \infty} 1000\left(1 + \frac{1}{m}\right)^{tm/20} = 1000\left[\lim_{m \to \infty}\left(1 + \frac{1}{m}\right)^m\right]^{t/20} = 1000e^{t/20}$$

Chapter 5, Section 2C

1 Take $K = 200$, $c' = 50/(200 - 50) = \frac{1}{3}$, $s = -(\frac{1}{7})\log(\frac{1}{3}) = 0.157$, so the equation becomes

$$x(t) = \frac{\frac{200}{3}}{\frac{1}{3} + e^{-0.157t}}$$

3 The solution $x(t)$ is given by (14). To show this is increasing, examine the derivative—this brings us back to (13)—and show it is positive for all $t \geq 0$. Assuming $s > 0$ (can you prove this?) we need only show the other two factors are > 0. Since we are assuming $x(0) < K$, $c' > 0$, which implies, using (14), that $x(t) > 0$. Now we show $x(t) < K$ for all t. In view of (14) this is equivalent to $c'/(c' + e^{-st}) < 1$, which is clearly true since $e^{-st} > 0$. Since we have shown $x(t) < K$, it follows that $1 - x(t)/K > 0$ and all factors are positive.

5 (a) $\lim_{x \to 0} [se^{-x/(K-x)}] = se^{-0/K} = s$

    (b) $\lim_{x \to K} [se^{-x/(K-x)}] = se^{-\infty} = 0$

7 $K = 109$ (or maybe 110), $c' = 10/(109 - 10) = 0.101$, $-st^* = \log(\frac{10}{99})$. To find t^*, the time at the inflection point, first calculate half the maximum population, about 54.5, and find the corresponding t. Thus $t^* = 5$, and so $s = -(\frac{1}{5})\log(\frac{10}{99}) = 0.459$.

Chapter 5, Section 2D

1 (a) Equilibrium: $x(t) = 0$. The solution is $x(t) = x(0)e^{-1.5t}$; $\lim_{t \to \infty} x(0)e^{-1.5t} = 0$, so we have stability around 0.

    (b) Equilibrium: $x(t) = 0$. The solution is $x(t) = (-\frac{1}{2})^t x(0)$ and we have $\lim_{t \to \infty} (-\frac{1}{2})^t x(0) = 0$. Therefore we have stability around 0.

2 The solution is $x(t) = (1 - r)^t x(0)$. When $0 < r < 2$, $\lim_{t \to \infty} (1 - r)^t = 0$ and 0 is a stable equilibrium. For $r \geq 2$ this is not the case.

3 The species could exist elsewhere and migrate to the area covered by the model.

Chapter 5, Section 2E

1 (*a*) We wish to show $p(t + 1) - p(t) > 0$. From (27) this equivalent to $(m - 1)p(t) + k > 0$. From (28) this is equivalent to

$$(m - 1)m^t p(0) + k(m^t - 1) + k > 0 \quad \text{or} \quad m^t[(m - 1)p(0) + k] > 0$$

We will show both factors in the first expression > 0. Since $m > 0$, $m^t > 0$. Since $p(0) < k/(1 - m)$, $(1 - m)p(0) < k$ and $0 < k + (m - 1)p(0)$.

    (*b*) Similar to part (*a*).

3 If more pumpernickel is needed, the ingredients can be diverted from other uses. Less white bread will be made so the wheat can be used in pumpernickel instead. There is very little time lag in doing this diversion.

5 a = elasticity of supply; c = elasticity of demand. $m = a/c$ is the ratio of elasticities. $|m| > 1$ means the supply elasticity is greater than the demand elasticity in absolute value. Therefore the corollary becomes: If the supply elasticity exceeds the demand elasticity in absolute value and $p(0) \neq k/(1 - m)$, then $\lim_{t \to \infty} |p(t)| = \infty$.

7 $s(t + 1) = ap(t) + a'p(t - 1) + b$; $d(t + 1) = cp(t + 1) + e$. Setting $d(t + 1) = s(t + 1)$ gives $p(t + 1) = (a/c)p(t) + (a'/c)p(t - 1) + (b - e)/c$.

9 If b were positive this would mean there would be a positive supply even if the price were 0. This is not reasonable. Thus b is either 0 or negative (probably negative). In Equation (26) if the price is 0 there should be positive demand. Thus $e > 0$.

Chapter 5, Section 2F

1 We need to show $b(1 - x)x > x$ if $0 < x < R$. This inequality is equivalent to the following:

$$-bx^2 + (b - 1)x > 0$$

$$x(b - 1 - bx) > 0$$

$$b - 1 - bx > 0$$

$$x < (b - 1)/b = R$$

3 (*a*) $|f'(R)| = |\cos(0)| = 1$, so the hypotheses of the theorem don't hold.

    (*b*) The values do approach 0. The graph of $y = \sin x$ lies between the equilibrium line and the x axis, so graphical iteration produces a downward-trending staircase.

4 (*a*) Substitute 2 for $x(t)$ and we obtain $x(t + 1) = 2$, so 2 is an equilibrium level. $d[2 + (x - 2)^3]/dx = 3(x - 2)^2$, which is 0 when $x = 2$, so Theorem 10 guarantees a neighborhood of stability.

    (*b*) If $x(0) = 3$, then $x(1) = 3$. Thus 3 is an equilibrium level also and can't be in a stable neighborhood of 2.

    (*c*) They are solutions of $x = 2 + (x - 2)^3$. Equivalently,

$$(x - 2)^3 - (x - 2) = 0$$

$$(x - 2)[(x - 2)^2 - 1] = 0$$

$$(x - 2)[x^2 - 4x + 4 - 1] = 0$$

$$(x - 2)(x - 3)(x - 1) = 0$$

The roots are 1, 2, 3.

7 (*a*) By the chain rule, $|d[f^2(u)]/du| = |f'[f(u)] \cdot f'(u)| = |f'(v)f'(u)| < 1$.

    (*b*) $f'(x) = b - 2bx$. Now substitute each of the period-2 values for x and multiply. Carry out the algebra and apply part (*a*).

10 If $0 \le x(t) \le 1$, then each factor of the right side of (31) is ≥ 0, so $x(t + 1) \ge 0$. To show $x(t + 1) \le 1$ we must show $b[1 - x(t)]x(t) \le 1$. Since $b \le 4$ this will be true if we can show $[1 - x(t)]x(t) \le \frac{1}{4}$. This can be established by finding the maximum of the function $y = (1 - x)x$ on the interval $0 \le x \le 1$ and showing it is $\le \frac{1}{4}$. This is a standard max-min problem from calculus.

Chapter 5, Section 3A

1 Relative frequency in cases (a), (b), (d); equal probabilities in case (f); subjective guess in cases (c), (e).

3 (a) $3x$ to x odds means probability $3x/(3x + x) = \frac{3}{4}$. The answer is independent of x.

   (b) We want r to 1 odds, where $r/(r + 1) = p$. Thus $r = rp + p$ and $r = p/(1 - p)$.

4 (a) $\frac{26}{52}$; (b) $\frac{8}{52}$; (c) $\frac{16}{52}$

7 (a) Values of the random variable are $1, 2, 3, 4, 5, 6$. Mean is $1(\frac{1}{6}) + 2(\frac{1}{6}) + 3(\frac{1}{6}) + 4(\frac{1}{6}) + 5(\frac{1}{6}) + 6(\frac{1}{6}) = \frac{21}{6} = 3.5$. The standard deviation is

$$\sqrt{(\tfrac{1}{6})[(1 - 3.5)^2 + (2 - 3.5)^2 + (3 - 3.5)^2 + (4 - 3.5)^2 + (5 - 3.5)^2 + (6 - 3.5)^2]} = 1.71$$

   (b) Mean is $(k + 1)(\frac{1}{6}) + (k + 2)(\frac{1}{6}) + (k + 3)(\frac{1}{6}) + (k + 4)(\frac{1}{6}) + (k + 5)(\frac{1}{6}) + (k + 6)(\frac{1}{6}) = 6k(\frac{1}{6}) + $ previous mean. When we set up the formula for the standard deviation, notice that k drops out.

10 Horse #1: mean $= (0.2)(10) + (0.3)(5) - (0.5)(2) = 2.5$
Horse #2: mean $= (0.3)(8) + (0.3)(4) - (0.4)(2) = 2.8$
Bet on horse #2.

Chapter 5, Section 3B

1 (a) $p = 10/(I + S)$; $\mu = IS/(I + S)$; $\sigma/\mu = \sqrt{1/I} = 1/\sqrt{0.1N} = 3.162/\sqrt{N}$

(c)

N	σ/μ
100	0.316
1,000	0.099
10,000	0.032
100,000	0.010
1,000,000	0.003

2 (a) $\dfrac{\sigma}{\mu} = \sqrt{\dfrac{0.8N - 0.2N}{2(0.2N)(0.8N)}} = \dfrac{1.369}{\sqrt{N}}$

9 The events described (the infectious contacts) are not all independent. Consider the events:

1. A, who is infective, infects C, who is initially susceptible.
2. B, who is infective, infects C, who is initially susceptible.

If one of these occurs, the other can't (has probability 0) since C will no longer be susceptible.

11 (a) $\mu = Ne^{-\lambda t}$; $\sigma = \sqrt{Ne^{-\lambda t}(1 - e^{-\lambda t})}$

   (c) t and N must both be very small.

Chapter 5, Section 3C

1 Since there are no cures, the pool of infectives never shrinks. $I(t)$ measures the size of this pool.

3 $S(t) + I(t) = N$, the constant population size. Thus $S(t) = N - I(t)$. Differentiate this.

5 $S(t) < 1$ is equivalent to $I(t) > N - 1$ or

$$\frac{NI(0)e^{\beta Nt}}{N - I(0) + I(0)e^{\beta Nt}} > N - 1$$

which is equivalent to $e^{\beta Nt} > (N - 1)[N - I(0)]/I(0)$. Taking logs gives $t > (1/\beta N)\log(N - 1) + (1/\beta N)t_{max}$.

6 $I(t)$ is asymptotic to N but never actually reaches it. Thus, $S(t)$ is asymptotic to 0 but never actually reaches it.

8 We need to show $dw/dt < 0$ if $I(0) > s(0)$. From (2) we have $dw/dt = d^2I/dt^2 = \beta(dI/dt)[N - 2I(t)]$. The sign of this depends only on the last factor since $\beta > 0$ and $dI/dt > 0$ for all t (because I is monotone increasing). Initially, when $t = 0$, $N - 2I(t) = N - 2I(0) = S(0) - I(0) < 0$. Since $I(t)$ can only increase as t increases, $N - 2I(t)$ can only decrease. Thus it starts negative and stays negative.

Chapter 5, Section 3D

2 (a) Substitute $i = N - 1$ in (6) to get

$$\frac{dP_{N-1}}{dt} = \beta(N - 2)(2)P_{N-2}(t) - \beta(N - 1)P_{N-1}(t)$$

Add Equation (7) to get the result.

(b) Substitute $i = N - 2$ in (6), then add the equation found in part (a)

4 Differentiate the formula $P_1(t) + \cdots + P_N(t) = 1$.

5 $M_K(t)$ is the probability the number of infectives is at least k at time t. Since there are no cures, for fixed k the probability of k infectives or more should be a nondecreasing function of t. A symbolic proof would involve showing that the derivative of $M_k(t)$ is nonnegative. This brings us back to the sorts of formulas dealt with in Exercise 2. The right sides of these formulas are always nonnegative.

6 The initial condition changes to

$$P_i(0) = \begin{cases} 1 & \text{for } i = 2 \\ 0 & \text{for } i = 1, 3, 4, \dots \end{cases}$$

The equations themselves don't change.

INDEX